Lecture Notes in Computer Science   10538

Commenced Publication in 1973
Founding and Former Series Editors:
Gerhard Goos, Juris Hartmanis, and Jan van Leeuwen

More information about this series at http://www.springer.com/series/7409

Zi Huang · Xiaokui Xiao · Xin Cao (Eds.)

Databases Theory and Applications

28th Australasian Database Conference, ADC 2017
Brisbane, QLD, Australia, September 25–28, 2017
Proceedings

Springer

Editors
Zi Huang
University of Queensland
Brisbane, QLD
Australia

Xin Cao
University of New South Wales
Sydney, NSW
Australia

Xiaokui Xiao
Nanyang Technological University
Singapore
Singapore

ISSN 0302-9743   ISSN 1611-3349   (electronic)
Lecture Notes in Computer Science
ISBN 978-3-319-68154-2   ISBN 978-3-319-68155-9   (eBook)
DOI 10.1007/978-3-319-68155-9

Library of Congress Control Number: 2017954904

LNCS Sublibrary: SL3 – Information Systems and Applications, incl. Internet/Web, and HCI

Printed on acid-free paper

This Springer imprint is published by Springer Nature
The registered company is Springer International Publishing AG
The registered company address is: Gewerbestrasse 11, 6330 Cham, Switzerland

Preface

It is our pleasure to present to you the proceedings of the 28th Australasian Database Conference (ADC 2017), which took place in Brisbane, Australia. The Australasian Database Conference is an annual international forum for sharing the latest research advancements and novel applications of database systems, data-driven applications, and data analytics between researchers and practitioners from around the globe, particularly Australia and New Zealand. The mission of ADC is to share novel research solutions to problems of today's information society that fulfill the needs of heterogeneous applications and environments and to identify new issues and directions for future research and development work. ADC seeks papers from academia and industry presenting research on all practical and theoretical aspects of advanced database theory and applications, as well as case studies and implementation experiences. All topics related to databases are of interest and within the scope of the conference. ADC gives researchers and practitioners a unique opportunity to share their perspectives with others interested in the various aspects of database systems.

As in previous years, the ADC 2017 Program Committee accepted those papers to be considered as being of ADC quality without setting any predefined quota. The conference received 32 submissions and accepted 22 papers, including 20 full research papers and two demo papers. Each paper was peer reviewed in full by at least three independent reviewers, and in some cases four referees produced independent reviews. A conscious decision was made to select the papers for which all reviews were positive and favorable. The Program Committee that selected the papers consists of 41 members from around the globe, including Australia, China, Finland, Japan, Korea, New Zealand, Singapore, Switzerland, the UK, and the USA, who were thorough and dedicated to the reviewing process.

We would like to thank all our colleagues who served on the Program Committee or acted as external reviewers. We would also like to thank all the authors who submitted their papers and all the attendees. This conference is held for you, and we hope that with these proceedings, you can have an overview of this vibrant research community and its activities. We encourage you to make submissions to the next ADC conference and contribute to this community.

August 2017

Zi Huang
Xiaokui Xiao
Xin Cao

General Chair's Welcome Message

Welcome to the proceedings of the 28th Australasian Database Conference (ADC 2017)! ADC is a leading Australia- and New Zealand-based international conference on research and applications of database systems, data-driven applications, and data analytics. In the past decade, ADC has been held in Sydney (2016), Melbourne (2015), Brisbane (2014), Adelaide (2013), Melbourne (2012), Perth (2011), Brisbane (2010), Wellington (2009), Wollongong (2008), and Ballarat (2007). This year, the ADC conference came to Brisbane.

In the past, the ADC conference series was held as a part of the Australasian Computer Science Week (ACSW). Starting from 2014, ADC conferences have departed from ACSW as the database research community in Australasia has grown significantly larger. Now the new ADC conference has an expanded research program and focuses on community building through a PhD School. ADC 2017 was the fourth of this new ADC conference series.

The conference this year had three eminent speakers to give keynote speeches: Divesh Srivastava from AT&T Labs-Research (USA), Masaru Kitsuregawa co-affiliated with the National Institute of Informatics and the University of Tokyo (Japan), and Mingsheng Ying from the University of Technology Sydney (Australia). In addition to 22 papers carefully selected by the Program Committee, we were also fortunate to have a distinguished lecture by Dr. Guoliang Li from Tsinghua University (China), and two invited talks presented by Gianluca Demartini from the University of Sheffield (UK) and Dacheng Tao from the University of Sydney (Australia). Furthermore, we had a PhD School program with great support from four invited speakers: Divesh Srivastava from AT&T Labs-Research (USA), Yu Zheng from Microsoft Research (China), Gianluca Demartini from the University of Sheffield (UK), and Rui Zhang from the University of Melbourne (Australia).

We wish to take this opportunity to thank all speakers, authors, and organizers. I would especially like to thank our Organizing Committee members, the Program Committee co-chairs Helen Huang and Xiaokui Xiao, for their dedication to ensuring a high-quality program, proceedings chair Xin Cao, for his efforts in delivering the conference proceedings, local organization co-chairs Hongzhi Yin and Sen Wang, for their efforts in covering every detail of the conference logistics, publicity and Web chair Jun Zhou for his efforts in maintaining the conference website, tutorial and distinguished lecture chair Sebastion Link for his efforts in selecting and inviting the tutorial/lecture speakers, panel chair Athman Bouguettaya for his efforts in choosing the topic of an inspiring panel discussion, and PhD School coordinator Junhao Gan for designing an exciting program for the PhD school. I would also like to thank the University of Queensland for the support that it gave to the conference. Without them, this year's ADC would not have been a success.

Brisbane is a multicultural city, and ADC 2017 was held on the St. Lucia campus of the University of Queensland. We trust that all ADC 2017 participants had a wonderful experience with the conference, the campus, and the city.

Yufei Tao

Organization

General Chair

Yufei Tao   University of Queensland, Australia

Program Committee Co-chairs

Zi Huang   University of Queensland, Australia
Xiaokui Xiao   Nanyang Technological University, Singapore

Local Organization Chairs

Hongzhi Yin   University of Queensland, Australia
Sen Wang   Griffith University, Australia

Panel Chair

Athman Bouguettaya   University of Sydney, Australia

Publication Chair

Xin Cao   University of New South Wales, Australia

Publicity and Web Chair

Jun Zhou   Griffith University, Australia

Tutorial and Distinguished Lecture Chair

Sebastion Link   University of Auckland, New Zealand

PhD School Coordinator

Junhao Gan   University of Queensland, Australia

Steering Committee

Rao Kotagiri   University of Melbourne, Australia
Timos Sellis   RMIT University, Australia
Gill Dobbie   University of Auckland, New Zealand
Alan Fekete   University of Sydney, Australia
Xuemin Lin   University of New South Wales, Australia

Yanchun Zhang   Victoria University, Australia
Xiaofang Zhou   University of Queensland, Australia

Program Committee

Zhifeng Bao   RMIT University, Australia
Runyao Duan   University of Technology Sydney, Australia
Junbin Gao   University of Sydney, Australia
Janusz Getta   University of Wollongong, Australia
Gabriel Ghinita   University of Massachusetts, Boston, USA
Yusuke Gotoh   Okayama University, Japan
Yu Gu   Northeastern University, China
Wook-Shin Han   POSTECH, Korea
Michael E. Houle   National Institute of Informatics, Japan
Guangyan Huang   Deakin University, Australia
Arijit Khan   Nanyang Technological University, Singapore
Guoliang Li   Tsinghua University, China
Feifei Li   University of Utah, USA
Jianxin Li   University of Western Australia, Australia
Xue Li   University of Qeensland, Australia
Zhixu Li   Soochow University, China
Xiang Lian   Kent State University, UK
Jiaheng Lu   University of Helsinki, Finland
Muhammad Naeem   University of Auckland, New Zealand
Lu Qin   University of Technology Sydney, Australia
Goce Ristanoski   Data61, Australia
Shazia Sadiq   University of Queensland, Australia
Timos Sellis   Swinburne University of Technology, Australia
Haichuan Shang   National Institute of Information and Communications,
                          Japan
Quan Z. Sheng   University of Adelaide, Australia
Jingkuan Song   University of Queensland, Australia
Bela Stantic   Griffith University, Australia
Farhan Tauheed   Oracle Labs Zurich, Switzerland
Anwaar Ulhaq   Victoria University, Australia
Junhu Wang   Griffith University, Australia
Yan Wang   Macquarie University, Australia
Yinghui Wu   University of California Santa Barbara, USA
Chuan Xiao   Nagoya University, Japan
Yang Yang   University of Electronic Science and Technology of China
Weiren Yu   Aston University, UK
Ying Zhang   University of Technology Sydney, Australia
Wenjie Zhang   University of New South Wales, Australia
Kai Zheng   Soochow University, China
Yi Zhou   University of Western Sydney, Australia

Lei Zou   Peking University, China
Feida Zhu   Singapore Management University, Singapore

External Reviewers

Sen Wang   Griffith University, Australia
Henry Nguyen   Griffith University, Australia
Saiful Islam   Griffith University, Australia
Binbin Gu   Griffith University, Australia
Jian Chen   Griffith University, Australia
Dong Wen   University of Technology Sydney, Australia
Xubo Wang   University of New South Wales, Australia
Seyed-Mohssen Ghafari   Macquarie University, Australia
Wei Emma Zhang   Macquarie University, Australia
Feng Zhu   Macquarie University, Australia

ADC Keynotes

Quantum Graph Reachability Problem

Mingsheng Ying

University of Technology Sydney, Australia

Abstract. Graph reachability is a fundamental problem in database theory and many other areas of computer science. In this talk, we consider quantum graph reachability problem, which originally arose in verification and analysis of quantum programs and model-checking quantum systems, but may also interest database community. We will discuss the following issues: 1. How we can naturally define a graph structure in the state Hilbert space of a quantum system from its (discrete-time) dynamics? 2. Why the approaches to classical graph reachability problem do not work for quantum reachability problem? 3. Strongly connected component decomposition theorem for quantum graphs. At the end of the talk, a series of open problems will be pointed out, including possible applications to database search in future quantum computers.

Short Biography. Mingsheng Ying was Cheung Kong Chair Professor at Tsinghua University and Director of the Scientific Committee, the National Key Laboratory of Intelligent Technology and Systems, China. Since 2008, he is Distinguished Professor and Research Director of the Centre for Quantum Software and Information, University of Technology Sydney, Australia. He is also Deputy Director for Research of the Institute of Software, Chinese Academy of Sciences.

Ying's research interests include quantum computation and quantum information, programming theory, and logical foundations of artificial intelligence. In particular, he developed Hoare logic for quantum programs and proved its (relative) completeness (TOPLAS'11). He defined the notion of invariants for quantum programs (POPL'17). He initiated the research line of model checking quantum Markov chains (CONCUR'12–14, TOCL'14). Ying is the author of Foundations of Quantum Programming (Morgan Kaufmann 2016).

Big Data Integration

Divesh Srivastava[1] and Masaru Kitsuregawa[2]

[1] AT&T Labs Research, USA
[2] National Institute of Informatics, University of Tokyo, Japan

Abstract. The Big Data era is upon us: data are being generated, collected and analyzed at an unprecedented scale, and data-driven decision making is sweeping through all aspects of society. Since the value of data explodes when it can be linked and fused with other data, addressing the big data integration (BDI) challenge is critical to realizing the promise of Big Data. BDI differs from traditional data integration in many dimensions: (i) the number of data sources, even for a single domain, has grown to be in the tens of thousands, (ii) many of the data sources are very dynamic, as a huge amount of newly collected data are continuously made available, (iii) the data sources are extremely heterogeneous in their structure, with considerable variety even for substantially similar entities, and (iv) the data sources are of widely differing qualities, with significant differences in the coverage, accuracy and timeliness of data provided. This talk presents techniques to address these novel challenges faced by big data integration, and identifies a range of open problems for the community.

Short Biography. Divesh Srivastava is the head of Database Research at AT&T Labs-Research. He is a Fellow of the Association for Computing Machinery (ACM) and the managing editor of the Proceedings of the VLDB Endowment (PVLDB). He has served as a trustee of the VLDB Endowment, as an associate editor of the ACM Transactions on Database Systems (TODS), as an associate Editor-in-Chief of the IEEE Transactions on Knowledge and Data Engineering (TKDE), and as a general or program committee co-chair of many conferences. He has presented keynote talks at several international conferences, and his research interests and publications span a variety of topics in data management. He received his Ph.D. from the University of Wisconsin, Madison, USA, and his Bachelor of Technology from the Indian Institute of Technology, Bombay, India.

Short Biography. Masaru Kitsuregawa received his Information Engineering Ph.D. degree from the University of Tokyo in 1983. Since then he joined the Institute of Industrial Science, the University of Tokyo, and is currently a professor. He is also a professor at Earth Observation Data Integration & Fusion Research Initiative of the University of Tokyo since 2010. He also serves Director General of National Institute of Informatics since 2013. Dr. Kitsuregawa's research interests include Database Engineering, and he had been principal researcher of Funding Program for World-Leading Innovative R&D on Science and Technology, MEXT Grant-in-Aids

Program for "Info-Plosion", and METI's Information Grand Voyage Project. He had served President of Information Processing Society of Japan from 2013 to 2015. He served as a committee member for a number of international conferences, including ICDE Steering Committee Chair. He is an IEEE Fellow, ACM Fellow, IEICE Fellow and IPSJ Fellow, and he won ACM SIGMOD E.F.Codd Contributions Award, Medal with Purple Ribbon, 21st Century Invention Award, and C&C Prize.

ADC Invited Talks

Human Computation
for Entity-Centric Information Access

Gianluca Demartini

The University of Sheffield, UK

Abstract. Human Computation is a novel approach used to obtain manual data processing at scale by means of crowdsourcing. In this talk we will start introducing the dynamics of crowdsourcing platforms and provide examples of their use to build hybrid human-machine information systems. We will then present ZenCrowd: a hybrid system for entity linking and data integration problems over linked data showing how the use of human intelligence at scale in combination with machine-based algorithms outperforms traditional systems. In this context, we will then discuss efficiency and effectiveness challenges of micro-task crowdsourcing platforms including spam, quality control, and job scheduling in crowdsourcing.

Short Biography. Dr. Gianluca Demartini is a Senior Lecturer in Data Science at the University of Sheffield, Information School. His research is currently supported by the UK Engineering and Physical Sciences Research Council (EPSRC) and by the EU H2020 framework program. His main research interests are Information Retrieval, Semantic Web, and Human Computation. He received the Best Paper Award at the European Conference on Information Retrieval (ECIR) in 2016 and the Best Demo Award at the International Semantic Web Conference (ISWC) in 2011. He has published more than 70 peer-reviewed scientific publications including papers at major venues such as WWW, ACM SIGIR, VLDBJ, ISWC, and ACM CHI. He has given several invited talks, tutorials, and keynotes at a number of academic conferences (e.g., ISWC, ICWSM, WebScience, and the RuSSIR Summer School), companies (e.g., Facebook), and Dagstuhl seminars. He is an ACM Distinguished Speaker since 2015. He serves as area editor for the Journal of Web Semantics, as Student Coordinator for ISWC 2017, and as Senior Program Committee member for the AAAI Conference on Human Computation and Crowdsourcing (HCOMP), the International Conference on Web Engineering (ICWE), and the ACM International Conference on Information and Knowledge Management (CIKM). He is Program Committee member for several conferences including WWW, SIGIR, KDD, IJCAI, ISWC, and ICWSM. He was co-chair for the Human Computation and Crowdsourcing Track at ESWC 2015. He co-organized the Entity Ranking Track at the Initiative for the Evaluation of XML Retrieval in 2008 and 2009. Before joining the University of Sheffield, he was post-doctoral researcher at the eXascale Infolab at the University of Fribourg in Switzerland, visiting researcher at UC Berkeley, junior researcher at the L3S Research Center in Germany, and intern at Yahoo! Research in Spain. In 2011, he obtained a Ph.D. in Computer Science at the Leibniz University of Hanover focusing on Semantic Search.

The Progress of AI

Dacheng Tao

University of Sydney, Australia

Abstract. Since the concept of Turing machine has been first proposed in 1936, the capability of machines to perform intelligent tasks went on growing exponentially. Artificial Intelligence (AI), as an essential accelerator, pursues the target of making machines as intelligent as human beings. It has already reformed how we live, work, learning, discover and communicate. In this talk, I will review our recent progress on AI by introducing some representative advancements from algorithms to applications, and illustrate the stairs for its realization from perceiving to learning, reasoning and behaving. To push AI from the narrow to the general, many challenges lie ahead. I will bring some examples out into the open, and shed lights on our future target. Today, we teach machines how to be intelligent as ourselves. Tomorrow, they will be our partners to get into our daily life.

Short Biography. Dacheng Tao is Professor of Computer Science and ARC Future Fellow in the School of Information Technologies and the Faculty of Engineering and Information Technologies at The University of Sydney. He was Professor of Computer Science and Director of the Centre for Artificial Intelligence in the University of Technology Sydney. He mainly applies statistics and mathematics to Artificial Intelligence and Data Science. His research interests spread across computer vision, data science, image processing, machine learning, and video surveillance. His research results have expounded in one monograph and 500+ publications at top journals and conferences, such as IEEE T-PAMI, T-NNLS, T-IP, JMLR, IJCV, IJCAI, AAAI, NIPS, ICML, CVPR, ICCV, ECCV, ICDM; and ACM SIGKDD, with several best paper awards, such as the best theory/algorithm paper runner up award in IEEE ICDM'07, the best student paper award in IEEE ICDM'13, and the 2014 ICDM 10-year highest-impact paper award. He received the 2015 Australian Scopus-Eureka Prize, the 2015 ACS Gold Disruptor Award and the 2015 UTS Vice-Chancellor's Medal for Exceptional Research. He is a Fellow of the IEEE, OSA, IAPR and SPIE.

Hybrid Human-Machine Big Data Integration (Distinguished Lecture)

Guoliang Li

Tsinghua University, China

Abstract. Data integration cannot be completely addressed by automated processes. We proposed a hybrid human-machine method that harnesses human ability to address this problem. The framework first uses machine algorithms to identify possible matching pairs and then utilizes the crowd to compute actual matching pairs from these candidate pairs. In this talk, I will introduce our two systems on hybrid human-machine big data integration. (1) DIMA: A distributed in-memory system on big-data integration that can use SQL to integrate heterogenous data. DIAM can be used to identify candidate matching pairs in big data integration. (2) CDB: A crowd-powered database system that provides declarative programming interfaces and allows users to utilize an SQL-like language for posing crowdsourced queries. CDB can be used to refine the candidate pairs in big data integration.

Short Biography. Guoliang Li is an Associate Professor of Department of Computer Science, Tsinghua University, Beijing, China. His research interests include crowd-sourced data management, large-scale data cleaning and integration, and big spatio-temporal data analytics. He has regularly served as the PC members of many premier conferences, such as SIGMOD, VLDB, KDD, ICDE, WWW, IJCAI, and AAAI. He was a PC co-chair of WAIM 2014, WebDB 2014, NDBC 2016, and an area chair of CIKM 2016-2017. He is an associated editor of some premier journals, such as TKDE, Big Data Research, and FCS. He has published more than 80 papers in premier conferences and journals, such as SIGMOD, VLDB, ICDE, SIGKDD, SIGIR, ACM TODS, VLDB Journal, and TKDE. His papers have been cited more than 3600 times. He received IEEE TCDE Early Career Award, and best paper awards/ nominations at DASFAA 2014 and APWeb 2014.

Contents

Spatial Databases

The Flexible Group Spatial Keyword Query 3
Sabbir Ahmad, Rafi Kamal, Mohammed Eunus Ali, Jianzhong Qi,
Peter Scheuermann, and Egemen Tanin

Surrounding Join Query Processing in Spatial Databases 17
Lingxiao Li, David Taniar, Maria Indrawan-Santiago, and Zhou Shao

Boosting Point-Based Trajectory Search with Quad-Tree 29
Munkh-Erdene Yadamjav, Sheng Wang, Zhifeng Bao, and Bang Zhang

Query Processing

Query Refinement for Correlation-Based Time Series Exploration 45
Abdullah M. Albarrak and Mohamed A. Sharaf

A Multi-way Semi-stream Join for a Near-Real-Time Data Warehouse 59
M. Asif Naeem, Kim Tung Nguyen, and Gerald Weber

Decomposition-Based Approximation of Time Series Data
with Max-Error Guarantees 71
Boyu Ruan, Wen Hua, Ruiyuan Zhang, and Xiaofang Zhou

Similarity Search

Searching k-Nearest Neighbor Trajectories on Road Networks 85
Pengcheng Yuan, Qinpei Zhao, Weixiong Rao, Mingxuan Yuan,
and Jia Zeng

Efficient Supervised Hashing via Exploring Local and Inner Data Structure . . . 98
Shiyuan He, Guo Ye, Mengqiu Hu, Yang Yang, Fumin Shen,
Heng Tao Shen, and Xuelong Li

Learning Robust Graph Hashing for Efficient Similarity Search 110
Luyao Liu, Lei Zhu, and Zhihui Li

Data Mining

Provenance-Based Rumor Detection 125
Chi Thang Duong, Quoc Viet Hung Nguyen, Sen Wang, and Bela Stantic

An Embedded Feature Selection Framework for Hybrid Data 138
    Forough Rezaei Boroujeni, Bela Stantic, and Sen Wang

A New Data Mining Scheme for Analysis of Big Brain Signal Data 151
    Siuly Siuly, Roozbeh Zarei, Hua Wang, and Yanchun Zhang

Extracting Keyphrases Using Heterogeneous Word Relations 165
    Wei Shi, Zheng Liu, Weiguo Zheng, and Jeffrey Xu Yu

Mining High-Dimensional CyTOF Data: Concurrent Gating,
Outlier Removal, and Dimension Reduction 178
    Sharon X. Lee

AI for Big Data

Locality-Constrained Transfer Coding for Heterogeneous
Domain Adaptation 193
    Jingjing Li, Ke Lu, Lei Zhu, and Zhihui Li

Training Deep Ranking Model with Weak Relevance Labels 205
    *Cheng Luo, Yukun Zheng, Jiaxin Mao, Yiqun Liu, Min Zhang,
    and Shaoping Ma*

A Deep Approach for Multi-modal User Attribute Modeling 217
    *Xiu Huang, Zihao Yang, Yang Yang, Fumin Shen, Ning Xie,
    and Heng Tao Shen*

Potpourri

Reversible Fragile Watermarking for Fine-Grained Tamper Localization
in Spatial Data 233
    Hai Lan and Yuwei Peng

Jointly Learning Attentions with Semantic Cross-Modal Correlation
for Visual Question Answering 248
    Liangfu Cao, Lianli Gao, Jingkuan Song, Xing Xu, and Heng Tao Shen

Deep Semantic Indexing Using Convolutional Localization Network
with Region-Based Visual Attention for Image Database 261
    *Mingxing Zhang, Yang Yang, Hanwang Zhang, Yanli Ji, Ning Xie,
    and Heng Tao Shen*

Demo Papers

Real-Time Popularity Prediction on Instagram 275
    Deming Chu, Zhitao Shen, Yu Zhang, Shiyu Yang, and Xuemin Lin

DSKQ: A System for Efficient Processing of Diversified
Spatial-Keyword Query   280
    Shanqing Jiang, Chengyuan Zhang, Ying Zhang, Wenjie Zhang,
    Xuemin Lin, Muhammad Aamir Cheema, and Xiaoyang Wang

Author Index   285

Spatial Databases

The Flexible Group Spatial Keyword Query

Sabbir Ahmad[1], Rafi Kamal[1], Mohammed Eunus Ali[1(✉)], Jianzhong Qi[2],
Peter Scheuermann[3], and Egemen Tanin[2]

[1] Department of CSE, Bangladesh University of Engineering and Technology,
Dhaka, Bangladesh
{ahmadsabbir,eunus}@cse.buet.ac.bd, rafikamalb@gmail.com
[2] School of CIS, University of Melbourne, Melbourne, Australia
{jianzhong.qi,etanin}@unimelb.edu.au
[3] Department of EECS, Northwestern University, Evanston, USA
peters@eecs.northwestern.edu

Abstract. We propose the flexible group spatial keyword query and
algorithms to process three variants of the query in the spatial textual
domain: (i) the group nearest neighbor with keywords query, which finds
the data object that optimizes the aggregate cost function for the whole
group Q of size n query objects, (ii) the subgroup nearest neighbor with
keywords query, which finds the optimal subgroup of query objects and
the data object that optimizes the aggregate cost function for a given
subgroup size m ($m \leq n$), and (iii) the multiple subgroup nearest neighbor with keywords query, which finds optimal subgroups and corresponding data objects for each of the subgroup sizes in the range $[m, n]$. We
design query processing algorithms based on branch-and-bound and best-
first paradigms. Finally, we conduct extensive experiments with two real
datasets to show the efficiency of the proposed algorithms.

1 Introduction

The *group nearest neighbor* (GNN) query [10] and its variants, the flexible aggregate nearest neighbor (FANN) [8] query and the consensus query [2] have been
previously studied in spatial database domain. Given a set Q of n queries and
a dataset D, a GNN query finds the data object that minimizes the aggregate
distance (e.g., sum or max) for the group, whereas an FANN query finds the optimal subgroup of query points and the data object that minimizes the aggregate
distance for a subgroup of size m, and a consensus query finds optimal subgroups
and the data objects for each of the subgroup sizes in the range $[n', n]$. In all
of these studies, the aggregate similarity is computed based on only spatial (or
Euclidean) distances between a data point and a group of query points. In this
paper, we address all the three variants of the above queries in the context of
spatial textual domain, where both spatial proximity and keyword similarity for
a *group or subgroups of users* to data points need to be considered. We call this
class of query as the *flexible group spatial keyword* query.

The flexible group spatial keyword query has many applications in spatial
and multimedia database domain. For example, a group of friends residing at

© Springer International Publishing AG 2017
Z. Huang et al. (Eds.): ADC 2017, LNCS 10538, pp. 3–16, 2017.
DOI: 10.1007/978-3-319-68155-9_1

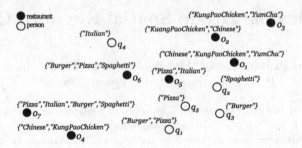

Fig. 1. An example of a group of users to find the optimal restaurant.

their homes or offices may share their locations as spatial coordinates and their preferences as sets of keywords with a location-based service provider, to find a Point of Interest (POI) (e.g., restaurant) that optimizes an aggregate cost function composed of spatial distances and keyword similarities for the group. Since finding a POI that suits all group members might be difficult due to the diverse nature of choices, the group might prefer a result that is not optimal for the entire group, but is optimal for subset of it. In such cases, we need to find optimal *subgroup* of users and a POI that minimizes the cost function for the subgroup.

Figure 1 shows an example, where a group of five friends $\{q_1, q_2, q_3, q_4, q_5\}$ is trying to decide a restaurant for a Sunday brunch. Each person provides his location and preferred type of food, represented by a set of keywords such as {"Burger", "Pizza"} or {"Italian"}, etc. There is a set of restaurants $\{o_1, o_2, ..., o_7\}$ to be selected from, and each restaurant is identified by its location and by a set of keywords describing the type of cuisine it offer, e.g., {"Pizza", "Italian"}. In general, it is preferred to find an answer that optimizes both spatial distance and keyword set dissimilarity at the same time, and o_7 is returned as the answer if we consider the whole group. However, if we allow leaving out a user, say q_4, then more answer candidates will become available. In particular, o_6 will now become the best choice of the subgroup $\{q_1, q_2, q_3, q_5\}$, as it covers all the keywords, and is the closest to members of the subgroup. In fact, leaving any other query user out (e.g., q_2) would not obtain a better cost function value. Thus $\{q_1, q_2, q_3, q_5\}$ is the optimal subgroup of size 4 with o_6 as the meeting point.

We observe that in many practical applications relaxing the requirement, i.e., not including all the query objects, has potential benefits in finding good quality answer. For example, a company may want to find a location for holding a marketing campaign, where it is often desired that the selected place optimizes for at least 60% of the customers as it may be difficult to find a place that suits all customers. Similarly, in a multimedia domain, one may want to find an image that matches with a subgroup of query images, where an object or query image is represented as a point (in a high-dimensional space) and a set of tag-words.

Generally, one may prefer the subgroup size to be maximized, and hence, it might be beneficial to explore the optimal solutions for different subgroup sizes.

The key challenge in processing the group spatial keyword queries is how to utilize both the spatial and keyword preferences and to efficiently prune the search space. Another major challenge is how to find the optimal subgroups of various sizes in one pass over the data set. Our contributions are as follows:

- We propose a new class of group queries in the spatial textual domain: (i) the group nearest neighbor with keywords (GNNK) query that finds the best POI with respect to our cost function for the whole group, (ii) the subgroup nearest neighbor with keywords (SGNNK) that finds the optimal subgroup and the corresponding best POI for a given subgroup size of size m (with $m \leq n$, the group size), and (iii) the multiple subgroup nearest neighbor with keywords (MSGNNK) that returns in one pass the optimal subgroups and corresponding POIs for all subgroups of size m, where $n' \leq m \leq n$ and n' being the minimum subgroup size.
- We propose pruning strategies based on branch and bound as well as best-first strategies for these three queries. The resultant algorithms can process the queries in a single pass over the dataset.
- We provide theoretical bounds for our algorithms, and evaluate them through an extensive experimental evaluation on real datasets.

2   Problem Statement

Let D be a geo-textual dataset. Each object $o \in D$ is defined as a pair $(o.\lambda, o.\psi)$, where $o.\lambda$ is a location point and $o.\psi$ is a set of keywords. A query object q is similarly defined as a pair $(q.\lambda, q.\psi)$. Let $dist(q.\lambda, o.\lambda)$ be the spatial distance between q and o, and $similarity\_key(q.\psi, o.\psi)$ be the similarity between their keyword sets. We normalize both $dist(q.\lambda, o.\lambda)$ and $similarity\_key(q.\psi, o.\psi)$ so that their value lie between 0 and 1 (inclusive). The cost of o with respect to q is expressed in terms of their spatial distance and keyword set distance:

$$cost(q, o) = \alpha \cdot dist(q.\lambda, o.\lambda) + (1 - \alpha) \cdot (1 - similarity\_key(q.\psi, o.\psi))$$

Here, α is a user-defined parameter to control the preference of spatial proximity over keyword set similarity. The spatial distance is normalized by the maximum spatial distance between any pair of objects in the dataset, d_{max}. Thus,

$$dist(q.\lambda, o.\lambda) = euclidean\_distance(q.\lambda, o.\lambda)/d_{max}$$

Each keyword in the dataset is associated with a weight. The weight of each keyword is normalized by the maximum keyword weight w_{max} present in the dataset. Let $y.w$ be the weight of keyword y. Then the text relevance between q and o is the normalized sum of the weights of the keywords shared by q and o:

$$similarity\_key(q.\psi, o.\psi) = \frac{1}{|q.\psi|} \sum_{y \in q.\psi \cap o.\psi} \frac{y.w}{w_{max}}$$

We formulate the GNNK, SGNNK and MSGNNK queries as follows.

Definition 1 (*GNNK*). *Given a set D of spatio-textual objects, a set Q of query objects $\{q_1, q_2, ..., q_n\}$, and an aggregate function f, the GNNK query finds an object $o_i \in D$ such that for any $o' \in D \setminus \{o_i\}$,*

$$f(cost(q_j, o_i) : q_j \in Q) \leq f(cost(q_j, o') : q_j \in Q)$$

Definition 2 (*SGNNK*). *Given a set D of spatio-textual objects, a set Q of query objects $\{q_1, q_2, ..., q_n\}$, an aggregate function f, a subgroup size m ($m \leq n$), and the set SG_m of all possible subgroups of size m, the SGNNK query finds a subgroup $sg_m \in SG_m$ and an object $o_i \in D$ such that for any $o' \in D \setminus \{o_i\}$,*

$$f(cost(q_j, o_i) : q_j \in sg_m) \leq f(cost(q_j, o') : q_j \in sg_m)$$

and for any subgroup $sg'_m \in SG_m \setminus \{sg_m\}$,

$$f(cost(q_j, o_i) : q_j \in sg_m) \leq f(cost(q', o') : q' \in sg'_m)$$

Definition 3 (*MSGNNK*). *Given a set D of spatio-textual objects, a set Q of query objects $\{q_1, q_2, ..., q_n\}$, an aggregate function f, and minimum subgroup size n' ($n' \leq n$), the MSGNNK query returns a set S of $(n - n' + 1)$ $\langle subgroup, data\ object \rangle$ pairs such that, each pair $\langle sg_m, o_m \rangle$ is the result of the SGNNK query with subgroup size m ($n' \leq m \leq n$).*

If the users are interested in the k-best POIs then the queries can be generalized as k-*GNNK*, k-*SGNNK* and k-*MSGNNK* queries. In this paper, we focus providing efficient solutions for the above queries for aggregate functions SUM ($\sum_{q_j \in Q} cost(q_j, o)$) and MAX ($\max_{q_j \in Q} cost(q_j, o)$).

3   Related Work

Group Nearest Neighbor Queries. The depth-first (DF) [11] and the best-first (BF) [7] algorithms are commonly used to process the k nearest neighbor (kNN) queries in spatial database. They assume the data objects to be indexed in a tree structure, e.g., the R-tree [6].

The group nearest neighbor (GNN) query finds a data point that minimizes the aggregate distance for a group of query locations. SUM, MAX and MIN are commonly used aggregate functions. The generalization of the GNN query is the kGNN query, where k best group nearest neighbors are to be found. Several methods for processing GNN queries have been presented in [10].

The flexible aggregate nearest neighbor (FANN) query [8] is a generalization of the GNN query. It returns the data object that minimizes the aggregate distance to any subset of ϕn query points, where n is the size of the query group and $0 < \phi \leq 1$. The query also returns the corresponding subset of query points.

A query similar to the FANN query called the *consensus query* [2] is the main motivation of our paper. Given a minimum subgroup size m and a set of n query points, the consensus query finds objects that minimize the aggregate distance for all subgroups with sizes in the range $[m, n]$. A BF algorithm was proposed to process the consensus query.

Spatial Keyword Queries. The spatial keyword query consists of a query location and a set of query keywords. A spatio-textual data object is returned based on its spatial proximity to the query location and textual similarity with the query keywords. A number of indexing structures for processing the spatial keyword query have been proposed [1,5,9,13]. Among them, the IR-tree [5,9] is shown to be a highly efficient one. The IR-tree augments each node of the R-tree with an inverted file corresponding to the keyword sets of the child nodes.

A variant of the spatial keyword query, called *spatial group keyword query* has been introduced [3,4]. It finds a group of objects that cover the keywords of a *single query* such that both the aggregate distance of the objects from the query location and the inter-object distances within the group are also minimized. Exact and approximate algorithms for three types of aggregate functions (SUM, MAX and MIN) have been presented in [3].

A work parallel to ours, the group top-k spatial keyword query (GLkT) has been proposed recently [12]. This paper presents a branch-and-bound technique to retrieve the top-k spatial keyword objects for a group. We show in our experimental evaluation (Sect. 6), our best-first technique always outperforms the branch-and-bound method substantially even for a single group query.

4 Our Approach

This section presents our algorithms to process the GNNK, SGNNK and MSGNNK queries. The key challenge is to utilize the spatial distance and keyword preference together to constrain the search space as much as possible, since the performance of the algorithms is directly proportional to the search space (in both running time and I/O). Another challenge in the SGNNK and MSGNNK queries is to find the optimal subgroup from all possible subgroups.

4.1 Preliminaries

We use the IR-tree [5] to index our geo-textual dataset D. Other extensions of the IR-tree, such as the CIR-tree, the DIR-tree or the CDIR-tree [5] can be used as well. The IR-tree is essentially an inverted file augmented R-tree [6]. The leaf nodes of the IR-tree contain references to the objects from dataset D. Each leaf node has also a pointer to an inverted file index corresponding to the keyword sets of the objects stored in that node. The inverted file index stores a mapping from the keywords to the objects where the keywords appear. Each node N of the IR-tree has the form $(N.\Lambda, N.\Psi)$, where $N.\Lambda$ is the minimum bounding rectangle (MBR) that bounds the child node entries, and $N.\Psi$ is the union of the keyword sets in the child node entries. The cost of an IR-tree node is defined similarly to the cost of a data object:

$$cost(q, N) = \alpha \ min\_dist(q.\lambda, N.\Lambda) + (1 - \alpha) \ (1 - similarity\_key(q.\psi, N.\Psi))$$

Here, $min\_dist(q.\lambda, N.\Lambda)$ is the minimum spatial distance between the query object location $q.\lambda$ and the MBR of N; $similarity\_key(q.\psi, N.\Psi)$ is the textual similarity between the query keywords and the keywords of the node. The cost of an IR-tree node gives a lower bound over the cost of its children, as formalized by the following lemma:

Lemma 1. *Let N be an IR-tree node and q be a query object. If N_c is a child of N, then $cost(q, N) \leq cost(q, N_c)$.*

Proof. The child N_c can either be a data object or an IR-tree node. In either case $min\_dist(q.\lambda, N.\Lambda)$ is smaller than or equal to that of N_c according to the R-tree structure. Meanwhile, the keyword set of N_c is a subset of the keyword set of N. Thus, N will have a higher (or equal) textual similarity value (and hence lower keyword set distance) with the query keywords. Overall, we have $cost(q, N) \leq cost(q, N_c)$. □

4.2 Branch and Bound Algorithms for GNNK and SGNNK

Traditional nearest neighbor algorithms access the data indexed in a spatial index (e.g., R-tree) and restricts its search space by pruning bounds [11]. We extend the idea to design two branch and bound algorithms for the GNNK and SGNNK queries. These algorithms work as the baseline in our experiments.

Branch and Bound Algorithm for GNNK. We use the following heuristic to prune the unnecessary nodes while searching the IR-tree for the best object with the minimum aggregate cost.

Heuristic 1. *A node N can be safely pruned if its aggregate cost with respect to the query set Q is greater than or equal to the smallest cost of any object retrieved so far.*

This heuristic is derived from Lemma 1. As f is a monotonic function and $cost(q, N) \leq cost(q, N_c)$ for any child N_c of N, $f(cost(Q, N)) \leq f(cost(Q, N_c))$. Let min_cost be the smallest cost of any data object retrieved so far. Then $f(cost(Q, N)) \geq min\_cost$ implies that the cost of any descendant of N is greater than or equal to min_cost, and we can safely prune N.

The branch and bound algorithm for GNNK is based on the heuristic and denoted by GNNK-BB. The algorithm keeps a stack and inserts the child nodes of the IR-tree into the stack, if the aggregate cost of the node is less than min_cost. After all the nodes are explored, the leaf node for the min_cost is returned.

Branch and Bound Algorithm for SGNNK. We design a similar branch and bound algorithm named SGNNK-BB for the SGNNK query. The following heuristic is used for pruning.

Heuristic 2. *Let N be an IR-tree node and m be the required subgroup size. If sg_m is the best subgroup of size m, and min_cost is the smallest cost of any size-m subgroup retrieved so far, we can safely prune N if $f(cost(sg_m, N)) \geq min\_cost$.*

This heuristic is derived from Lemma 1. Let N_c be a child of N and sg'_m be the best subgroup corresponding to N_c. Then we have $f(cost(sg'_m, N)) \leq f(cost(sg'_m, N_c))$. Meanwhile sg_m is the best subgroup for N among all possible subgroups of size m. Thus, $f(cost(sg_m, N)) \leq f(cost(sg'_m, N))$.

The above two inequalities imply that $f(cost(sg_m, N)) \leq f(cost(sg'_m, N_c))$, i.e., the aggregate cost for the best size-m subgroup of N is lower than or equal to that of the best size-m subgroup of any of its children. Therefore, if $f(cost(sg_m, N)) \geq min\_cost$, $f(cost(sg_m, N_c))$ will also be greater than or equal to min_cost, and we should prune N. The overall tree traversal procedure is similar to that of the GNNK-BB algorithm. The difference is in the calculation of the optimization function, where the optimization function value is computed based on the top-m queries with the lowest costs.

4.3 Best-First Algorithms for GNNK and SGNNK

Branch and bound techniques may access unnecessary nodes during query processing. To improve the query efficiency by reducing disk accesses, we propose in this section best-first search techniques that only access the necessary nodes.

Best-First Algorithm for GNNK. The best-first procedure for the GNNK query is denoted by GNNK-BF. This algorithm uses a minimum priority queue to maintain the nodes/objects to be visited according to their aggregate costs. If an intermediate node (leaf node) is popped, all the child nodes (child objects) are pushed into the queue. When an object is first popped from the queue, it denotes the minimum cost object and is returned as the query result. The algorithm is not shown due to space limitation.

Best-First Algorithm for SGNNK. The best-first algorithm for the SGNNK query, denoted by SGNNK-BF, is similar to GNNK-BF algorithm. Here, the

---

Algorithm 1. SGNNK-BF (R, Q, m, f)

INPUT: IR-tree index R, n query points $Q = \{q_1, q_2, ..., q_n\}$, subgroup size m, f.
OUTPUT: A data object o and a set of m query points sg_m that minimize $f(cost(sg_m, o))$
1: Initialize a new min priority queue P and $P.push(root, 0)$
2: **repeat**
3:      $E \leftarrow P.pop()$
4:      **if** E is an intermediate node N **then**
5:          **for all** N_c in $N.children$ **do**
6:              Compute $cost(q_1, N_c), ..., cost(q_n, N_c)$
7:              $sg_m \leftarrow$ first m query points with the lowest cost values
8:              $P.push(N_c, f(cost(sg_m, N_c)))$
9:      **else if** E is a leaf node N **then**
10:         **for all** o in $N.children$ **do**
11:             Compute $cost(q_1, o), ..., cost(q_n, o)$
12:             $sg_m \leftarrow$ first m query points with the lowest cost values
13:             $o.best\_subgroup = sg_m$
14:             $P.push(o, f(cost(sg_m, o))$
15:     **else if** E is a data object o **then**
16:         return $(o, o.best\_subgroup)$
17: **until** P is empty
18: return $null$

---

optimization function is computed for top-m queries. Best subgroup is chosen from the lowest m query points, and pushed into the priority queue. For an intermediate node, aggregate costs and best subgroup are calculated for all the child nodes of the node. For a leaf node, it is done for all the children objects, and then pushed into the priority queue. When an object is first popped, it is returned as the result. The pseudocode is shown in Algorithm 1.

4.4 Algorithms for MSGNNK

To process the MSGNNK query with a minimum subgroup size m, we can run SGNNK-BF $n - m + 1$ times (for subgroup sizes $m, m + 1, ..., n$) and return the combined results. We call this the MSGNNK-N algorithm. However, MSGNNK-N requires accessing the dataset $n - m + 1$ times, which is too expensive. To avoid this repeated data access, we design an algorithm based on best-first method that can find the best data objects for all subgroup sizes between m and n in a single pass over the dataset. Algorithm 2 summarizes the proposed procedure, denoted as MSGNNK-BF. The algorithm is based on the following heuristic.

Algorithm 2. MSGNNK-BF (R, Q, m, f)

INPUT: Index R of objects, query set Q, min subgroup size $m (m \leq n)$, and f.
OUTPUT: A set of $\langle data\_object, subgroup \rangle$ pairs $\langle o_k^*, sg_k^* \rangle$ for all subgroup sizes between m and
    n (inclusive), where $\langle o_k^*, sg_k^* \rangle$ minimizes $f(cost(sg_k, o))$.
1: Initialize a new min priority queue P and $P.push(root, 0)$
2: $min\_costs[i] \leftarrow \infty$ and $root.query\_costs[i] \leftarrow 0$ for $m \leq i \leq n$
3: **repeat**
4:     $E \leftarrow P.pop()$
5:     **if** $\exists i \in [m, n]: E.query\_costs[i] < min\_costs[i]$ **then**
6:         **if** E is an intermediate node **then**
7:             **for all** N_c in $E.children$ **do**
8:                 Compute $cost(q_1, N_c), ..., cost(q_n, N_c)$
9:                 $total\_cost \leftarrow 0$
10:                 **for** $i = m \rightarrow n$ **do**
11:                     $sg_i \leftarrow$ top i lowest cost query points
12:                     $total\_cost += f(cost(sg_i, N_c))$
13:                     $N_c.query\_costs[i] = f(cost(sg_i, N_c))$
14:                 **if** $f(cost(sg_i, N_c)) < min\_costs[i]$ for any subgroup size $i \in [m, n]$ **then**
15:                     $P.push(N_c, total\_cost)$
16:         **else if** E is a leaf node **then**
17:             **for all** o in $N.children$ **do**
18:                 Compute $cost(q_1, o), ..., cost(q_n, o)$
19:                 **for** $i = m \rightarrow n$ **do**
20:                     $sg_i \leftarrow$ top i lowest cost query points
21:                     **if** $f(cost(sg_i, o)) < min\_costs[i]$ **then**
22:                       $min\_costs[i] \leftarrow f(cost(sg_i, o))$
23:                       $best\_objects[i] \leftarrow o$
24:                       $best\_subgroups[i] \leftarrow sg_i$
25: **until** P is empty
26: **return** $best\_objects, best\_subgroups$

Heuristic 3. *Let N be an IR-tree node and m be the minimum subgroup size. Let sg_i be the best subgroup of size i $(m \leq i \leq n)$, and min_cost_i be the smallest cost for subgroup size i from any object retrieved so far. We can safely prune N if $f(cost(sg_i, N)) \geq min\_cost_i$ for any i.*

The proof of correctness is straightforward based on Heuristics 1 and 2, and is omitted due to space limit.

A Relaxed Pruning Bound. A possible simplification of Heuristic 3 is to only test whether $f(cost(sg_m, N)) \geq min\_cost_n$, i.e., whether the best subgroup of size m corresponding to N has a cost lower than the min_cost for the whole group of size n found so far. If this holds, then N can be safely pruned, as formalized by the following heuristic.

Heuristic 4. *Let N be an IR-tree node and m be the minimum subgroup size. Let sg_m be the best subgroup of size m corresponding to N, and min_cost_n be the smallest cost for the whole group of size n from any object retrieved so far. We can safely prune N if $f(cost(sg_m, N)) \geq min\_cost_n$.*

The proof is straightforward and thus omitted. Note that, while this heuristic simplifies the node pruning computation (Lines 12 to 17 in Algorithm 2), it also relaxes the pruning bound, which may cause more nodes to be processed.

5 Cost Analysis

We analytically compare the I/O and CPU cost of the algorithms. Let C_m be the maximum number of entries in a disk block, C_e be the effective capacity of the IR-tree used to index the dataset D, and $|D|$ be the size of D. We assume that an IR-tree node size equals a disk block. I/O cost and CPU cost of the preloading is denoted by io_i and cpu_i, respectively. We quantify the percentage of pruned nodes in the tree traversal as the pruning power, denoted by w and is represented by w_{gb}, w_{gf}, w_{sb}, w_{sf}, and w_{mb} for GNNK-BB, GNNK-BF, SGNNK-BB, SGNNK-BF, and MSGNNK-BF, respectively. We denote the I/O cost of accessing the inverted index by io_l, and the associated CPU cost by cpu_l. Per node CPU cost of GNNK-BB and GNNK-BF is denoted by cpu_g, SGNNK-BB and SGNNK-BF by cpu_s and MSGNNK-BF by cpu_m. Table 1 summarizes the analytical results. Calculation of cost analysis is omitted due to page limitation.

Table 1. Summary of costs

Algorithm	I/O	CPU
GNNK-BB	$io_i + (1-w_{gb})(\frac{|D|}{C_e-1} + \frac{|D|}{C_e} \cdot io_l)$	$cpu_i + (1-w_{gb})(\frac{|D|}{C_e-1} \cdot cpu_g + \frac{|D|}{C_e} \cdot cpu_l)$
GNNK-BF	$io_i + (1-w_{gf})(\frac{|D|}{C_e-1} + \frac{|D|}{C_e} \cdot io_l)$	$cpu_i + (1-w_{gf})(\frac{|D|}{C_e-1} \cdot cpu_g + \frac{|D|}{C_e} \cdot cpu_l)$
SGNNK-BB	$io_i + (1-w_{sb})(\frac{|D|}{C_e-1} + \frac{|D|}{C_e} \cdot io_l)$	$cpu_i + (1-w_{sb})(\frac{|D|}{C_e-1} \cdot cpu_s + \frac{|D|}{C_e} \cdot cpu_l)$
SGNNK-BF	$io_i + (1-w_{sf})(\frac{|D|}{C_e-1} + \frac{|D|}{C_e} \cdot io_l)$	$cpu_i + (1-w_{sf})(\frac{|D|}{C_e-1} \cdot cpu_s + \frac{|D|}{C_e} \cdot cpu_l)$
MSGNNK-N	$io_i + (n-m+1)(1-w_{sf})(\frac{|D|}{C_e-1} + \frac{|D|}{C_e} \cdot io_l)$	$cpu_i + (n-m+1)(1-w_{sf})(\frac{|D|}{C_e-1} \cdot cpu_s + \frac{|D|}{C_e} \cdot cpu_l)$
MSGNNK-BF	$io_i + (1-w_{mb})(\frac{|D|}{C_e-1} + \frac{|D|}{C_e} \cdot io_l)$	$cpu_i + (1-w_{mb})(\frac{|D|}{C_e-1} \cdot cpu_m + \frac{|D|}{C_e} \cdot cpu_l)$

6 Experimental Evaluation

6.1 Experimental Settings

We evaluate the performance of the proposed algorithms. The BB algorithms are used as the baseline to compare with the BF algorithms for the GNNK and SGNNK queries. We use the MSGNNK-N algorithm as the baseline algorithm for the MSGNNK queries, and compare it with the MSGNNK-BF algorithm.

Table 2. Dataset properties

Parameter	Flickr	Yelp
Dataset size	1,500,000	60,667
Number of unique keywords	566,432	783
Total number of keywords	11,579,622	176,697
Avg. number of keywords per object	7.72	2.91

We use two real datasets from Yahoo! Flickr and Yelp[1] (Table 2). We generate 20 groups of spatio textual query objects for each experiment and average the results. The parameters that are varied are shown in Table 3.

We use the IR-tree to index the datasets, which is disk resident. The fanout of the IR-tree is chosen to be 50, and the page size is 4 KB. All the algorithms are implemented in Java and the experiments are conducted on a Core i7-4790 CPU @ 3.60 GHz with 4 GB of RAM. We measure the running time and the I/O cost (number of disk page accesses), where the running time includes the computation and I/O time. We use Flickr as our default dataset.

Table 3. Query parameters

Parameter name	Values	Default value
Number of queried data points (k)	1, 10, 20, 30, 40, 50	10
Query group size (n)	10, 20, 40, 60, 80	10
Subgroup size (m, $\%n$)	40%, 50%, 60%, 70%, 80%	60%
Number of query keywords	1, 2, 4, 6, 8, 10	4
Size of the query space	.001%, .01%, .02%, .03%, .04%, .05%	0.01%
Size of the query keyword set	1%, 2%, 3%, 4%, 5%	3%
Spatial vs. textual preference (α)	0.1, 0.3, 0.5, 0.7, 1.0	0.5
Dataset size (Flickr)	1M, 1.5M, 2M, 2.5M	1.5M

---

[1] webscope.sandbox.yahoo.com, www.yelp.com/academic_dataset.

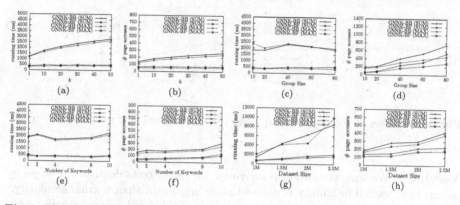

Fig. 2. The effect of varying k (a–b), query group size (c–d), number of query keywords (e–f), query keyword set size (g–h) and dataset size (i–j) in running time and I/O

6.2   The GNNK Query Algorithms

We conduct seven sets of experiments to evaluate the performance of GNNK-BB and GNNK-BF. In each set of experiments, one parameter is varied while all other parameters are set to their default values. GNNK-BF outperforms GNNK-BB in all experiments both in terms of running time and I/O cost.

Varying k. Figure 2(a–b) shows that for both GNNNK-BB and GNNK-BF, the processing time and the I/O cost increase with the increase of k. For both SUM and MAX, on average GNNK-BF runs 3.5 times faster than GNNK-BB. The I/O cost of GNNK-BF is much less than that of GNNK-BB as GNNK-BF only accesses the necessary nodes.

Varying Query Group Size. The query processing costs of both algorithms increase as the value of n increases (Fig. 2(c–d)). On average, GNNK-BF runs approximately 4 times faster than GNNK-BB.

Varying Number of Query Keywords. Figure 2(e–f) shows the effect of the number of keywords in each query object. GNNK-BF again outruns GNNK-BB in all the experiments. Also, the query processing costs of both algorithms increase as the number of keywords in each query object increases. This can be explained by that a larger set of query keywords takes more time to compute the aggregate cost function.

Varying Query Space Size. We observe that the running time of our algorithms remains almost constant with the change of the query space area (not shown in graphs). Since varied query space areas are insignificant in compared to the data space, we do not observe any significant change in this experiment.

Varying Query Keyword Set Size. We see that the running time of our algorithms do not follow any regular pattern with the change of the query keyword set size and remains relatively stable for varying of query keyword set size (the subset of keywords from where the query keywords are generated).

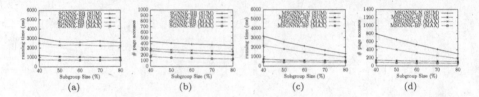

Fig. 3. The effect of varying subgroup size m (a–b) and minimum subgroup size (c–d)

Varying α. We observe that, as α increases, the query costs decrease. A larger α means that spatial proximity is deemed more important than textual similarity. When α increases, the impact of the keyword similarity becomes smaller and algorithms converge faster (not shown in graphs).

Varying Dataset Size. Figure 2(g–h) shows the effect of varying number of objects. Both running time and I/O cost of our proposed algorithms increase at a lower rate than the baseline algorithms.

6.3 The SGNNK Query Algorithms

We performed experiments on SGNNK-BB and SGNNK-BF, by varying query group size, subgroup size, number of query keywords, query space size, query keyword set size, k, dataset size, and α. SGNNK-BF outperforms SGNNK-BB in all the experiments. For space constraints, we only show the effect of varying the subgroup size (in % n) in Fig. 3(a–b). On average, SGNNK-BF runs 3.5 times faster and takes 40% less I/O than SGNNK-BB.

6.4 The MSGNNK Query Algorithms

In similar experiments, MSGNNK-BF significantly outperforms MSGNNK-N. Due to space constraints, we only show the effect of varying the minimum subgroup size (in percentage of n) in Fig. 3(c–d). When the minimum subgroup size increases, the running time of both algorithms decrease as expected. The smaller change in cost of MSGNNK-BF demonstrates its the better scalability. On average, MSGNNK-BF runs about 4 times faster than MSGNNK-N.

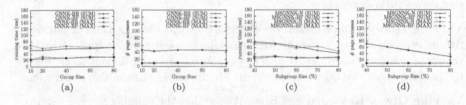

Fig. 4. The effect of varying group size (a–b) and minimum subgroup size (c–d)

6.5   Experiments on Yelp Dataset

We have run the same set of experiments as mentioned above on the Yelp dataset. All of our experimental results show similar trends in both datasets. Due to page limitations, we only present the experimental results for varying group size for GNNK queries and minimum subgroup size for MSGNNK queries with Yelp dataset in Fig. 4(a–b) and (c–d), respectively.

7   Conclusion

In this paper, we have presented a new type of group spatial keyword query suitable for a collaborative environment. This query aims to find the best POI that minimizes the aggregate distance and maximizes the text relevancy for a group of users. We have studied three instances of this query, which return (i) the best POI for the whole group, (ii) the optimal subgroup with the best POI given a subgroup size m, and (iii) the optimal subgroups and the corresponding best POIs of different subgroup sizes in $m, m + 1, ..., n$. In all these queries, our proposed best-first approach runs approximately 4 times faster (on average) than the branch and bound approach for both real datasets.

Acknowledgment. This research is partially supported by the ICT Division, Government of the People's Republic of Bangladesh. Jianzhong Qi is supported by The University of Melbourne Early Career Researcher Grant (project number 603049).

References

1. Ali, M.E., Khan, S.-u.-I., Khan, S.M.S., Nasim, M.: Spatio-temporal keyword search for nearest neighbour queries. J. Locat. Based Serv. **9**(2), 113–137 (2015)
2. Ali, M.E., Tanin, E., Scheuermann, P., Nutanong, S., Kulik, L.: Spatial consensus queries in a collaborative environment. TSAS **2**(1), 3:1–3:37 (2016)
3. Cao, X., Cong, G., Guo, T., Jensen, C.S., Ooi, B.C.: Efficient processing of spatial group keyword queries. TODS **40**(2), 13 (2015)
4. Cao, X., Cong, G., Jensen, C.S., Ooi, B.C.: Collective spatial keyword querying. In: SIGMOD, pp. 373–384 (2011)
5. Cong, G., Jensen, C.S., Wu, D.: Efficient retrieval of the top-k most relevant spatial web objects. PVLDB **2**(1), 337–348 (2009)
6. Guttman, A.: R-trees: a dynamic index structure for spatial searching. In: SIGMOD, pp. 47–57 (1984)
7. Hjaltason, G.R., Samet, H.: Distance browsing in spatial databases. TODS **24**(2), 265–318 (1999)
8. Li, Y., Li, F., Yi, K., Yao, B., Wang, M.: Flexible aggregate similarity search. In: SIGMOD, pp. 1009–1020 (2011)
9. Li, Z., Lee, K.C., Zheng, B., Lee, W.-C., Lee, D., Wang, X.: IR-tree: an efficient index for geographic document search. TKDE **23**(4), 585–599 (2011)
10. Papadias, D., Tao, Y., Mouratidis, K., Hui, C.K.: Aggregate nearest neighbor queries in spatial databases. TODS **30**(2), 529–576 (2005)

11. Roussopoulos, N., Kelley, S., Vincent, F.: Nearest neighbor queries. ACM SIGMOD Rec. **24**, 71–79 (1995)
12. Yao, K., Li, J., Li, G., Luo, C.: Efficient group top-k spatial keyword query processing. In: Li, F., Shim, K., Zheng, K., Liu, G. (eds.) APWeb 2016. LNCS, vol. 9931, pp. 153–165. Springer, Cham (2016). doi:10.1007/978-3-319-45814-4_13
13. Zhang, D., Chee, Y.M., Mondal, A., Tung, A.K., Kitsuregawa, M.: Keyword search in spatial databases: Towards searching by document. In: ICDE, pp. 688–699 (2009)

Surrounding Join Query Processing in Spatial Databases

Lingxiao Li[(⊠)], David Taniar, Maria Indrawan-Santiago, and Zhou Shao

Monash University, Melbourne, Australia
lli278@student.monash.edu,
{david.taniar,maria.indrawan,joe.shao}@monash.edu

Abstract. Spatial join queries play an important role in spatial data-base, and mostly all the distance-based join queries are based on the *range* search and nearest neighbour (NN), namely *range join query* and *kNN join query*. In this paper, we propose a new join query which is called *surrounding* join query. Given two point datasets Q and P of mul-tidimensional objects, the *surrounding* query retrieves for each point in Q its all surrounding points in P. As a new spatial join query, we propose algorithms that are able to process such query efficiently. Evaluation on multiple real world datasets illustrate that our approach achieves high performance.

Keywords: Spatial join · Spatial indexing · Spatial database · Nearest neighbour

1 Introduction

The spatial join query involves two datasets Q and P retrieves the object pairs from the Cartesian Product $Q \times P$ which satisfy a spatial predicate, From the theoretical point of view, the spatial join is similar as join that in the traditional database system domain. The main difference is join predicate, which can be intersection, topological, directional or distance, rather than simply equijoin. The intersection and distance-based join queries have been widely studied. A typical example of an intersection join is *"find all suburbs that are crossed by Southern Link Highway (M1), Western Link (M2) and East Link Highway (M3) in the city of Melbourne"*. In the example, we regard highway and suburb as spatial objects, line and polygon respectively. On the other hand, a case of distance-based join could be *"find all pairs of hotels and restaurants within 1 km apart"*. Both hotel and restaurant denote spatial point.

In this paper, we only focus on the distance-based join queries, most common join queries in this category are range join and kNN join. More specifically, the range join is a query for each query point finds all the target points that within the pre-specified range ϵ. In contrast, the kNN join query retrieves k nearest neighbours for each query points. However, both above queries have some main problems. For range join query, the result set cardinality is difficult to control.

© Springer International Publishing AG 2017
Z. Huang et al. (Eds.): ADC 2017, LNCS 10538, pp. 17–28, 2017.
DOI: 10.1007/978-3-319-68155-9_2

If the distance is defined too small or too large, the size of result set will change enormously in some situations. The problem of range join can be overcome by kNN join, which make sure each point in one dataset exactly combined with its k closest neighbours in the other dataset. However, if there is a cluster of points near all the query points, then the kNN join result is restricted to that scope [15]. Besides, before the query processing, we have to specify the range distance ϵ for range join and value of k for kNN join.

Inspired by these limitations, we introduce a new join query, called the Surrounding Join (SJ) query and propose efficient query processing techniques. This new join query is based on the surrounding query. A surrounding query is a query to retrieve all the nearest objects that surround the query object. Figure 1 shows an example of surrounding query. In the figure, the blue point X denotes a user's position, and the black dots (A to R) are all groceries in this suburb. From the perspective of this user, the query of surrounding groceries are the points {G, I, J, L, K, M}. If this user picks a surrounding grocery (for example I), which means that she doesn't need to know the groceries (A, B, H) behind the grocery I. In this case, A, B and H are dominated by I.

Fig. 1. An example of surrounding query (Color figure online)

In summary, the contributions of this paper are summarized as follows:

- We introduce surrounding join (SJ) query in spatial databases, which belongs to the distance-based join queries and involves spatial point data type.
- To solve the SJ queries, We propose two approaches; the first one is a straightforward algorithm that relies on a Voronoi diagram; The second approach is a hierarchical algorithm which prunes unnecessary nodes for obtaining the surrounding points. Meanwhile, it has higher performance.
- We have conducted extensive experimental studies on two real datasets that demonstrate the efficiency of our algorithms.

2   Related Work

In geospatial domain, the nearest neighbour (NN) query is to retrieve the points in the target dataset P that has shortest distance to a query point q. It has been widely used in many different type of queries, such as k Nearest Neighbor (kNN) [1,6,12], Reverse k Nearest Neighbor (RkNN) [7,14,16,17] and skyline query [2,9,13]. The existing NN algorithms always assume that the target dataset is indexed by an R-tree. In the R-tree index, the data point is completely and properly enclosed by a minimum bounding rectangle (MBR). In [11], Roussopoulos et al. proposed a algorithm to find the nearest neighbour object to a point, which is called *branch and bound R-tree traversal*. The metrics *MINDIST* denotes the minimum distance of point p_i to q, $p_i \in P$. The algorithm access the R-tree in a depth first (DF) manner. Starting from root node, and the entry with the smallest *MINDIST* is accessed first. The process is repeated recursively until the leaf node is visited where a potential NN is found. An optimal NN algorithm has been introduced in [10]. Consider query point q is center point and radius equals to the distance from q to its NN, then a *query circle* is created. In this case, the algorithm only traverses nodes whose MBR intersect with *query circle*.

The Voronoi diagram (VD) of a given set G of k points $\{g_1, g_2,...,g_m\}$ in a Euclidean plane partitions the space $\mathbb{R}^d$ into k regions. Each region contains a point g_i ($g_i \in G$) that is regard as generator point. The Euclidean distance form any other point in its region to g_i is smaller than to any other generator. Two generator points shares a Voronoi edge and three generator points form a Voronoi vertex. Existing algorithms for generating VD can be briefly divided into tree categories. The first category of algorithms are incremental algorithms, which create the VD by inserting a point at a time [5]. The second are divide and conquer algorithms. The set of points is divided into multiple parts, and VD of each part constructed recursively [19] The last category of algorithms compute VD by implementing the sweepline technique [4].

3   Problem Definition

The surrounding join (SJ) query is defined as below:

Definition 1 (*SJ Queries*). *Given a set Q of m query points q_1, q_2,...,q_m and a set P of n target points p_1, p_2,...,p_n, a SJ query $Q \underset{SJ}{\bowtie} P$ returns for each query point $q_i \in Q$, a sub-set $P\prime \subseteq P$. In terms of the sub-set $P\prime$, $\forall p_j\prime \in P\prime$ to the query point q_i has the shortest Euclidean distance in a particular direction. Meanwhile, $\forall p_j\prime \in P\prime$ is not dominated by the other points in $P\prime$.*

For a surrounding join query, we are going to find all the nearest target points that just surround each query point. As depicted in Fig. 2, three query points are respectively connected to its surrounding points.

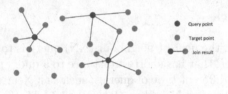

Fig. 2. Surrounding join

4   A Sketch-First Approach

Inspired by the Voronoi Diagram (VD), we can instantly get an idea that all the surrounding target points of each query point look like the adjacent vertexes in the VD [18]. Towards addressing this idea, a possible solution could be like this: Firstly, create a VD based on all the query and target points, and then for each query point q_i retrieves all the adjacent vertexes which surround this query point. Here, we assume that the adjacent vertexes are same as the surrounding points, which are what we need for the join query. To creating a VD, we apply the Fortune's sweep line algorithm [4], which guarantees the $O(n \log n)$ worst-case running time and uses $O(n)$ space. However, there are some limitations of this straightforward approach: (i) the data from two datasets need to be merged first and then sorted as the input of the algorithm. If two datasets are very big, which contains millions of spatial points; The sorting phase will take a substantial amount of computing resources and very inefficient. (ii) When we add, update, delete points in any dataset, the VD will be changed accordingly. It means that we have to create a new VD for the join queries.

Figure 3 depicts an example of the VD processing. The target points are denoted as A, B, C..., H. For simplicity and clarity, we only specify one query point P that is represented as red square. Obviously, the surrounding points of point P should be all adjacent vertexes, namely points $\{D, E, F, G, H\}$. In Fig. 3, we observe that each of these green points share a VD *edge* with query point P.

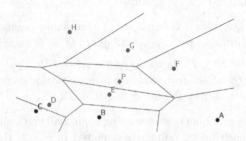

Fig. 3. VD approach (Color figure online)

5   Our Proposed Approach

In this section, we present our approach for the *surrounding join* queries computation, which is mainly composed of two parts: *Filter* and *Refinement*. More specifically, in *Filter* phase, we implement the global Branch and Bound skyline (GBBS) to prune all dominated points in target dataset. GBBS is an enhanced customization of the original *Branch and Bound Skyline* BBS algorithm [8]. Then, based on the skyline points from phase 1, a VD is created. The purpose the VD here is to help us retrieve all the adjacent points which surround the query point. We start by introducing the skyline and its variation *global-skyline*, and then we continue with a description of algorithms in *Filter* phase and *Refinement* phase.

5.1   Skyline

Give a set T of d-dimensional points, the original skyline operation returns all points in T are not dominated by any other point. More specifically, assume a point t_j is dominated by another point t_i, the condition is that coordinate of t_i on any axis is not greater than the corresponding coordinate of t_j, and strictly smaller in at least one axis. Informally, this implies that point t_i is preferable to t_j based on any real scenario. Figure 4(a) shows an example of original skyline in a two-dimensional space. Three solid dots A, B, C are skyline points which are dominating all the other points. If we refer to x and y axis to distance and price attribute, retroactively, and assume all the dots denote different restaurants. For instance, because point B dominates the point F, we can say restaurant B is better than restaurant F. The reason is that restaurant B is cheaper and closer than restaurant F. In short, the skyline of a multi-dimensional dataset encloses the *best* points according to any preference function that is monotone in each dimension [3].

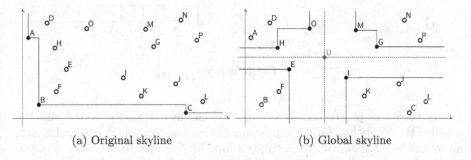

(a) Original skyline   (b) Global skyline

Fig. 4. Skyline

Original skyline considers the static attribute values of each data point in the dimensional space, and only examines one direction. Meanwhile, the query point is not involved into the operation. Since our aim is surrounding join, which means we need to consider all the directions rather than one direction. Besides, both query and target points should be taken into account. Accordingly, we apply *global*-skyline to solve our problem. As a variations of original skyline, *global*-skyline concerns about the potential targets points for each a query point, and returns all the points that are not *globally* dominated by other points. In other words, the *global*-skyline considers the directions of the processing, and the minimization of the coordinate distances between a query point and the target point is taken into account.

Figure 4(b) illustrates the *global*-skyline of query point U contains six points, H, O, M, G, I, E, which dominates the other points on all directions. Notice that these dominating points surround the query point U, which means no other target point is better than one of them with regard to U. Actually, these dominating points are the initial result of the surrounding join query. In the following section, we present the algorithm to generate these *global*-skyline points.

5.2   Filter Phase: GBBS Process

Same as NN and BBS, the GBBS join algorithm is also based on the nearest neighbour [11] search. Although all of these algorithms could be implemented by using data partition method, in this paper we use R-tree as index for target dataset due to its simplicity and popularity in spatial area. The set of 2-dimensional data points are used in Fig. 3, which is organized in the R-tree of Fig. 5.

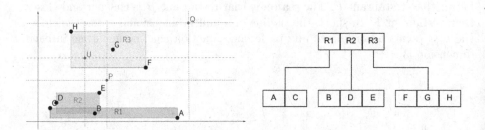

Fig. 5. R-tree

For the query processing, we take point P as the query point to describe the detail of the algorithm. GBBS starts from the root node of the R-tree and inserts all its entries (R_1, R_2, R_3) in a empty heap. The element in the heap is sorted by the Euclidean distance from point P as ascending order. Then, the entry with the minimum distance R_3 is expanded. This expansion prunes R_3 from the heap and add its children (F, G, H). Currently, the elements in the heap are (R_2, R_1, F, G, H). Then, the entry R_2 with minimum distance is expanded, and insert its children (E, B, D), in which the first nearest data point E appears.

Algorithm 1. GBBS Join Algorithm

Input: query points Q, R-tree R of all target points
Output: S. *list of target points*
1 initialization: $i \leftarrow \{\}$, $S \leftarrow \{\}$, insert all entries of R in H;
2 **while** $H.size() \neq 0$ **do**
3      $\quad e \leftarrow$ poll first element of H;
4      $\quad$ **if** e *is not leaf* **then**   // MBR, intermediate node
5          $\quad\quad$ **for** *each child* e_i *of* e **do**
6              $\quad\quad\quad$ **if** e_i *is not dominate by any item in* S **then**
7                  $\quad\quad\quad\quad |$ insert e_i to H;
8                  $\quad\quad\quad$ **end**
9          $\quad\quad$ **end**
10     $\quad$ **else if** e *is leaf* **then**   // leaf node
11         $\quad\quad$ **for** *each child* e_i *of* e **do**
12             $\quad\quad\quad$ **if** e_i *is not dominate by any item in* S **then**
13                 $\quad\quad\quad\quad |$ insert e_i to H;
14                 $\quad\quad\quad$ **end**
15         $\quad\quad$ **end**
16     $\quad$ **else**
17         $\quad\quad |$ insert e_i to S;
18     $\quad$ **end**
19 **end**
20 Return result list S

Point E belongs to the global skyline and is added to result list S. After we moved E from the heap to S. The first element in the heap is $R1$, which still is an intermediate node, not the real data node. GBBS proceeds with the $R1$ and inserts its children (A, C). The heap now becomes (G, B, F, D, C, H, A), and $S = \{E\}$. The algorithm processes in the same way until the heap becomes empty thus all global skyline points are added in the result list S. The join result of P in phase 1 is a list $\{E, F, G, H, A\}$. The join operation between the other two query point U, Q and target points will be processed as the same manner, and the result about U, Q are the list $\{B, E, F, G, H, D\}$ and $\{A, F, G, H\}$, respectively. The pseudo-code of GBBS is shown in Algorithm 1.

5.3   Refinement Phase: VD Process

In this phase, we consider the join result from step 1 is candidate result which needs further process. Therefore, we create VD based on the query point and skyline points which come from step 1. This means, as soon as we find the skyline S of query point q, we check if the point t in S is the adjacent of the q. If this is the case, we add this t to the final join result. Otherwise, we can safely prune point t. Note that the number of target points is much smaller than the number of the original dataset and consequently, the VD can be created much faster. For generating VD, we still use Fortune's sweep line algorithm [4].

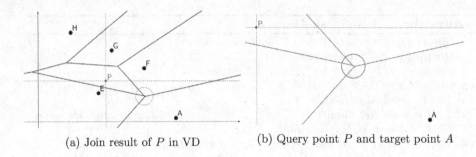

(a) Join result of P in VD   (b) Query point P and target point A

Fig. 6. Candidate result in VD

In the following discussion, we continue to use query point P to describe the detail of the refinement algorithm. So for, we have already got the skyline points of P is a list $\{E, F, G, H, A\}$. Therefore, we merge p and five skyline points together and create a VD. The VD is shown in the Fig. 6(a). Next, we retrieve the points around the cell of P. If a point share a same edge with P, then we add this point to reset list. For example, each of point E, F, G and H shares a VD edge with P, we can say the surrounding join result of P is the list of target points $\{E, F, G, H\}$. Note that, the point A is pruned in this phase. The reason is obvious, because A is not the adjacent of P. In Fig. 6(b), We can see there is an edge between P and A. For the other query points, the same process can be followed to get the join result. The pseudo-code of refinement is shown in Algorithm 2.

Algorithm 2. Refinement Algorithm

Input: query points Q, Skyline points S
Output: Result set R
1 initialization: Merger Q and Skyline points, generate VD;
2 **for** *each edge e of VD.edges* **do**
3   **if** *e.left is equal to Q* **then**
4   insert *e.right* into R;
5   **else if** *e.right equals Q* **then**
6   insert *e.left* into R;
7 **end**
8 Return result set R

At this point, we get the join result of query points U, P, Q and target points A, B, C, D, E, F, G, H. For Point P, if implement sketch-first approach, which is a pure VD approach, then we get the join result $P \rightarrow \{D, E, F, G, H\}$. The detail is shown in Sect. 4, Fig. 3. In contrast, in our improved approach, the join result of P is the $\{E, F, G, H\}$. If we compare these two lists, we can find that the first approach contains extra point D. Consider this case based on the perspective of the global-Skyline; we can see D is dominated by E. Note that, E is a surrounding point of P. Since there are no other existing approaches to answer the surrounding join query, and we are the first to propose these two possible solutions. Therefore, if we only consider these approaches, the second approach is more accurate than sketch-first approach.

6 Experiments

6.1 Experimental Setup

According to our literature research, in addition to our two approaches, there are no prior methods to process surrounding join queries. Therefore, we compare two proposed algorithms with each other to evaluate their performance. We refer to our sketch-first approach, improved approach as the *VDS* and *SVDS* in the following evaluation report, respectively. The experiments are performed on the real datasets which are road network of San Francisco and California. Both datasets are retrieved from the website[1]. For the input data, we randomly obtain 2000 points and set them as query points. Then, get rid of those 2000 points, we randomly generate five target datasets which contain 2000, 5000, 8000, 11000 and 14000 points, respectively. The experiments are repeated 100 times, and the average value is reported. All algorithms were implemented in JAVA and experiments were conducted on a Linux PC with 16 Intel Xeon E312xx 2 GHz CPUs and 64 GB main memory.

6.2 Experimental Results

In this section, we will evaluate two approaches of surrounding join query from four different perspectives, namely index construction, a diverse number of target points, a diverse number of query points and the detail of the second approach.

Evaluation on Index Construction: In the first approach *VDS*, we create a Voronoi Diagram first, then conduct the join operation on this VD. We assume the VD that stored in the main memory is a kind of index. Besides, The second approach *SVDS* is based on the R-Tree. Therefore, we compare the time and space consumption of the index construction of two approaches. Figure 7(a) illustrates the runtime of two index construction on same datasets. We specify the number of query point equals to 2000 and gradually increase the number of target points. The index creation time of *SVDS* is slighter faster than *VDS*. However, the VD-index need more memory space to store index as shown in Fig. 7(b). Specifically, the size of VD-index is about seven times that of SVDS-index, which dues to that VD-index involves both query and target datasets, and the structure is not good as R-Tree.

Evaluation on the Varied Number of Query and Target Points: We evaluate the overall performance of *VDS* and *SVDS* from two perspectives,

---

[1] http://www.cs.utah.edu/~lifeifei/SpatialDataset.htm.

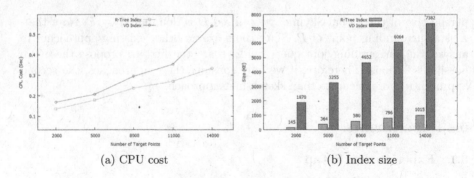

(a) CPU cost

(b) Index size

Fig. 7. Index construction

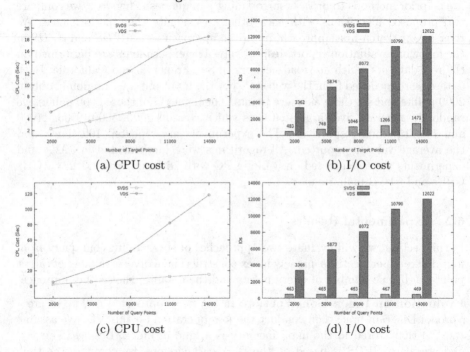

(a) CPU cost

(b) I/O cost

(c) CPU cost

(d) I/O cost

Fig. 8. Effect of varying number of points

CPU cost and average I/O cost. Figure 8(a) shows the CPU cost and Fig. 8(b) shows the I/O cost of each method for increasing the number of the target point. The number of query point is specified as 2000. When the number of target point increases, the CPU and I/O cost of both two approaches increase correspondingly. However, with the increment of the number of target points, the cost of *VDS* rises rapidly because it has to access all the points. On the other hand, we set target point equals to 2000 and increase the number of the query point. For this case, Fig. 8(c) and (d) show the processing time and I/O cost, respectively. The experimental results still indicate that *SVDS* is more efficient

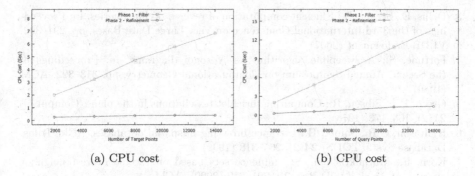

(a) CPU cost   (b) CPU cost

Fig. 9. Performance comparison on two phases of SVDS

than *VDS*. The main reason behind this is *SVDS* prunes all the dominated points in the filter phase. Nevertheless, *VDS* always access all the points.

Evaluation of *SVDS*: Based on the above evaluation, we understand the performance of *SVDS* is much better than *VDS*. Evaluating the *filter* and *refinement* phase in SVDS is necessary. In Fig. 9(a), we observe that the running time of *filter* phase increases gradually with the increase of the number of the target point. In contrast, the *refinement* phase remains almost constant during the whole experiment. The CPU cost of two phases is displayed in Fig. 9(b), which roughly illustrates the similar characteristics as Fig. 9(a). The reason is obvious. The candidate join result as the input for refinement phase which comes from filter phase is much smaller than the original size of target points.

7   Conclusion

In this paper, we introduced a new type of query, namely *Surrounding Join Query* that enables for each query point to identify the surrounding target points. It enriches the semantics of the conventional distance-based spatial join query. To efficiently process a *Surrounding Join Query*, we proposed two approaches. The first one, *VDS*, relies on the Voronoi Diagram. In contrast, the second approach, *SVDS* that combines the skyline and Voronoi Diagram to answer the query. Our experiments also illustrate that our algorithm has the capability to process the query efficiently. In the future, we are going to implement the surrounding for range join query and other spatial data types.

References

1. Berchtold, S., Ertl, B., Keim, D.A., Kriegel, H.-P., Seidl, T.: Fast nearest neighbor search in high-dimensional space. In: Proceedings of the 14th International Conference on Data Engineering, pp. 209–218. IEEE (1998)
2. Borzsony, S., Kossmann, D., Stocker, K.: The skyline operator. In: Proceedings of the 17th International Conference on Data Engineering, pp. 421–430. IEEE (2001)

3. Dellis, E., Seeger, B.: Efficient computation of reverse skyline queries. In: Proceedings of the 33rd International Conference on Very Large Data Bases, pp. 291–302. VLDB Endowment (2007)
4. Fortune, S.: A sweepline algorithm for Voronoi diagrams. In: Proceedings of the Second Annual Symposium on Computational Geometry, pp. 313–322. ACM (1986)
5. Green, P.J., Sibson, R.: Computing dirichlet tessellations in the plane. Comput. J. 21(2), 168–173 (1978)
6. Hjaltason, G.R., Samet, H.: Distance browsing in spatial databases. ACM Trans. Database Syst. (TODS) 24(2), 265–318 (1999)
7. Korn, F., Muthukrishnan, S.: Influence sets based on reverse nearest neighbor queries. ACM SIGMOD Rec. 29, 201–212 (2000). ACM
8. Papadias, D., Tao, Y., Fu, G., Seeger, B.: An optimal and progressive algorithm for skyline queries. In: Proceedings of the 2003 ACM SIGMOD International Conference on Management of Data, pp. 467–478. ACM (2003)
9. Papadias, D., Tao, Y., Greg, F., Seeger, B.: Progressive skyline computation in database systems. ACM Trans. Database Syst. (TODS) 30(1), 41–82 (2005)
10. Papadopoulos, A., Manolopoulos, Y.: Performance of nearest neighbor queries in R-trees. In: Afrati, F., Kolaitis, P. (eds.) ICDT 1997. LNCS, vol. 1186, pp. 394–408. Springer, Heidelberg (1997). doi:10.1007/3-540-62222-5_59
11. Roussopoulos, N., Kelley, S., Vincent, F.: Nearest neighbor queries. ACM SIGMOD Rec. 24, 71–79 (1995). ACM
12. Shao, Z., Cheema, M.A., Taniar, D., Lu, H.: Vip-tree: an effective index for indoor spatial queries. PVLDB 10(4), 325–336 (2016)
13. Sharifzadeh, M., Shahabi, C.: The spatial skyline queries. In: Proceedings of the 32nd International Conference on Very Large Data Bases, pp. 751–762. VLDB Endowment (2006)
14. Stanoi, I., Agrawal, D., El Abbadi, A.: Reverse nearest neighbor queries for dynamic databases. In: ACM SIGMOD Workshop on Research Issues in Data Mining and Knowledge Discovery, pp. 44–53 (2000)
15. Taniar, D., Rahayu, W.: A taxonomy for nearest neighbour queries in spatial databases. J. Comput. Syst. Sci. 79(7), 1017–1039 (2013)
16. Taniar, D., Safar, M., Tran, Q.T., Rahayu, W., Park, J.H.: Spatial network RNN queries in GIS. Comput. J. 54(4), 617–627 (2011)
17. Tao, Y., Papadias, D., Lian, X.: Reverse KNN search in arbitrary dimensionality. In: Proceedings of the Thirtieth International Conference on Very Large Data Bases, vol. 30, pp. 744–755. VLDB Endowment (2004)
18. Xuan, K., Zhao, G., Taniar, D., Safar, M., Srinivasan, B.: Voronoi-based multi-level range search in mobile navigation. Multimedia Tools Appl. 53(2), 459–479 (2011)
19. Yap, C.K.: Ano (n logn) algorithm for the Voronoi diagram of a set of simple curve segments. Discrete Comput. Geom. 2(1), 365–393 (1987)

Boosting Point-Based Trajectory Search
with Quad-Tree

Munkh-Erdene Yadamjav[1]($^{(\boxtimes)}$), Sheng Wang[1], Zhifeng Bao[1], and Bang Zhang[2]

[1] RMIT University, Melbourne, Australia
{munkh-erdene.yadamjav,sheng.wang,zhifeng.bao}@rmit.edu.au
[2] DATA61 — CSIRO, Canberra, Australia
mattbang.zhang@data61.csiro.au

Abstract. The availability of spatial data generated by objects enables people to search for a similar pattern using a set of query points. In this paper, we focus on point-based trajectory search problem which returns top-k results to a set of query points. The primary purpose of this work is to revisit state-of-the-art search algorithms on various indices and find the best choice of spatial index while giving a reason behind it. Furthermore, we propose an optimization on the search method, which is able to find the initial upper bound for the query points, leading to further performance improvement. Lastly, extensive experiments on real dataset verified the choice of the index and our proposed search method.

Keywords: Spatial database · Trajectory search · Indexing technique

1 Introduction

Mobile devices equipped with GPS, such as smart phones, have generated a large amount of geo-location data and *trajectory* data which is composed of multiple points from a same device. The availability of trajectory data enables us to develop useful applications such as route planning and trip recommendation [5, 10]. Finding similar trajectories for the given query also contributes significantly to users explore valuable resources nearby [12–15].

Specifically, Chen et al. [1] first proposed a new search called *k-Best Connected Trajectories* (k-BCT), which searches k trajectories for the query composed by a set of locations. k-BCT searches for trajectories that best connect given set of query points, the distance between the trajectory and query points is calculated using the distance between each query point and its nearest point of the trajectory. Figure 1a shows a set of query points and trajectories. k-BCT chooses (q_1, p_{31}), (q_2, p_{32}), (q_3, p_{33}) as the shortest matching points to calculate the distance between trajectory T_3 and the query points.

IKNN(Incremental k-NN) [1] is first proposed to answer k-BCT query by searching the nearest trajectory points to each query point independently. It checks k-BCT using the trajectory points retrieved in each iteration and continues until reaching terminating condition. After the candidate generation

© Springer International Publishing AG 2017
Z. Huang et al. (Eds.): ADC 2017, LNCS 10538, pp. 29–41, 2017.
DOI: 10.1007/978-3-319-68155-9_3

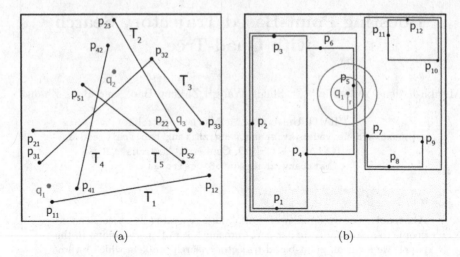

Fig. 1. (a) Trajectory and a set of query points (b) Expansion of query on R-Tree

ends, the candidate refinement stage iterates over partially matched trajectories to compute actual distance to query points. Further, GH/QE (Global Heap/Qualifier Expectation) [11] and SRA (Spatial Range-based Approach) [8] algorithms are proposed to tackle the problem of multiple visit of tree nodes, which causes performance degradation (more details in Sect. 2).

Motivation. A number of studies have been done for answering k-BCT query efficiently by proposing different candidate generation and refinement techniques. However, algorithms proposed by [1,8,11] all work on top of R-Tree [4], while R-tree does not work well in the range-based or nearest neighbor-based expansion for candidates. This motivates us to study how much the performance can be improved by using other indexing structures, such as Grid [6], Quad-tree [3]. These indexing techniques perform better when doing the expansion to get new candidates (more details on Sect. 3). Moreover, existing approaches for k-BCT all choose a random query to evaluate the algorithms. This motivates us to test the scalability of each method by various queries in substantially crowded areas, such as using crowded area more than 10K points in 1 Km radius around the query point.

Contributions. In this paper, we study the performance of well-known trajectory search algorithms in dense populated dataset which has more points around each query point in a real dataset called Geolife [16]. Based on the observations, we study how the best algorithm performs on top of different indexing structure which are R-Tree, Grid, and Quad-Tree. We propose how to compute the initial upper bound of the search region for each query point to avoid poor performance caused by outlier location. Extensive experiments show that our algorithm performs better than other state-of-the-art algorithms. Our contributions can be summarized as follows:

- We employ a labeled Quad-tree to avoid the drawbacks of using R-tree, i.e., expansion without candidates and time consuming in the traversal from root to leaf.
- We optimize the initial search upper bound to save computational cost.
- We conduct extensive experiments to verify our newly proposed index and optimizations.

Outline. The rest of the paper is organized as follows. We discuss the problem formulation and our motivation in detail in Sect. 2. In Sect. 3, we presented our algorithm on top of Quad-Tree index. We present our experiment results in Sect. 4 and conclude in Sect. 5.

2 Background and Problem Formulation

2.1 Problem Definition

We formally define the problem of k-BCT search. Table 1 summarizes the notation used throughout this paper.

Table 1. Summary of notations

Notation	Definition
D	Trajectory database
T	Trajectory
Q	A set of query points
C	A set of candidate trajectories
L	A set if labels assigned to Quad-Tree leafs
q_i	Query point
p_i, p_j	Trajectory points
$Euclidean(p_i, q_j)$	Euclidean distance between p_i and q_j points
$d(T, Q)$	Distance between trajectory T and query Q
UB_k	Distance of kth-best connected trajectory
r_i	Currently explored search radius for q_i
r_{total}	Total explored search radius for all query points

Definition 1 (Trajectory). *A Trajectory T in dataset D is defined as a sequence of n spatial points $\{p_1, p_2, \ldots, p_n\}$ where each point is represented by [latitude, longitude].*

Definition 2 (Query). *The input query Q (of size m) of k-BCT search over the collection D is defined as a set of query points $Q = \{q_1, \ldots, q_m\}$.*

Definition 3 (Matching Pair). *If* $\forall p_k \neq p_i, Euclidian(p_i, q_j) <$ *Euclidian*(p_k, q_j) *we say* $[p_i, q_j]$ *is the shortest matching pair of* q_j *and* T. *The distance between the trajectory and the query point is calculated using the matching pairs of the trajectory and query points.*

Definition 4 (Trajectory to Query Distance). *We define the distance between trajectory* T *and query* Q *as the aggregated sum of the shortest matching pair of each query point* q_j. $d(T, Q) = \sum_{q_j \in Q} d(T, q_j) = \sum_{q_j \in Q} Euclidian(p_i, q_j)$.

Definition 5 (k-Best Connected Trajectory). *k-BCT query finds the best result set* $S_k = \{T_1, T_2, \ldots, T_k\}$ *trajectories from the dataset* D *where* $\forall T_i \in S_k, \forall T_j \in D - S_k, \ d(T_i, Q) \leq d(T_j, Q)$.

The goal of k-BCT query is to find k trajectories that have the shortest aggregated distance to a set of query points. For instance, the distance between the trajectory T_3 and the query points in Fig. 1a can be computed as: $d(T_3, Q) = d(p_{31}, q_1) + d(p_{32}, q_2) + d(p_{33}, q_3) = 10 + 32 + 12 = 54$.

2.2 Preliminary

There have been three algorithms [1,8,11] to answer k-BCT query over spatial database. These methods follow the procedure of conducting *expansions* based on basic spatial index to scan candidate points from closest to furthest, until all valid candidates have been found. Then, the scanned candidate trajectories are *refined* by computing real distance.

Table 2 summarizes three methods that we compare the performance with our solution. We briefly compare the solutions for the nearest neighbor-based algorithms which are IKNN [1] and GH/QE [11] and the spatial range-based approach which is SRA [8]. Specifically, we further distinguish them in terms of expansion and terminating condition.

Expansion Method. Chen et al. [1] uses round robin technique to generate candidates for each query point in the query set Q by conducting k-NN [9] search. Candidate set is incrementally updated based on the result set of k-NN search, where k increases in each iteration. Since a large number of points are scanned multiple times due to several rounds of expansions with k-NN [1,11],

Table 2. Comparison of related works [1,8,11] to our work.

Method	Expansion	Termination	Index	Query generation
IKNN [1]	Repetitive	Bound	R-Tree	Random
GH/QE [11]	Repetitive	k Full-match	R-Tree	Random
SRA [8]	Non-repetitive	Bound	R-Tree	Random
Our solution	Non-repetitive	Bound	Quad-tree	Density-sensitive

Qi et al. [8] proposed a range-based expansion method based on R-tree instead of k-NN search. It avoids repetitive scanning by only visiting new candidates based on range search over R-tree.

However, we find that R-tree is not the best index structure for range-based expansion in terms of reducing I/O operation and the number of candidates. Through analysis and experiments in Sect. 4, we found that Quad-tree can perform well in reducing I/O operation and the number of candidates.

Terminating Condition. Chen et al., Qi et al. [1,8] employed the bound computation, i.e., comparing the lower bound of current results based on expansions to the upper bound of unscanned trajectories, which is similar to [2]. The scanning stops when the lower bound becomes bigger than the upper bound. For the bound computation, we further describe it in Sect. 3.

Tang et al. [11] takes another way to terminate, it checks whether k full matched trajectories exist which have been scanned all matching points. Once the algorithm finds k full matches, the candidate generation terminates as the lower bounds of all unseen trajectories are always bigger than k^{th} full match.

3   Quad-Tree Based Approach

The main idea of most query processing methods can be divided into two parts [1,8,11]: (1) one is filtering the unscanned trajectories which are impossible to be the results, this part is done by comparing the upper bound distance of existing result and the lower bound distance of unscanned trajectories. (2) it refines the scanned candidates by computing their real distances to the query.

To generate candidates in the filtering stage, existing state-of-the-art algorithms conduct a range-based expansion over R-tree [4]. A circle-range query with a given radius finds new leaf nodes that are inside the range as we can see in the Fig. 1b. However, there are two drawbacks of using R-tree to conduct the incremental expansions.

- It depends on the radius of a range. In some cases, no candidate can be returned due to the fixed incremental radius as shown in Fig. 1b.
- It needs to traverse from the root to all leaf nodes inside the range to retrieve, which requires extra time on I/O.

Thus, we propose to use *Label-Tree* structure which expands the search range dynamically based on the density of POIs.

3.1   Label-Tree

Quad-Tree index divides the whole space into four equal-sized square regions. The capacity to store maximum number of points is defined before the creation of Quad-Tree. If the number of points to be indexed in given region exceeds the capacity, that particular region is divided further into four equal-sized square regions.

However, the main challenge here is that existing Quad-tree does not support accessing the neighbor leaf nodes of the current node. To access efficiently to the neighbors for a given leaf node, we build a *Label-Tree* for all the leaf nodes inside a Quad-tree, which can address the problems of using R-Tree on a range-based search.

Everytime we split the region, top-left, top-right, bottom-left and bottom right sub regions are assigned 1, 2, 3, 4 values respectively as shown in Fig. 2a. The label length increases as we further divide a region into sub regions. This labeling structure requires much less space compared to the actual dataset. Neighbor nodes can also guarantee an upper bound for the unseen trajectories. Based on the *Label-Tree*, we can expand the search radius dynamically. This can avoid an expansion without candidates and finding too many candidate points like a range-based search in [8].

3.2 Our Algorithm

The pseudocode of our proposed algorithm is shown in Algorithm 1. We call our algorithm *Quad-Tree based Spatial Range Approach* (SQRA). Candidate set and total radius are initialized in Line 1. The median point for all query points is computed in Line 2. Line 3 computes initial k^{th} upper bound distance using k-th nearest point to the median point while SRA algorithm uses sum-ANN approach [7]. Each radius is initialized ξ which is the half length of the smallest square in Quad-Tree index in Line 4–6. Line 7 indicates stop condition of the candidate generation. We choose query point with the smallest number of candidates points in Line 8. We only retrieve points of labeled Quad-Tree nodes where the query point resides in first iteration.

Starting from second iteration, we search for *Label-Tree* nodes that intersect with the query region of length $2 * r_c$ centred at q_c. This condition is represented in Line 9–14. Points are retrieved from the disk using labels that have not been checked in previous iteration in Line 15. Line 16 updates search radius in next iteration for query point q_c based on the length of labels found in last iteration. Candidate trajectory set is updated in Line 17.

A new upper bound is calculated using current candidate set C according to Eq. 1 in Line 18. Line 19 computes total search range using current search range of each query point according to Eq. 2. Algorithm refines results in Line 21 and returns k-BCT in line 22. The main differences of our algorithm and the state-of-the-art algorithm lie in three aspects: (1) The initial upper bound calculation in Line 2–3, (2) Retrieve candidate points that intersect with the search region using *Label-Tree* in Line 9–15, and (3) Lower bound calculation using search radius in line 16.

Bound Computation. The upper bound distance between T_i and Q:

$$UB(T_i, Q) = \sum_{q_j \in Q_i} d(p_{ij}, q_j) + \sum_{q_j \in Q \setminus Q_i} d(T_i.p_j, q_j) \tag{1}$$

Algorithm 1. Quad-Tree based Search Range Approach (SQRA)

Input: Trajector database D, set of query points Q, number of results k
Parameters: candidate set C, label set L, k-th distance upper bound UB_k,
               current r_i search radius for each $q_i \in Q$, total search radius r_{total},
               set of labeled Quad-tree nodes N
Output: top-k list of trajectories R

1   initialize $C \leftarrow \emptyset, L \leftarrow \emptyset, r_{total} \leftarrow 0$;
2   $q_{median} \leftarrow$ computeMedian(Q);
3   compute UB_k invoking a k-NN(q_{median}, N);
4   **foreach** $q_j \in Q$ **do**
5 | initialize $r_j \leftarrow \xi$
6   **end**
7   **while** $r_{total} \leq UB_k$ **do**
8   select q_c with minimum candidate points;
9   **if** $r_c = \xi$ **then**
10 | $L \leftarrow$ find the label where the region contains q_c;
11   **end**
12   **else**
13 | $L \leftarrow$ find labels where the region intersects with the region of r_c
         | centered at q_c ;
14   **end**
15      $S \leftarrow retrievePoints(L)$;
16      $r_c \leftarrow$ update radius using the minimum length of intersecting region ;
17   Update C with S;
18   Update UB_k;
19   Update r_{total} with r_c;
20   **end**
21   $R \leftarrow Refine(k, T, Q, C)$;
22   return R;

---

where p_{ij} denotes a trajectory point that already matched and $T_i.p_j$ denotes the closest trajectory point to q_j found so far. Then, the lower bound of unseen trajectories is computed as:

$$LB = \sum_{q_i \in Q} r_i \qquad (2)$$

where r_i denotes the search range for query point q_i.

Figure 2a shows the algorithm execution. We search for Top-1 result with three given query points. Each candidate trajectory contains same number of slots with query points. We fill each slot of candidate trajectory based on the points we have retrieved so far. Point p_{22} of trajectory T_2 is the nearest point from the computed median point of all query points. Thus, the initial $UB_k = 112 + 68 + 10 = 190$.

We first choose query point q_1 as current search point. We compute Quad-Tree label 3.3 using q_1 location and find point p_{11}. Trajectory T_1 is added to candidate set. The slots are filled as follows $T_1(p_{11}, p_{11}, p_{11})$. Aggregated distance

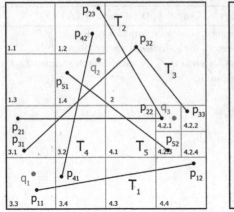 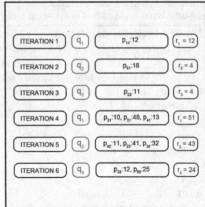

(a) Quad-tree Index (Node capacity of 2)   (b) Running example using SQRA

Fig. 2. An example of Index and Search method

of T_1 to query Q is $d(T_1, Q) = 12 + 112 + 115 = 239$. The search radius is set to $r_{q_1} = 12$. $r_{total} = 12 + 0 + 0 = 12$. We continue as $r_{total} < UB_k$. Algorithm continues until the sixth iteration. Iterations from 2 to 5 can be seen from Fig. 2b.

Sixth iteration chooses query point q_3 and finds two points p_{33}, p_{52} using labels 4.2.2 and 4.2.3. The search radius is set to $r_{q_1} = 24$. The slots of the corresponding candidates are updated as follows $T_3(p_{31}, p_{32}, p_{33}), T_5(p_{51}, p_{51}, p_{52})$. Aggregated distances of the trajectories to query Q are: $T_3 = 10 + 32 + 12 = 54, T_5 = 83 + 18 + 25 = 126$. As we found trajectory with shorter distance, $UB_k = 54$. Total radius of search area is $r_{total} = 51 + 43 + 24 = 118$. We stop candidate generation as $r_{total} > UB_k$. Our Top-1 result will be T_3 where $d(T_3, Q) = 54$.

4   Experiment

4.1   Setup

Dataset. We conducted our analysis using GPS trajectory dataset that was collected in (Microsoft Research Asia) Geolife project. GeoLife [16] dataset contains 17,620 trajectories and 23,642,416 points. We filtered the search space by the latitude value of 39.0 and 41.0 and the longitude value of 116.0 and 118.0 to generate densely populated new dataset. This filtered dataset contains 19,476,949 points of 16438 unique trajectories. The latitude and longitude of each point is normalized into [0,1] range to build a Grid index.

Parameters. Experimental parameters are shown in the following table. Default values are shown in bold (Table 3).

Table 3. Parameter setting

Description	Parameter	Values
Number of results	k	1, 5, **10**, 50, 100
Number of query points	$|Q|$	2, 4, **6**, 8, 10
Search range	r	0.01
Nearest neighbor increment	λ	1000
R-Tree node capacity	Fanout	200
Quad-Tree node capacity	Fanout	1000
Grid dimension	d	128×128

Table 4. Grid density

Number of unique POIs	Color
Less than 10	
Between 10 and 10K	
Between 10K and 50K	
More than 50K	

Table 5. Index footprint

Index	Size/MB/	Height
R-Tree	918	4
GRID	767	N/A
Quad-Tree	968	16

Query Generation. We choose query points randomly from the grid cells that have more than 10K points to evaluate how algorithms scale with respect to the number of candidates to process. Grid presentation can be seen from Fig. 3 and Table 4. Each query execution, I/O cost and the number of candidates have been averaged for all 100 queries.

Comparisons. We test the following methods to answer k-BCT queries on Geo-Life dataset. (1) **IKNN:** the nearest neighbor based algorithm on top of R-Tree indexing structure (2) **GH:** the nearest neighbor based algorithm on top of R-Tree indexing structure (3) **SRA:** the baseline algorithm that we optimized to increase the performance. (4) **SRA-G:** the baseline algorithm SRA on top of Grid indexing structure. (5) **SRA-Q:** the baseline algorithm SRA on top of Quad-Tree index. (6) **SQRA:** the proposed algorithm on top of Quad-Tree index structure. All algorithms are implemented in JAVA and experiment is done on a machine with Intel Core i7-2630 2.0 GHz CPU and 8 GB main memory running Windows 10.

4.2   Data Profiling

Figure 3 shows that our dataset is non-uniformly distributed. We show grid cells with different colors based on the number of unique points in the cell. Table 4 shows the color representation of the grid density.

(a)   (b)

Fig. 3. POI distribution in Geolife dataset according to Table 4

4.3 Evaluation

Overview. In summary, our main observations are (Table 5):

1. Our proposed algorithm performs up to 3 times faster than the best performing algorithm.
2. Grid index can save more time than R-Tree index when we search for small number of results.
3. Quad-Tree based index always save I/O cost by applying a dynamic range-based search compared to R-Tree index. Moreover, Quad-Tree based index generates less candidates than other two indices.
4. The initial upper bound calculation will slow down the algorithm performance when the query area is big or the density is high.

Effect of Indexing Structure. Figure 4 shows SRA algorithm's execution, I/O cost, the number of results on different indexing structure. It can be seen that I/O cost directly influences on overall query performance. SRA-Q algorithm generates less candidates as this method expands the search region dynamically

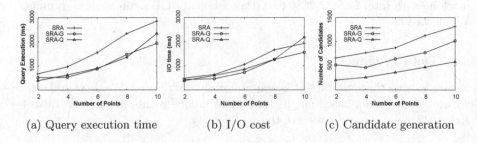

(a) Query execution time (b) I/O cost (c) Candidate generation

Fig. 4. Scalability test of SRA on varying number of query points

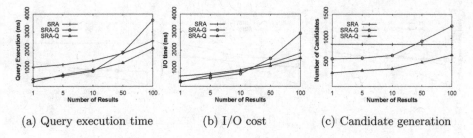

(a) Query execution time   (b) I/O cost   (c) Candidate generation

Fig. 5. Scalability test of SRA on varying number of results

depending on the point density as shown in Fig. 4c. Grid index performs more I/O operation when the number of query points increases. The reason is that Grid index retrieves all points in the cell without considering the distance to the query point. Quad-tree based index always performs better than R-Tree in terms of I/O cost, computation cost and the number of candidates as shown in Fig. 4 and 5.

Effect of $|Q|$. Figure 6 shows the query execution time, I/O cost and the number of candidates for different numbers of query points. Query execution and I/O cost show an increasing trend as we conduct queries with more points for all algorithms. However, our proposed algorithm based on Quad-Tree always performs better than other algorithms in terms of query execution and I/O cost as shown in Fig. 6a and b. The nearest neighbor based approach generates less candidates than the range-based approach as shown in Fig. 6c.

Effect of k. Figure 7 shows how the different number of results influences on query execution time and I/O cost. Fixed range for SRA algorithm retrieves more points than other algorithms with few iterations. Query execution time of algorithm highly depends on I/O cost as you see the correlation between Fig. 7a and b. More candidates increase the overall query execution time. Our proposed algorithm performs better than other algorithms as shown in Fig. 7a.

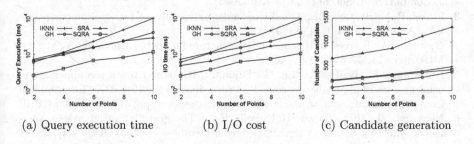

(a) Query execution time   (b) I/O cost   (c) Candidate generation

Fig. 6. Scalability test on varying number of query points

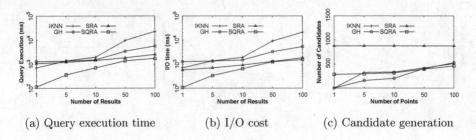

(a) Query execution time   (b) I/O cost   (c) Candidate generation

Fig. 7. Scalability test on varying number of results

5   Conclusion

We studied the problem of efficiently answering k-BCT query on spatial database with low computational and I/O cost in this paper. By utilizing the characteristics of a Quad-Tree, we modified the indexing structure used by state-of-the-art algorithms and constructed a *Label-Tree*. We proposed SQRA algorithm based on the *Label-Tree* and further optimized query processing. Experiment showed that our algorithm performs up to three times faster than the best performing algorithm. As future work, we plan to investigate how to compute tighter bounds to accelerate candidate generation phase.

Acknowledgment. Zhifeng Bao is partially supported by ARC DP170102726 and Google Faculty Research Award. Munkh-Erdene Yadamjav is a recipient of Data61 PhD Scholarship.

References

1. Chen, Z., Shen, H.T., Zhou, X., Zheng, Y., Xie, X.: Searching trajectories by locations: an efficiency study. In: Proceedings of the 2010 ACM SIGMOD International Conference on Management of Data, pp. 255–266. ACM (2010)
2. Fagin, R., Lotem, A., Naor, M.: Optimal aggregation algorithms for middleware. J. Comput. Syst. Sci. **66**(4), 614–656 (2003)
3. Finkel, R.A., Bentley, J.L.: Quad trees a data structure for retrieval on composite keys. Acta Inform. **4**(1), 1–9 (1974)
4. Guttman, A.: R-trees: a dynamic index structure for spatial searching. ACM SIGMOD Rec. **14**, 47–57 (1984). ACM
5. Kurashima, T., Iwata, T., Irie, G., Fujimura, K.: Travel route recommendation using geotags in photo sharing sites. In: Proceedings of the 19th ACM International Conference on Information and Knowledge Management, pp. 579–588. ACM (2010)
6. Nievergelt, J., Hinterberger, H., Sevcik, K.C.: The grid file: an adaptable, symmetric multikey file structure. ACM Trans. Database Syst. (TODS) **9**(1), 38–71 (1984)
7. Papadias, D., Tao, Y., Mouratidis, K., Hui, C.K.: Aggregate nearest neighbor queries in spatial databases. ACM Trans. Database Syst. (TODS) **30**(2), 529–576 (2005)

8. Qi, S., Bouros, P., Sacharidis, D., Mamoulis, N.: Efficient point-based trajectory search. In: Claramunt, C., Schneider, M., Wong, R.C.-W., Xiong, L., Loh, W.-K., Shahabi, C., Li, K.-J. (eds.) SSTD 2015. LNCS, vol. 9239, pp. 179–196. Springer, Cham (2015). doi:10.1007/978-3-319-22363-6_10
9. Roussopoulos, N., Kelley, S., Vincent, F.: Nearest neighbor queries. ACM SIGMOD Rec. **24**, 71–79 (1995). ACM
10. Shang, S., Ding, R., Yuan, B., Xie, K., Zheng, K., Kalnis, P.: User oriented trajectory search for trip recommendation. In: Proceedings of the 15th International Conference on Extending Database Technology, pp. 156–167. ACM (2012)
11. Tang, L.-A., Zheng, Y., Xie, X., Yuan, J., Yu, X., Han, J.: Retrieving k-nearest neighboring trajectories by a set of point locations. In: Pfoser, D., Tao, Y., Mouratidis, K., Nascimento, M.A., Mokbel, M., Shekhar, S., Huang, Y. (eds.) SSTD 2011. LNCS, vol. 6849, pp. 223–241. Springer, Heidelberg (2011). doi:10.1007/978-3-642-22922-0_14
12. Vlachos, M., Kollios, G., Gunopulos, D.: Discovering similar multidimensional trajectories. In: Proceedings of the 18th International Conference on Data Engineering, pp. 673–684. IEEE (2002)
13. Wang, S., Bao, Z., Culpepper, J.S., Sellis, T., Sanderson, M., Yadamjav, M.-E.: Interactive trip planning using activity trajectories. In: ADCS, pp. 77–80 (2016)
14. Wang, S., Bao, Z., Culpepper, J.S., Sellis, T., Sanderson, M., Qin, X.: Answering top-k exemplar trajectory queries. In: IEEE 33rd International Conference on Data Engineering (ICDE), pp. 597–608. IEEE (2017)
15. Zheng, K., Shang, S., Yuan, N.J., Yang, Y.: Towards efficient search for activity trajectories. In: IEEE 29th International Conference on Data Engineering (ICDE), pp. 230–241. IEEE (2013)
16. Zheng, Y., Xie, X., Ma, W.-Y.: Geolife: a collaborative social networking service among user, location and trajectory. IEEE Data Eng. Bull. **33**(2), 32–39 (2010)

Query Processing

Query Refinement for Correlation-Based Time Series Exploration

Abdullah M. Albarrak$^{(\boxtimes)}$ and Mohamed A. Sharaf

University of Queensland, Brisbane, Australia
{a.albarrak,m.sharaf}@uq.edu.au

Abstract. In this paper, we focus on the problem of exploring sequential data to discover time sub-intervals that satisfy certain pairwise correlation constraints. Differently than most existing works, we use the deviation from targeted pairwise correlation constraints as an objective to minimize in our problem. Moreover, we include users preferences as an objective in the form of maximizing similarity to users' initial sub-intervals. The combination of these two objectives are prevalent in applications where users explore time series data to locate time sub-intervals in which targeted patterns exist. Discovering these sub-intervals among time series data is extremely useful in various application areas such as network and environment monitoring.

Towards finding the optimal sub-interval (i.e., optimal query) satisfying these objectives, we propose applying query refinement techniques to enable efficient processing of candidate queries. Specifically, we propose QFind, an efficient algorithm which refines a user's initial query to discover the optimal query by applying novel pruning techniques. QFind applies two-level pruning techniques to safely skip processing unqualified candidate queries, and early abandon the computations of correlation for some pairs based on a monotonic property. We experimentally validate the efficiency of our proposed algorithm against state-of-the-art algorithm under different settings using real and synthetic data.

1 Introduction

Exploration of time series data [10,16,17] is present in many domains (e.g., for environment monitoring, network traffic analysis, etc.) and is a key ingredient of various analysis tasks such as detecting patterns or anomalies among multiple time series [6,8]. Pearson's correlation [3,4,11,12] is widely considered to be a powerful tool for performing such analysis tasks, as it reveals the true pairwise similarity between any time series pairs.

However, computing correlation of time series based on the whole time interval [9,13] (rather than sub-intervals) is vulnerable to the classical Yule-Simpson effect [1]. At the same time, computing correlation for all sub-intervals is a much harder problem for users, since the number of sub-intervals increases quadratically with the length of time series. The following toy example shows the usefulness and the challenges of computing correlation of time series data based on

© Springer International Publishing AG 2017
Z. Huang et al. (Eds.): ADC 2017, LNCS 10538, pp. 45–58, 2017.
DOI: 10.1007/978-3-319-68155-9_4

	timestamp														
	6:00	7:00	8:00	9:00	10:00	11:00	12:00	13:00	14:00	15:00	16:00	17:00	18:00	19:00	20:00
T_1	0.2	0.3	0.3	0.5	0.8	0.9	0.8	0.9	0.7	0.8	0.4	0.3	0.3	0.2	0.1
T_2	0.15	0.2	0.15	0.22	0.41	0.54	0.48	0.49	0.36	0.43	0.19	0.17	0.16	0.11	0.06
T_3	0.05	0.1	0.15	0.28	0.21	0.13	0.32	0.41	0.34	0.37	0.21	0.13	0.14	0.09	0.04

Fig. 1. Relation R stores hourly CPU load readings of three connected servers T_1, T_2 and T_3 in a hypothetical data center.

	T_1	T_2	T_3		T_1	T_2	T_3
T_1		1	0.8	T_1		0.98	-0.55
T_2			0.7	T_2			-0.47
T_3				T_3			
		M_{Q1}				M_{Q2}	

Fig. 2. Correlation matrices M_{Q_1} and M_{Q_2} computed from Q_1 and Q_2 outputs

sub-intervals. This example is prevalent in data center management systems [7] where users analyze servers' loads collectively (e.g., Query 3 and 4 in [13]) using the pairwise correlation of all servers' loads.

Example 1 (**Data Center Monitoring System**): Assume a hypothetical data center with 3 connected servers T_1, T_2, T_3, where T_1 is responsible for forwarding incoming requests to T_2 and T_3 as evenly as possible. The hourly CPU load readings of these servers are stored in a database table R, as shown in Fig. 1.

Further, assume an admin who wants to detect any abnormal behavior based on the pairwise correlation of the servers' loads. Let this abnormal behaviour be: T_1's load increases but T_2's or T_3's loads simultaneously decreases.

Consequently, a correlation matrix M is created to assist in the automatic detection of this abnormal behavior. To create M, a selection query is executed $Q_1 : \sigma_{6 \leq timestamp \leq 20}(R)$; then its output is used to compute the pairwise correlation of the pairs $(T_1, T_2), (T_1, T_3)$ and (T_2, T_3) in M. The correlation matrix M_{Q_1} for Q_1 is shown in Fig. 2 for the whole time interval of the series.

It appears that no abnormal behavior exists within the results of Q_1: the loads of T_2 and T_3 do follow the same pattern as their parent T_1, which the high pairwise correlation values in M_{Q_1} confirm. Nonetheless, between 9 and 12 there is somehow an abnormal behavior: T_3's load breaks the pattern and decreases while T_1's load increases. This abnormal pattern is captured by the following query $Q_2 : \sigma_{9 \leq timestamp \leq 12}(R)$; and its matrix M_{Q_2} is shown in Fig. 2. ∎

In Example 1, it is assumed that the abnormal behaviour (i.e., matrix M_{Q_2}) is well-known by the admin. However, the time sub-interval for which M_{Q_2} was produced is what the admin explores for. Hence, she would have to submit queries for all possible sub time intervals and manually examine the correlation matrices to find Q_2. Specifically, there is a total of $\frac{m(m-1)}{2}$ possible queries, and this number increases quadratically with the length of time series m. Manually examining these queries is a labor intensive task which leads to users frustration and adds unnecessary overload to the database system [18].

A more suitable solution is for the user to define the target correlation matrix (e.g., M_{Q_2} in Example 1) that represents an abnormal behavior and a hint of where this abnormal behavior might be at (e.g., Query Q_1 in Example 1), and the system automatically finds the query (i.e., sub-interval) that outputs this matrix. This is an instance of the Query Refinement problems [2], where the goal is to automatically refine a user's query until its result satisfies her expectations. That is, based on Example 1, the goal is to refine Q_1 (i.e., its time interval) until its result produce M_{Q_2}. This simple yet computationally challenging problem is the focus of this paper.

Challenges: The task of automatically finding a query with a target correlation matrix is computationally challenging because: (1) an algorithm has to go through all possible candidate queries, which increase quadratically with the length of time series, (2) the number of pairs in the correlation matrix increases quadratically as well with the number of time series, (3) computing correlation from scratch hinder the exploration process, while caching some results to boost correlation computations is limited by the amount of available memory.

Optimization Opportunities: We propose to take advantage of a monotonic propriety to avoid processing some of the candidate queries by applying two-level pruning techniques, and to cache some of the computed results which might be helpful for incrementally computing correlation of later queries.

Our contributions are as follows:

- We formally defined the problem of Query Refinement based on Targeted Correlation Matrix (Sect. 2).
- Then, we proposed the QFind algorithm based on the classical graph traversal method BFS to be as solutions for the problem (Sect. 3).
- Further, we optimized our algorithm by applying two-level pruning techniques, which utilize the monotonicity property to avoid processing of unqualified queries and to early abandon the computations of pairwise correlation for some of the pairs in the correlation matrix (Sect. 4).
- Finally, we conducted extensive experiments to show the efficiency of our proposed algorithms on synthetic and real datasets, and compared the results to state-of-the-art algorithm (Sect. 5).

2   Problem Formulation

In this section we formally define the Query Refinement based on Targeted Correlation Matrix (**QuReLat**) problem. We assume the presence of n synchronized, equal length time series stored in a flat relational table $R : \{T_1, T_2, ..., T_n\}$ where each $T_i \in R$ contains m real values $\{v_1^i, v_2^i, ..., v_m^i\}$ such that v_j^i is the j-th value with time stamp j in T_i.

Users explore R by submitting SQL range queries on the `timestamp` attribute to select a time interval $[s, e]$ of the time series in R, then further analyze the results based on the pairwise correlation of all time series in R, i.e., correlation matrix M. Next, we define the basic notions of this problem, then later give the formal problem definition.

Definition 1. *Q is a range selection query on the* `timestamp` *attribute:*

$$\sigma_{s \leq timestamp \leq e}(R)$$

such that $1 \leq s \leq e \leq m$.

For ease of readability, a query will be denoted as $Q[s, e]$, or Q if the time interval $[s, e]$ can be omitted. Refining $Q[s, e]$ implies modifying its time interval. Hence, the number of possible refined queries is $\frac{m(m-1)}{2}$, which increases quadratically with the length of time series m. This observation renders the problem at hand to be computationally hard because the correlation matrix M will be computed for each one of these queries.

Definition 2. *Correlation matrix M is a symmetric matrix of size $n \times n$. Each entry $M[i][j] \in M$ is precisely the pairwise correlation of T_i and T_j: $\rho(T_i, T_j)$.*

Definition 3. *The pairwise correlation $\rho(T_i, T_j)$ of length l is measured by the Pearson's coefficient ρ:*

$$\rho(T_i, T_j) = \frac{l \sum_{k=1}^{l} v_k^i v_k^j - \sum_{k=1}^{l} v_k^i \sum_{k=1}^{l} v_k^j}{\sqrt{l \sum_{k=1}^{l} (v_k^i)^2 - (\sum_{k=1}^{l} v_k^i)^2} \sqrt{l \sum_{k=1}^{l} (v_k^j)^2 - (\sum_{k=1}^{l} v_k^j)^2}} \tag{1}$$

We are now in place to formally define the problem at hand:

Definition 4. *Query Refinement based on Targeted Correlation Matrix Problem (QuReLat):* *Given an input query Q_I and a target correlation matrix M_t. QuReLat's goal is to automatically refine Q_I to Q^* such that $f(Q^*)$ is maximized.*

$$f(Q^*) = f(Q_I, Q^*, M_t) = \lambda P(Q_I, Q^*) + (1 - \lambda)C(Q^*, M_t) \tag{2}$$

$$P(Q_I, Q^*) = 1 - \frac{1}{1 + e^{-d(Q_I, Q^*)}} \tag{3}$$

$$C(Q^*, M_t) = 1 - (\frac{1}{z} \sum_{i=0}^{n-1} \sum_{j=i+1}^{n} (M_t[i][j] - M_{Q^*}[i][j])^2) \tag{4}$$

where z is a normalization factor, and:

$$d(Q_I, Q^*) = |Q_I.s - Q^*.s| + |Q_I.e - Q^*.e| \tag{5}$$

As stated in Definition 4, the optimal solution Q^* is the one with the maximum preference $P()$ and the maximum closeness to the target correlation matrix $C()$, balanced by a user parameter λ. Ensuring maximum preference of Q^* to Q_I is useful when users are interested for a time interval that is close from Q_I's interval. Similarly, ensuring maximum closeness to M_t is important to maximally achieve the target.

Modeling the user preference $P()$ as a Sigmoid function on the `timestamp` attribute has two advantages: it is a parameter-free function, and it expresses users interests to the input query: at the beginnings everything close by Q_I seems interesting, though once moving away from Q_I (i.e., $d(Q_I, Q^*)$ increases), all other queries seem rather unrelated. As for $C()$, we use the Sum of Square Errors (SSE) since it indicates the tightness of M_t to a matrix M_Q of a candidate query Q. Its normalized value ranges between [0–1], where a small value denotes a tight fit of M_Q to the target M_t.

Towards finding the optimal solution Q^*, we propose an efficient search algorithm called **QFind**. QFind adopts the classical graph traversal strategy: Breadth First (BFS) which allows for innovative optimization techniques to be incorporated in to efficiently find Q^* without a compromise on the solution accuracy. Details are in the following section.

3   Methodology

We propose an efficient search algorithm called (**QFind**) as a solution for the **QuReLat** problem. In short, QFind starts by the input query Q_I then recursively *refine* it to obtain the next candidate queries (Sect. 3.1). The order in which QFind visits the next query is determined by the traditional traversal strategy Breadth First Strategy (BFS). Employing BFS enables QFind to incrementally compute M (Sect. 3.2), i.e., incrementally computing the pairwise correlation of every candidate query, which leads to considerable cost savings (Sect. 3.3). Further, QFind applies two simple yet powerful pruning techniques (Sect. 4) to enable far more efficient processing of the search space. These techniques enable QFind to avoid processing unqualified queries and to early abandon the correlation computations of unpromising pairs in M.

As shown in Fig. 3, QFind starts by the input query Q_I then recursively applies four refinement operations on the current query to obtain the next set of candidate queries. Next section explains how and on what those refinement operations are applied.

3.1   Queries Refinement

To refine a query Q with a selection predicate on the `timestamp` attribute, two operations are applied on that time interval $[s, e]$:

1. **Expansion:** to expand $[s, e]$ from either sides s or e by δ. For instance, $[\hat{s}, e]$ is expanded from s side by δ such that $\hat{s} = s - \delta$ while $[s, \hat{e}]$ is expanded from e side by δ such that $\hat{e} = e + \delta$. We encode those two operations as $\mathcal{LE}$ (left expansion) and $\mathcal{RE}$ (right expansion).

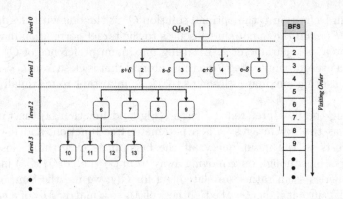

Fig. 3. QFind employs the classical traversal strategy: Breadth First (BFS) to decide the visiting order of the candidate queries in the search space starting from the input query Q_I.

2. **Contraction:** to contract $[s, e]$ from either sides s or e by δ. For instance, $[\hat{s}, e]$ is contracted from s side by δ such that $\hat{s} = s + \delta$ while $[s, \hat{e}]$ is contracted from e side by δ such that $\hat{e} = e - \delta$. Similarly, we encode those two operations as $\mathcal{LC}$ (left contraction) and $\mathcal{RC}$ (right contraction).

With those two refinement operations, QFind is able to recursively generate all possible $m(m - 1)/2$ combinations of candidate queries. Specifically, QFind applies $\mathcal{LE}, \mathcal{RE}, \mathcal{LC}, \mathcal{LE}$ on $[s, e]$ to generate the offspring ($[s + \delta, e]$, $[s, e + \delta]$, $[s - \delta, e]$ and $[s, e - \delta]$) then iteratively apply them again on the offspring and so on. To remove any approximation and to ensure no possible candidate queries are missed, we set $\delta = 1$.

For each candidate query Q, an algorithm has to compute the correlation of all pairs (i.e., M) within Q's interval, which can be done incrementally as explained next.

3.2 Caching Essential Arrays

Based on the observation that Eq. 1 can be computed incrementally [15], we propose to cache the essential arrays: $\sum x$, $\sum x^2$ and $\sum xy$ of a query after its correlation matrix has been evaluated and only if it happened to have offspring. This enables computations reusing when computing M for Q's offspring later on, and lead to cost savings as we experimentally show later in Sect. 5.1.

The essential arrays of a query Q are added to memory by storing them into a simple data structure (e.g., a hash table) called $\mathcal{H}$, indexed by Q's time interval. Each component is a 1-dimensional array of size n, except $\sum xy$ which is a 2-dimensional array of size $n(n - 1)/2$.

3.3 Cost Model Analysis

Similar to [9,15], we focus on the computational bound costs involved when searching for Q^* because computing M is extremely expensive even if we completely ignore the I/O cost.

1. **Number of Operations (OP):** OP is the number of operations to compute a correlation matrix M, and it depends on the length ℓ and the number n of the time series in Q. Specifically, for each pair of time series T_i, T_j of length ℓ in M, the five summation components in $\rho(T_i, T_j)$ will require exactly $(\ell-1) + (\ell-1) + 2(\ell-1) + 2(\ell-1) + 2(\ell-1) = 8(\ell-1)$ operations. Hence, M requires a total of $\frac{n(n-1)}{2} \times 8(\ell-1)$ operations.
2. **Maximum Size of Memory (MaxMemory):** After evaluating a query Q, its essential arrays $\sum x$, $\sum x^2$ and $\sum xy$ are cached in memory $\mathcal{H}$. With the assumption of a limited space for caching results, it is crucial for an algorithm to minimize the size of $\mathcal{H}$. Hence, an algorithm should release a query from $\mathcal{H}$ once it is expired, i.e., its cached arrays will not be needed anymore. Thus, we consider the maximum size of $\mathcal{H}$ as a cost of refinement.

3.4 QFind Algorithm

Algorithm 1 illustrates the main steps followed by QFind when searching for the optimal solution Q^*. Using a queue $\mathcal{Q}$, QFind orders the candidate queries to be visited following a BFS strategy. QFind stops the search when $\mathcal{Q}$ becomes empty, i.e., no more candidate queries to be evaluated (line 3).

For each candidate query Q, QFind looks up $\mathcal{H}$ for a suitable overlap to incrementally compute the matrix M_Q. If no such overlap exists in $\mathcal{H}$, then M_Q is computed from scratch (line 7). While traversing the search space, QFind remembers the best solution Q^* and its f_b (lines 12–13). Once QFind finishes evaluating the current query, it evokes the auxiliary function `refine()` (line 14) to generate the offspring of Q (Sect. 3.1) and inserts them into the queue $\mathcal{Q}$ as long as they have never been visited nor already in the queue.

Adding Arrays to $\mathcal{H}$ and Searching for Overlap: QFind adds the essential arrays of a query Q to $\mathcal{H}$ after fully evaluating Q, provided that Q has offspring. When QFind picks unvisited query Q to be evaluated, it looks for the best overlapping query Q_o of Q in $\mathcal{H}$ based on their time intervals, then merges Q_o's essential arrays with the new (and hopefully few) read values of Q. The best overlapping query Q_o of Q is the one with the minimum number of steps away from Q (ideally, the parent of Q).

Remove Arrays from $\mathcal{H}$: QFind removes the essential arrays of a query Q from $\mathcal{H}$ once it has no benefit in future steps. The removal is triggered when the current query's parent is different from the previous query's parent.

Algorithm 1. QFind

Require: Input query $Q_I[s, e]$, preference weight λ, target correlation M_t
Ensure: Q^*, f_b
 1: $f_b = -\infty; Q^* = \phi; \mathcal{Q} = \phi; \mathcal{H} = \phi;$
 2: $\mathcal{Q}.\text{add}(Q_I);$
 3: **while** ($\mathcal{Q} \neq \phi$) **do**
 4:     $Q = \mathcal{Q}.\text{pop}();$
 5:     $Q_o = \mathcal{H}.\text{findOverlap}(Q);$
 6:     **if** ($Q_o = \phi$) **then**
 7:         $Q = \text{probe}(Q);$
 8:     **else**
 9:         $Q = \text{merge}(Q_o, Q);$
10:     update $\mathcal{H};$
11:     $f = \lambda P(Q_I, Q) + (1 - \lambda)C(M_Q, M_t);$
12:     **if** ($f > f_b$) **then**
13:         $f_b = f; Q^* = Q;$
14:     $C = \text{refine}(Q);$
15:     **for all** $Q \in C$ **do**
16:         $\mathcal{Q}.\text{push}(Q);$
17:     release($\mathcal{H}$);
18: **return** $Q^*, f_b;$

---

4   Two-Level Pruning

We further extend QFind to utilize a monotonic property at two levels: preference level, and pairwise correlation level. The former enables efficient pruning of the search space with no false dismissal, which addresses the quadratic search space $\mathcal{O}(m^2)$, while the latter addresses the quadratic number of pairs in a correlation matrix instance $\mathcal{O}(n^2)$ by abandoning the computation of correlation for pairs as early as possible.

4.1   Preference-Aware Pruning Technique

QFind applies a simple yet powerful pruning technique to avoid visiting unpromising queries. This technique makes use of the monotonic property of the preference function $P()$, i.e., Eq. 3.

Lemma 1. $P()$ *is a monotonic decreasing function.*

Proof. It is easy to see from Fig. 3 that all candidate queries at level i have the same $P()$ since they are at the same distance from the root, i.e. $d(Q_I, Q_i)$. Also, it is clear that candidate queries at level $i + 1$ are at one extra step from the queries in the previous level, hence, their $P()$ is lower. This pattern continues through out the whole tree. Hence, $P()$ is a monotonic decreasing function in terms of levels.

With Lemma 1, QFind is able to early terminate the search and abandon all candidate queries that are yet to be explored once the current query's estimated $f()$ (i.e., f_e) is worse than the best solution found so far f_b. Specifically, if the current query's estimated objective $f_e = \lambda P() + (1 - \lambda) \times 1$ is lower than f_b, then this query, its offspring, the remaining queries in the queue, and all other unvisited queries will definitely have lower f than f_b and can no longer increase f_b further, thus they can be abandoned.

4.2   Pairwise Correlation Pruning Technique

Even with an arbitrary ordering of candidates, QFind is able to utilize a monotonic decreasing property of the correlation function $C()$, i.e., Eq. 4, to early abandon the correlation computations of some pairs in M.

Lemma 2. *$C()$ is a monotonic decreasing function.*

Proof. Recall that $C(M_t, M_Q)$ is the normalized sum of all absolute differences of correlation values between a target matrix M_t and a given one M_Q. By assuming M_Q to be an exact replica of M_t (which returns the maximum value of $C(M_t, M_Q)$), and iterating over all pairs in M_Q and inserting their real values, $C(M_t, M_Q)$ will gradually decrease. Hence, $C(M_t, M_Q)$ is a monotonic decreasing function in terms of number of pairs.

For a candidate instance Q, QFind assumes that M_Q is an exact replica of M_t, then whenever the algorithm inserts a real correlation value for a pair in M_Q, it simultaneously checks if $f_e = \lambda P() + (1 - \lambda) C(M_t, M_Q)$ is lower than f_b. If so, the rest of unevaluated pairs are skipped and the algorithm moves on to the next candidate query.

The order which QFind follows in examining the pairs in M_Q is crucial. QFind should follow an ordering that enables more pruning of pairs in M_Q.

Systematic Ordering: The default ordering of pairs in QFind is a systematic ordering (SYS). As its name suggests, SYS ordering is as follows: $\{M_Q[i][j] | i = 1, 2, ..., n - 1; j = i + 1, i + 2, ..., n; i < j\}$. Hence, QFind examines the pairs for all candidate queries in the same exact order.

Greedy Ordering: Another more intuitive ordering is to rearrange the pairs in M_Q in an ascending order based on their scores. The score of a pair $M_Q[i][j]$ is its distance d to the corresponding pair in M_t. Hence, the first pair to be examined under this greedy ordering is the one with the maximum distance to the target M_t. The intuition behind this ordering is to increase the chances of hitting the threshold f_b to early abandon the computations of correlation of the pairs in M_Q, as explained above in Sect. 4.2. However, since M_Q is not known until it is examined, QFind utilizes the history of the computed correlation values to estimates d. That is, QFind exploits a reference matrix M_r that approximates the current one M_Q, and computes the distance d from that reference.

This reference matrix M_r can be either static or dynamic. In the static case, M_r is an exact copy of the input query's correlation matrix (M_{Q_I}), while in the dynamic case M_r is an exact copy of the current query's parent matrix. Consequently, the static case (REF) implies QFind to compute the order only once (after computing M_{Q_I}), then reusing it for all remaining candidate queries. However, the dynamic case (REF-DY) implies QFind to re-compute the order every time the parent of a current query changes, which entails further cost overhead when compared to REF and SYS.

5   Experiments

We have performed extensive experiments to evaluate the performance of our proposed algorithms. Before discussing the results in Sect. 5.1, we explain the experiments setup.

Evaluated Algorithms: We experimented with the variants of QFind (Algorithm 1): QFind-PFP (Preference pruning, Sect. 4.1), QFind-PWC (pairwise pruning, Sect. 4.2), and QFind-PFP-PWC, under the default SYS ordering. Note that state-of-the-art algorithm ZES [5] is extended to cater for more than one pair of time series. The algorithms were implemented using Java SDK and run on a Windows machine with 16 GB RAM and Intel i7 CPU 3.0 GHz.

Datasets: In our experiments, we used two datasets: synthetic and a real dataset. The synthetic dataset was generated according to a Random Walk model, while the real dataset was extracted from Google Cluster Usage Data [14]. There are a total of $n = 1000$ time series, and their maximum length is $m = 1000$.

Workload: a workload consists of a set of runs. Each run is a trio: input query, target correlation matrix and a user preference (Q_I, M_t, λ). A query's time interval $[s, e]$ length is either: short, medium or long. The interval is also either on the left hand side, right hand side, or in the middle of the original time series interval. Q_I is generated by random from these $2^3 + 1$ classifications. M_t is arbitrarily chosen from a query that is generated from the above classifications to guarantee an exact solution existence. Hence, the size of the workload is $(2^3 + 1)^2$.

We report the cost components defined in Sect. 3.3 and vary the parameters n, m and λ in the experiments. Their default values are 30, 500 and 0.05 respectively.

5.1   Results

We test the scalability of our proposed algorithms in terms of n, their sensitivity to the user preference weight λ, and the ordering of pairs effects on cost.

Scalability Results: Figure 4 shows the results from the Google Cluster Dataset experiments. Figure 4a and b indicates that measuring the number of operations is in fact a suitable performance indicator since it is correlated with the execution time. Our proposed algorithm QFind-PFP-PWC is able to reduce the computational costs by almost %50 than state-of-the-art, as shown in

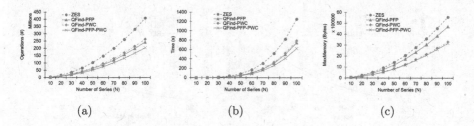

(a) (b) (c)

Fig. 4. Scalability results: $\lambda = 0.005$ and $m = 200$

Fig. 4a. As for the maximum size required to cache the essential arrays, QFind-PFP-PWC also reduces it by almost %42, as shown in Fig. 4c. This shows how powerful the two level pruning techniques are in avoiding unnecessary computations. Specifically, QFind-PFP reduces the computational costs by almost %40 by pruning unqualified candidate queries using Lemma 1. Similarly, QFind-PWC utilizes Lemma 2 to early abandon the computation of some pairs in the correlation matrix, reducing its cost by %35.

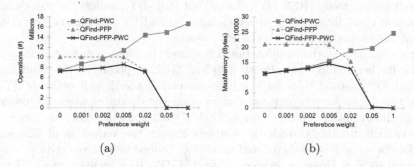

(a) (b)

Fig. 5. Sensitivity results: $n = 20$ and $m = 200$

Sensitivity to λ Results: Figures 5a and b show an interesting relationship between the user preference weight λ and the cost components. As mentioned previously, $P()$ is a monotonic decreasing function, i.e., it decreases as the algorithm moves away from the root (Q_I), which is a simple yet powerful property that QFind utilizes to prune the unqualified queries. This is apparent in both figures, as more weight is assigned to $P()$, i.e., increasing λ, QFind-PFP prunes more unqualified queries, resulting into reducing the number of operations and the amount of memory required. However, QFind-PWC becomes less efficient in its correlation level pruning technique, since the threshold f_b becomes more loose due to the less weight $C()$ gets as $P()$'s weight is increased. QFind-PFP-PWC which combines the two techniques achieves the best of these two techniques, and is more (or at least) as efficient as the other two versions QFind-PFP and QFind-PWC.

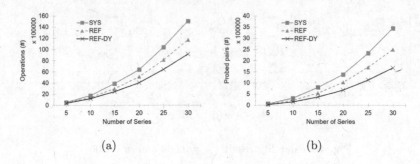

(a)   (b)

Fig. 6. Ordering of pairs results: $\lambda = 0$ and $m = 200$

Ordering of Pairs Results: We further examined how ordering of pairs can effect the cost components of our algorithm QFind-PFP-PWC using a smaller dataset. As briefly mentioned in Sect. 4.2, the order which QFind follows in examining the pairs in M can have an effect on the overall performance. Figures 6a and b show the computational cost (operations) and number of probed pairs for three different approaches of ordering: systematic (SYS), greedy (REF) and a dynamic greedy (REF-DY). Recall that REF-DY reorders the pairs when the current candidate query's parent changes, while REF performs this ordering once at the beginning based on the input query.

From the figures, the computational cost and the number of examined pairs can be further reduced by almost %40 and %52, respectively, if QFind uses the REF-DY method to order the pairs instead of the default ordering SYS. While this seems very promising, recomputing the distances when the parent of a candidate query changes entails further computational cost, as shown in Table 1. REF ordering provides a relatively competitive reduction of %22 and %30 for the computational cost and number of probed pairs, respectively, when compared to SYS. However, in contrast to REF-DY, REF incurs zero additional cost.

Table 1. Additional computational costs of REF-DY

Number of pairs	10	45	105	190	300	435
Operations (#)	33788	153264	370228	680697	1131465	1671644

6 Conclusions

Motivated by the prevalent need to support exploring the continuously growing time series data, we proposed the Query Refinement based on Targeted Correlation Matrix problem. Then, we proposed the QFind algorithm as an efficient solution. QFind extends state-of-the-art and applies innovative pruning techniques to avoid processing unqualified candidate solutions. We also showed the performance gains of QFind under different experimental settings. The functionality of QFind can be integrated with the correlation engines in TSDBMS such as Gorilla [11] to enable far more exploration capabilities.

References

1. Blyth, C.R.: On simpson's paradox and the sure-thing principle. J. Am. Stat. Assoc. **67**(338), 364–366 (1972)
2. Chaudhuri, S.: Generalization and a framework for query modification. In: Proceedings of the Sixth International Conference on Data Engineering, Los Angeles, California, USA, 5–9 February 1990, pp. 138–145 (1990)
3. Gavrilov, M., Anguelov, D., Indyk, P., Motwani, R.: Mining the stock market (extended abstract): which measure is best? In: Proceedings of the Sixth ACM SIGKDD International Conference on Knowledge Discovery and Data Mining, Boston, MA, USA, 20–23 August 2000, pp. 487–496 (2000)
4. Guo, T., Sathe, S., Aberer, K.: Fast distributed correlation discovery over streaming time-series data. In: Proceedings of the 24th ACM International Conference on Information and Knowledge Management, CIKM 2015, Melbourne, VIC, Australia, 19–23 October 2015, pp. 1161–1170 (2015)
5. Li, Y., Huo U, L., Yiu, M.L., Gong, Z.: Efficient discovery of longest-lasting correlation in sequence databases. VLDB J. **25**(6), 767–790 (2016)
6. Lin, J., Keogh, E.J., Lonardi, S., Lankford, J.P., Nystrom, D.M.: Visually mining and monitoring massive time series. In: Proceedings of the Tenth ACM SIGKDD International Conference on Knowledge Discovery and Data Mining, Seattle, Washington, USA, 22–25 August 2004, pp. 460–469 (2004)
7. Liu, J., Terzis, A.: Sensing data centres for energy efficiency. Philos. Trans. R. Soc. Lond. A Math. Phys. Eng. Sci. **370**(1958), 136–157 (2012)
8. Matsubara, Y., Sakurai, Y., Ueda, N., Yoshikawa, M.: Fast and exact monitoring of co-evolving data streams. In: 2014 IEEE International Conference on Data Mining, ICDM 2014, Shenzhen, China, 14–17 December 2014, pp. 390–399 (2014)
9. Mueen, A., Nath, S., Liu, J.: Fast approximate correlation for massive time-series data. In: Proceedings of the ACM SIGMOD International Conference on Management of Data, SIGMOD 2010, Indianapolis, Indiana, USA, 6–10 June 2010, pp. 171–182 (2010)
10. Palpanas, T.: Data series management: the road to big sequence analytics. SIGMOD Rec. **44**(2), 47–52 (2015)
11. Pelkonen, T., Franklin, S., Cavallaro, P., Huang, Q., Meza, J., Teller, J., Veeraraghavan, K.: Gorilla: a fast, scalable, in-memory time series database. PVLDB **8**(12), 1816–1827 (2015)
12. Rakthanmanon, T., Campana, B.J.L., Mueen, A., Batista, G.E.A.P.A., Westover, M.B., Zhu, Q., Zakaria, J., Keogh, E.J.: Searching and mining trillions of time series subsequences under dynamic time warping. In: The 18th ACM SIGKDD International Conference on Knowledge Discovery and Data Mining, KDD 2012, Beijing, China, 12–16 August 2012, pp. 262–270 (2012)
13. Reeves, G., Liu, J., Nath, S., Zhao, F.: Managing massive time series streams with multiscale compressed trickles. PVLDB **2**(1), 97–108 (2009)
14. Reiss, C., Wilkes, J., Hellerstein, J.L.: Google cluster-usage traces: format + schema. Technical report, Google Inc., Mountain View, CA, USA, November 2011

15. Sakurai, Y., Papadimitriou, S., Faloutsos, C.: BRAID: stream mining through group lag correlations. In: Proceedings of the ACM SIGMOD International Conference on Management of Data, Baltimore, Maryland, USA, 14–16 June 2005, pp. 599–610 (2005)
16. Tabachnick, B.G., Fidell, L.S.: Using Multivariate Statistics, 5th edn. Allyn & Bacon Inc., Needham Heights (2006)
17. Utomo, C., Li, X., Wang, S.: Classification based on compressive multivariate time series. In: Cheema, M.A., Zhang, W., Chang, L. (eds.) ADC 2016. LNCS, vol. 9877, pp. 204–214. Springer, Cham (2016). doi:10.1007/978-3-319-46922-5_16
18. Vartak, M., Raghavan, V., Rundensteiner, E.A.: Qrelx: generating meaningful queries that provide cardinality assurance. In: Proceedings of the ACM SIGMOD International Conference on Management of Data, SIGMOD 2010, Indianapolis, Indiana, USA, 6–10 June 2010, pp. 1215–1218 (2010)

A Multi-way Semi-stream Join
for a Near-Real-Time Data Warehouse

M. Asif Naeem[1], Kim Tung Nguyen[1(✉)], and Gerald Weber[2]

[1] School of Engineering, Computing and Mathematical Sciences,
Auckland University of Technology, Auckland, New Zealand
`mnaeem@aut.ac.nz, jvc0109@autuni.ac.nz`
[2] Department of Computer Science, The University of Auckland,
Auckland, New Zealand
`gerald@cs.auckland.ac.nz`

Abstract. Semi-stream processing, the operation of joining a stream of data with non-stream disk-based master data, is a crucial component of near real-time data warehousing. The requirements for semi-stream joins are fast, accurate processing and the ability to function well with limited memory. Currently, semi-stream algorithms presented in the literature such as MeshJoin, Semi-Stream Index Join and CacheJoin can join only one foreign key in the stream data with one table in the master data. However, it is quite likely that stream data have multiple foreign keys that need to join with multiple tables in the master data. We extend CacheJoin to form three new possibilities for multi-way semi-stream joins, namely Sequential, Semi-concurrent, and Concurrent joins. Initially, the new algorithms can join two foreign keys in the stream data with two tables in the master data. However, these algorithms can be easily generalized to join with any number of tables in the master data. We evaluated the performance of all three algorithms, and our results show that the semi-concurrent architecture performs best under the same scenario.

Keywords: Multi-way stream processing · Join operator · Near-real-time data warehouse

1 Introduction

Near-real-time data warehousing (RDW), with its ability to process and analyze data nearly instantly, is increasingly adopted by the business world. Among several approaches to RDW, Data Stream Processing is a crucial component, handling the continuous incoming information - a stream of data, from multiple sources [1]. One method of stream processing is a join operation which combines the streaming data with the slowly changing disk-based master data (denoted as R) [1,2]. As the join deals with two sources, one being a stream, and the other being fairly stable data stored in a disk, such as master data, the join is considered "semi-stream".

Z. Huang et al. (Eds.): ADC 2017, LNCS 10538, pp. 59–70, 2017.
DOI: 10.1007/978-3-319-68155-9_5

With the rapid development of new technologies, the large capacity of current main memories as well as the availability of powerful cloud computing platforms can be utilized to execute stream-based operations [3]. However, to enable the efficient use of ICT infrastructure, semi-stream joins that can process the streaming data in a near-real-time manner while requiring minimum resource consumption, are still of interest. Several semi-stream joining methods have been proposed so far. The authors of the MeshJoin algorithm [8] argued for the need to support streaming updates in RDW [4]. Since then, many other join operators have been developed by improving or adding more features to MeshJoin, such as R-MeshJoin [5], Partition-based Join [6], HybridJoin [7], Semi-Stream Index Join (SSIJ) [2] and CacheJoin [3], to name a few. The authors of the MeshJoin operator suggest that one of the most important research topics in the field that need to be examined next is multi-way semi-stream joins between a stream (whose tuples have two or more foreign keys) and many relations [8]. Indeed, it would be quite practical to process stream data with multiple foreign keys to join with multiple tables in R.

In this paper, we address the problem by developing a multi-way semi-stream join. We propose three different approaches to the joins namely Sequential, Semi-concurrent and Concurrent. The joins are developed by extending CacheJoin (CJ), one of the most advanced semi-stream joins proposed in the field [3]. The advantage of CJ is that it requires very little in the way of computing resources while its service rate is higher than other joins such as MeshJoin, R-MeshJoin and HybridJoin [3]. As extended versions of CJ, the new multi-way joins inherit the main characteristics of their precursor. For example, as CJ performs well with skewed, non-uniformly distributed data, such as the Zipfian distribution of foreign keys in the stream data [3], the newly developed multi-way joins are expected to have the same characteristics.

In this paper, we first develop new multi-way joins which can match a stream data having two foreign keys with two tables in R. The joins then can be generalized to join more tables. To test the new algorithms, we apply them to a scenario where a stream tuple includes customer and product foreign keys which need to join with customer and product tables in R. In the Sequential approach, there are two CJs running concurrently where the first CJ joins customer keys and produces output as the input for the second CJ. After this, the second CJ processes the product foreign key and produces output for the whole multi-way join. In Semi-concurrent, only part of the stream tuples are processed in sequence, and the rest are processed concurrently by two separate CJs. In Concurrent, there are also two CJs running concurrently, but they match the two foreign keys of a tuple at the same time, and the tuple will be sent to output only when both keys are matched. After testing the new joins with different datasets, results show that Semi-concurrent performs best under the same memory setting.

The rest of this paper is organized as follows. Section 2 presents a review of the available semi-stream joins in the academic literature, which focuses on the architecture of the CJ algorithm. This is expected to provide the background theory required to comprehend the new multi-way join algorithms. Section 3 describes the architectures of the Sequential, Semi-Concurrent and Concurrent joins in detail. In Sect. 4 we present a cost model to measure the performance of the new joins. Section 5 presents the performance evaluation and, from the experimental data, it is concluded that the Semi-concurrent performs best while Concurrent performs worst among the three. In Sect. 6 we explain our explanation for this order. Finally Sect. 7 concludes the paper.

2   Related Work

This section presents an overview of some of the semi-stream joins available and then examines in detail the architecture and characteristics of CJ, which is the antecedent of our multi-way semi-stream joins.

In the past, the algorithm MeshJoin was proposed for joining a data stream with a slowly changing table under limited main memory conditions [8,9]. The two fundamental features of MeshJoin are: (1) accessing the disk-based R with fast sequential scans, and (2) armotizing the cost of I/O operations over a large number of stream tuples. The features, therefore, can help MeshJoin reduce costly disk access. Other advantages of MeshJoin are: (1) it can work well with limited main memory and, (2) the organization of R has hardly any effect on its performance. However, the join operation has some limitations. The first limitation is caused by the fact that MeshJoin does not consider the distribution of the incoming stream data as well as the organization of R. Therefore its performance on skewed data is inferior [10]. Also, the performance of MeshJoin is inversely proportional to the size of R. Thus this algorithm does not perform well with large Rs [3].

To improve the MeshJoin algorithm, R-MeshJoin (Reduced MeshJoin) was developed in 2010 [5]. R-MeshJoin improves the MeshJoin operator by clearly defining the dependent relationships between its antecedent's components. Therefore, R-MeshJoin is simpler and obtains slightly better performance than MeshJoin.

We presented another improved version named HybridJoin in the past [7]. The main goals of HybridJoin are: (1) to amortize the fast-coming data stream with slow disk access using limited computer memory, and (2), to deal with an input data stream sent in small and sporadic groups [10]. The main technique used by HybridJoin to amortize the fast-coming data stream is an index-based approach to access R, which is quite efficient. However, like MeshJoin, Hybrid-Join does not take data distribution of the streaming data into consideration.

CJ is an improved HybridJoin operator that inherits the advantages and solves the limitation of its former algorithms [3]. The architecture of CJ is presented in Fig. 1. The main improvement of CJ is an additional hash table stored in computer memory, which stores the most frequent tuples coming from the stream (denoted as H_R). When tuples from the stream arrive, they enter the cache phase first where they are matched with H_R. In this way more frequent tuples can be processed faster as memory access is faster than disk access. If a tuple is not matched in the cache phase, it will be sent to disk phase which

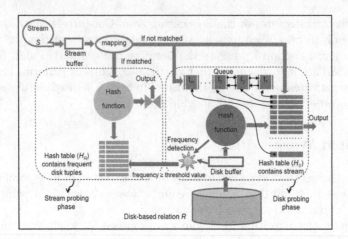

Fig. 1. CACHEJOIN architecture

is basically a HybridJoin. In disk phase, stream tuples are stored in a hash table named H_S and their foreign keys are also added to a queue. To minimize expensive disk access, a few disk pages of R are loaded to a Disk-Buffer (DB) whenever the join conducts a database query. The oldest tuple in the queue is used to determine the partition of R which will be loaded in each probing itera-tion. More specifically, after probing the foreign key of a tuple into R, a few disk pages starting from the matching page in R will be put into DB. All these tuples will be matched with H_S, in order to amortize the seek and disk access time. Thus, the higher the number of tuples in H_S, the higher the probability that some tuples in H_S can be matched with DB, which leads to the faster service rate of CJ's Disk Phase. Another important component of CJ is the frequency detector whose algorithm is as follows: In each matching iteration between DB and H_S, rows that have the number of matches above a certain threshold will be considered as frequent tuples and added to H_R.

A comparison between CJ and MeshJoin shows that CJ out performs MeshJoin in many cases, such as with different settings of R, under different memory conditions and when stream data is skewed [3]. The only situation where CJ processes slower than MeshJoin is when the distribution of stream data is completely uniform, which hardly ever happens in practice. Of the algo-rithms described above, CJ is the only one which considers the distribution of the stream data, while still including the positive features of the others.

All of the above algorithms can join only one-foreign-key stream data with a single table in the master data, but in business there is a need of joining multiple foreign keys with multiple tables in R. The review of current literature shows that not much research has been carried out in this direction.

3   Multi-way Semi-stream Joins

In this paper, we developed three different multi-way semi-stream joins extended from CJ and named them Sequential, Semi-concurrent and Concurrent. As mentioned above, in our experiment presented here, the joins were applied to match two foreign keys of stream tuples with two tables in R. As there are two keys that need to be joined with two tables, our approach is to process each key using a CJ. Thus, we need to organize the process of the two CJs in a suitable order to optimize the multi-way joins' performance in regard to both service rate and resource consumption. The first decision was whether we should create two CJ threads executing the two keys concurrently, or only one thread which processes one key at a time. The two-thread approach was our preference for the following reasons:

- Both approaches require the same level of memory: as both of them contain two CJs, they have similar objects.
- The two-thread approach is feasible. Although running two threads concurrently means doubling the CPU calculation, this approach is still feasible as CJ consumes few resources [3].
- Utilizing multiple threads may improve the applications' performance [11].

Another advantage of the two-thread approach is that it reduces the idle time of the join operator. In CJ, after sending a SQL query to a Database Management System (DBMS) such as MySQL, the join is idle as it waits for the DBMS to execute the query and return the results. Similarly, the DBMS sleeps when the CJ is processing the data returned from the queries. By running two CJs concurrently, the idle time of the both systems (CJs and DBMS) will be reduced as one thread may be working while the other is idling. Therefore, we expect that the time required to process two keys will be less than double the time required to match only one key of the stream tuples.

3.1   Sequential Multi-way Semi-stream Join

Figure 2 presents the simplified architecture of the sequential join, which abstracts CJ to the cache and disk phase level (The cache phase and disk phase boxes are referred to in more detail in Fig. 1). Basically, the sequential join contains two CJs running in sequence, i.e. a tuple is firstly joined with the Customer

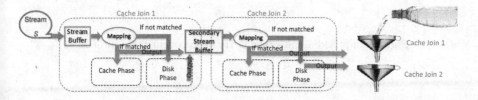

Fig. 2. Simplified architecture of the sequential approach

table by the Customer CJ. Then the matched tuple taken from the Customer table is attached to the stream tuple to form the input for the second CJ, which is the secondary stream buffer (SB). The second CJ takes tuples from SB and processes the other key of the tuples (Product key) and adds the probed product tuple to the stream tuple to form the final join's output. It is worth noting that, although the two keys of a stream tuple are processed in sequence, the two CJs are running concurrently. In Fig. 2, we use a visual metaphor where the water is the stream of data, and the two funnels depict the two CJs running concurrently while the tuples are processed in sequence. With this architecture, we may expect that, although Sequential matches two keys of a tuple, its service rate is equal to the service rate of the slower of the two CJs.

3.2 The Semi-concurrent Approach

Figure 3 presents the simplified architecture of the semi-concurrent algorithm. Similar to CJ, the semi-concurrent join has a cache phase and a disk phase. When a tuple first enters Semi-concurrent, both of its keys are matched with two hash tables H_{R_C} and H_{R_P}, which retain the most frequent tuples of the customer and product tables in cache respectively. If both keys of a stream tuple are matched, the tuple will be ready for output. In all other cases, the tuple will be sent to the disk phase. Semi-concurrent's disk phase has two CJ disk phases running concurrently, processing tuples in sequence, which is quite similar to the sequential process. In Fig. 3, we use the same visual metaphor as the sequential join, but the two funnels are only disk phases instead of complete CJs. If only one key of a tuple is matched in the cache phase, the tuple will be sent to the relevant CJ disk phase to be joined with the other key, e.g. if the product key of a tuple is matched in cache, the tuple will be sent to the customer disk phase. After the second key is processed, the disk phases will produce the final output for the join. With this architecture, only tuples having both keys unmatched within cache go through both customer and product disk phases.

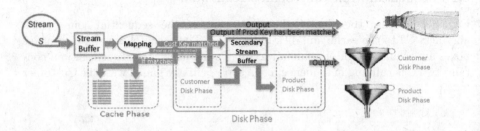

Fig. 3. Simplified architecture of the semi-concurrent approach

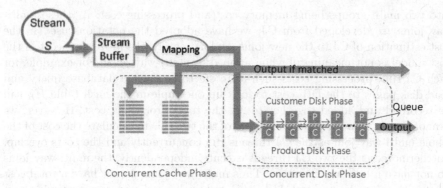

Fig. 4. Abstract level architecture of the concurrent approach

3.3 The Concurrent Approach

Figure 4 presents the simplified architecture of the concurrent join. The cache phase of the concurrent system is very similar to that of the semi-concurrent system, while its disk phase has a new processing method. The concurrent system stores stream, customer and product tuples in its queue, which makes the queue the largest component of the join with regard to memory consumption. At each queue node, its customer/product tuple will be set to null if its customer/product key has not been matched, otherwise the customer/product tuple will store the matched item. During the disk phase, there are two CJ disk phase threads simultaneously executing the unmatched customer and product keys in the queue, which are called customer and product disk phases. The disk phases are supported by two hash tables H_{S_C} and H_{S_P}, whose key/value pairs are an unmatched key and its associated queue node. In this architecture, a queue node will be sent to output only when both customer and product tuples are not null.

For example, if the customer key of a stream tuple is matched in the cache phase, this partially matched stream tuple will be added to a queue where the product key will be matched. At the same time, as only its product key has not been found in cache, the product key is put to H_{S_P}. In another instance, if neither key of a tuple is matched with the cache, the tuple will be added to a queue node where both customer and product items are null, and the customer and product keys are put to H_{S_C} and H_{S_P} respectively.

As opposed to its predecessor CJ, the number of queue nodes in the concurrent system is not equal to the numbers of tuples in H_{S_C} and H_{S_P}. Rather, the numbers of unmatched customer and product keys in the queue are equal to the sizes of H_{S_C} and H_{S_P}, respectively.

4 Cost Model

To evaluate the new multi-way joins, we developed a cost model to measure critical factors of their performance. In the case of CJ, the factors are classified into two main groups being memory cost and processing cost. As these multi-way joins are developed from CJ, we have adopted the notations used in the cost estimation of CJ to the new joins. Unfortunately, the processing cost in the cost model is not meaningful when applied to multi-way joins. For example, for each CJ run, processing costs such as; costs to conduct a database query and read disk pages to the DB, cost to look up one tuple in the hash table H_R can be recorded and added together to get the total processing cost. However, we cannot simply sum the processing costs of the two CJs to calculate the cost of the whole multi-way join as the CJ threads run concurrently and the costs overlap. Furthermore, as the two CJ threads execute independently, the multi-way joins do not have a common iteration. Thus multi-way joins do not have a total cost for one loop iteration as in CJ. To this end we have chosen one processing cost factor for evaluating multi-way joins, which is service rate. The service rates of the new joins are calculated as follows:

$$SR = \frac{total\_processing\_time}{total\_number\_of\_tuples\_processed} \qquad (1)$$

In regard to memory cost, we used the total runtime memory required by the Java programs to operate the multi-way joins in order to compare their performance. The runtime memory of a Java program includes both used and free memory, which are the memory allocated for currently used objects and possible new objects respectively [12]. In this way runtime memory may best reflect the memory cost of each semi-stream join. In our research, we use the memory cost objects adopted from CJ to calculate the memory required by all objects of the joins, but it is only an estimation because sizes of some objects change overtime. For example, the size of Concurrent's queue depends on the number of matched tuples in its nodes, but the number changes overtime. Another example is the secondary SB of Sequential and Concurrent, whose memory size is also not stable. By having the estimations, we adjust the setting of each multi-way join, so that the three joins have the same memory setting.

5 Evaluation

5.1 Experimental Setup

Testing Environment. We ran our experiments on an Core i3-2310 CPU@ 2.10 GHz with Solid State Drive (SSD). We implemented our experiments in Java, using *Eclipse Java Neon 4.6.3*. Measurements were taken with Apache plug ins and *nanoTime()* from *Java API*. The R is stored on a disk using a *MySQL* database, the fetch size for the result set was set to be equal to the disk buffer size. Synthetic data, the stream data, was generated with a Zipfian distribution of the foreign key. The detailed specifications of the data set used for analysis are shown in Table 1.

Table 1. Data specifications

Object	Value
Stream tuple size	*20 bytes*
Size of customer disk tuple	*120 bytes*
Size of product disk tuple	*120 bytes*
Data set	based on Zipf's law (exponent is set to 1)
	Case 1: Both customer and product tables have 1 million tuples
	Case 2: Customer table: 1 million tuples, product table: 300,000 tuples

Memory Setting. In the concurrent join, the largest component in terms of memory use is the queue. Indeed, each node of the queue stores the stream, product and customer objects, where customer and product objects are null if the objects have not been matched. To avoid memory consumption of the join becoming too high, there is a fixed maximum number of queue nodes. The memory size of the queue, therefore, will reach its maximum when all nodes are half-matched (either the customer or the product object is matched). We used N_{CQ} to denote the number of nodes in the concurrent queue.

In the Sequential and Semi-concurrent joins, the largest components in term of memory use are their two hash tables H_{S_C} and H_{S_P} and these hash tables' sizes also need to be fixed. In both joins, we set the same size for both Customer and Product Hash tables, and used N_{SQ} and N_{SCQ} to denote the size of the sequential and semi-sequential hash tables respectively.

To test the performance of each join, we attempted to allocate the same amount of memory for each multi-way join. For our test dataset, the size of each customer and product object are the same (120 bytes), and the size of a stream object is 20 bytes. With this setting, to allocate the same amount of memory to all the joins, N_{SQ} and N_{SCQ} are set to equal to around 2/3 of N_{CQ}.

5.2 Comparison of the Three Multi-way Joins

Figures 5 and 6 show comparisons between the three approaches and a single CJ in two different cases as stated in Table 1. It must be remembered that, while the multi-way joins match two keys of a stream tuple with two tables in R, CJ joins only one. It can be observed that the time required to join two keys in the newly developed multi-way joins is less than double the time of a single CJ to process one key. In regard to the memory cost, these three new joins consume a similar level of memory, around 600 MB and is three times more than CJ.

In both cases, the semi-concurrent join is the best performer, and Concurrent is the slowest multi-way join. The average time Semi-concurrent requires to process 1000 tuples in Case 1 is 7.5 s, while the single CJ requires 5.5 s, and, in Case 2, the difference is only one second.

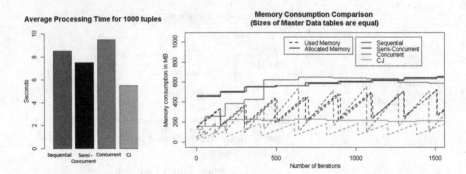

Fig. 5. Comparison of the three multi-way joins with the original CJ (Customer table: 1 million tuples, Product table: 1 million tuples)

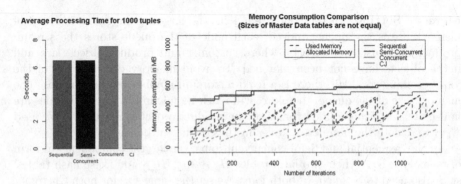

Fig. 6. Comparison of the three multi-way joins and the original CJ (Customer table: 1 million tuples, Product table: 300,000 tuples)

6　Discussion

The reason why Concurrent is slower than Semi-concurrent is as follows. Since both joins have the same cache phase, their disk phases cause the difference. As mentioned above, a node in concurrent process's queue will be moved to output only when both its customer and product keys are matched. While the concurrent process progresses, the number of half-matched nodes increases, which leads to the numbers of unmatched customer and product keys decreasing (because the total number of queue nodes is fixed to N_{CQ}). However, a characteristic of the CJ algorithm mentioned above is that the fewer unmatched items there are, the slower the join performs. In our experiment, after the join runs for a while, the number of unmatched customer and product keys is around 60% of N_{CQ}, which is smaller than in N_{SCQ} (which is equal to 2/3 of N_{CQ}). As a result, the concurrent system becomes slower than the semi-concurrent system because the number of unmatched keys in the semi-concurrent join is always fixed at N_{SCQ}. Figure 7 simulates the concurrent join's queue status while the join is in operation.

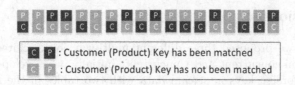

| C | P | : Customer (Product) Key has been matched |
| C | P | : Customer (Product) Key has not been matched |

Fig. 7. Simulation of a concurrent queue at a given time

There are several reasons to explain why the sequential join's service rate is lower than that of the semi-concurrent join. First, in a case where both keys of a tuple are recognized as frequent keys, the semi-concurrent join will send the tuple direct to output after matching it with the cache phase. However, the sequential join requires more steps in processing the tuple as, after matching the first key, the sequential join puts the tuple into a secondary stream buffer, and the tuple must wait for the second CJ to be executed. Second, if the product key of a tuple is matched with the semi-concurrent join's cache, the tuple will go to the customer disk phase, and this phase may directly send the tuple to output. However, after processing this tuple's customer key, sequential join also needs to put it in the secondary stream buffer, and again the tuple must wait for the second CJ to be executed.

Although Sequential has some weaknesses when compared with Semi-concurrent, the two joins have quite similar architecture. Basically, the two joins have two CJ disk phase threads running concurrently and processing tuples in sequence, and this architecture has been proved to be more effective than the concurrent architecture. This provides an answer as to why the concurrent join performs the least well of the three.

Another advantage of the semi-concurrent architecture is that the join is quite flexible. Depending on the case we can adjust its components to achieve better performance. For example, if the size of the product disk tuple is smaller than the customer disk tuple, we can put the product disk phase first in the architecture to save memory. In the semi-concurrent join, after we match a stream tuple with the second CJ, the tuple will be sent to output. Therefore, we do not keep the disk tuple of the second CJ in memory. However, after matching a stream tuple with the first CJ, we need to put keep the matched tuples for the other key to be matched. Hence, by putting the disk tuple which has a smaller memory first in the processing order, the memory required to store the tuples will decrease.

The semi-concurrent join can also be generalized to match more keys by adding more H_R tables to its cache phase and more CJ disk phase threads to its disk phase. The main problem with generalization is that the more keys the join needs to match, the more memory the join requires. Even so, the multi-way join is still expected to be more efficient than other approaches.

7 Conclusion

In this paper, we proposed three different multi-way join architectures called Sequential, Semi-concurrent and Concurrent. Initially, we developed new joins to match two-foreign keys in stream data with two tables in the master data. We also developed a cost model to measure the joins' performance. We compared the performance of the all three newly developed joins with the original CJ. Our results show that Semi-concurrent performed best among the three approaches. In future we aim to generalize our multi-way semi-concurrent approach to join with n number of tables in the master data. Also we will optimize Semi-concurrent by making some adjustments on the algorithm such as the frequency detector and allocating different memories to different CJs in accordance with the distributions of each streaming tuple's foreign key.

References

1. Naeem, M.A., Jamil, N.: An efficient stream-based join to process end user transactions in real-time data warehousing. J. Digit. Inf. Manag. **12**(3), 201–215 (2014)
2. Naeem, M.A., Weber, G., Dobbie, G., Lutteroth, C.: SSCJ: A semi-stream cache join using a front-stage cache module. In: Bellatreche, L., Mohania, M.K. (eds.) DaWaK 2013. LNCS, vol. 8057, pp. 236–247. Springer, Heidelberg (2013). doi:10. 1007/978-3-642-40131-2_20
3. Naeem, M.A., Dobbie, G., Weber, G.: A lightweight stream-based join with limited resource consumption. In: Cuzzocrea, A., Dayal, U. (eds.) DaWaK 2012. LNCS, vol. 7448, pp. 431–442. Springer, Heidelberg (2012). doi:10.1007/978-3-642-32584-7_35
4. Bornea, M., Deligiannakis, A., Kotidis, Y., Vassalos, V.: Semi-streamed index join for near-real time execution of etl transformations. In: 27th International Conference on IEEE, pp. 159–170 (2011)
5. Naeem, M.A., Dobbie, G., Weber, G., Alam, S.: R-MESHJOIN for near-real-time data warehousing. In: Proceedings of the ACM 13th International Workshop on Data Warehousing and OLAP, DOLAP 2010. ACM, Toronto, Canada (2010)
6. Chakraborty, A., Singh, A.: A partition-based approach to support streaming updates over persistent data in an active datawarehouse. In: Proceedings of the 2009 IEEE International Symposium on Parallel & Distributed Processing, IPDPS 2009, pp. 1–11. IEEE Computer Society, Washington, DC (2009)
7. Naeem, M.A., Dobbie, G., Weber, G.: HybridJoin for near-real-time data warehousing. Int. J. Data Warehous. Min. **7**(4), 24–43 (2011)
8. Polyzotis, N., Skiadopoulos, S., Vassiliadis, P., Simitsis, A., Frantzell, N.: Meshing streaming updates with persistent data in an active data warehouse. IEEE Trans. Knowl. Data Eng. **20**(7), 976–991 (2008)
9. Polyzotis, N., Skiadopoulos, S., Vassiliadis, P.: Supporting streaming updates in an active data warehouse. In: ICDE 2007 Proceedings of the 23rd International Conference on Data Engineering, Istanbul, pp. 476–485 (2007)
10. Naeem, M., Dobbie, G., Lutteroth, C., Weber, G.: Skewed distributions in semi-stream joins: How much can caching help? Inf. Syst. **64**, 63–74 (2017)
11. Oracle. Advantages and disadvantages of a multithreaded/multicontexted application. https://docs.oracle.com/cd/E13203_01/tuxedo/tux71/html/pgthr5.htm
12. Shirazi, J.: Java Performance Tuning. O'Reilly Media Inc., California (2003)

Decomposition-Based Approximation of Time Series Data with Max-Error Guarantees

Boyu Ruan, Wen Hua$^{(\boxtimes)}$, Ruiyuan Zhang, and Xiaofang Zhou

The University of Queensland, Brisbane, QLD 4067, Australia
{b.ruan,w.hua,ruiyuan.zhang}@uq.edu.au, zxf@itee.uq.edu.au

Abstract. With the growing popularity of IoT nowadays, tremendous amount of time series data at high resolution is being generated, transmitted, stored, and processed by modern sensor networks in different application domains, which naturally incurs extensive storage and computation cost in practice. Data compression is the key to resolve such challenge, and various compression techniques, either lossless or lossy, have been proposed and widely adopted in industry and academia. Although existing approaches are generally successful, we observe a unique characteristic in certain time series data, i.e., significant periodicity and strong randomness, which leads to poor compression performance using existing methods and hence calls for a specifically designed compression mechanism that can utilise the periodic and stochastic patterns at the same time. To this end, we propose a decomposition-based compression algorithm which divides the original time series into several components reflecting periodicity and randomness respectively, and then approximates each component accordingly to guarantee overall compression ratio and maximum error. We conduct extensive evaluation on a real world dataset, and the experimental results verify the superiority of our proposals compared with current state-of-the-art methods.

Keywords: Time series compression · High periodicity · Strong randomness · Decomposition-based algorithm · Max-error guarantee

1 Introduction

With the growing popularity of IoT (Internet of Things), tremendous amount of time series data is being generated nowadays by modern sensor networks in various domains [1], such as power grids, manufacturing networks, and medical care systems. For example, due to the upgrading process from conventional power grid to "smart grid", all levels of components in a grid, ranging from power plants, substations, transformers, distributors to smart home appliances, are being monitored simultaneously. Hence, the central controller needs to receive and process massive high resolution time series data from the smart electricity meters and other sorts of measurement devices [2]. Apparently, this will cause problems in terms of data storage, transmission, and processing. First, the expense on data

© Springer International Publishing AG 2017
Z. Huang et al. (Eds.): ADC 2017, LNCS 10538, pp. 71–82, 2017.
DOI: 10.1007/978-3-319-68155-9_6

storage is always a crucial financial concern for any organisation who owns millions of customers and measurement equipments. It is natural for the size of time series data to reach petabyte level, which requires enormous amount of storage space. Second, since massive data is being generated at real time, it will cause congestion during transmission.

Data compression is the key to overcome the aforementioned problems. Existing data compression techniques can be roughly classified into two categories: lossless compression and lossy compression [3,4]. As its name implies, lossless compression can fully guarantee the compression accuracy, and the original data can be reconstructed flawlessly from the compressed data. Nevertheless, the compression ratio of lossless techniques is generally much smaller than that of the lossy techniques. As a result, challenges caused by the high volume of time series data cannot be easily resolved by lossless methods. Lossy compression techniques, on the contrary, achieve a higher compression ratio at a slight sacrifice of the quality of the compressed data, and hence have attracted extensive attention from both industry and academia during recent decades.

So far, many model-based lossy compression techniques [5] have been proposed for various application domains where either the time series data is strictly stationary or it is compressed without any error guarantee. However, in practice, many time series data, such as traffic volume, illumination, solar energy generation, etc., demonstrate a common characteristic: they contain both *significant global periodicity* and *high local randomness*. As can be observed from Fig. 1(a) which illustrates solar energy generation for three consecutive days, although data of each day follows an approximately same distribution (periodicity), they differ from each other to various extent because of natural weather conditions (randomness). In this case, it is hardly possible for existing model-based compression methods to detect an appropriate model to approximate the original data due to frequent local fluctuations. Furthermore, the repeated patterns in each period can be utilised as well to improve compression ratio.

In this work, we propose a decomposition-based compression technique which divides the original time series into periodic component and random component (e.g., STL decomposition in Fig. 1(b)), and then approximates each component respectively using different methods to achieve the best performance. Our contributions can be summarised as below.

- We observe a common characteristic in certain time series data, i.e., high global periodicity and strong local randomness, where existing data compression techniques cannot be easily applied.
- We design a novel DBA (Decomposition-Based Approximation) algorithm to approximate such time series with max-error guarantee, based on which an optimised algorithm DBAfS (Decomposition-Based Approximation for Streams) is further proposed for online stream data compression.
- Experimental results on a real world dataset verify the superiority of our proposals compared with state-of-the-art compression techniques.

The rest of this paper is organised as follows: in Sect. 2, we briefly discuss related work for time series compression; then we formally define the research

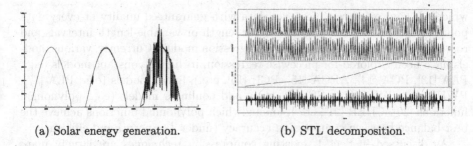

(a) Solar energy generation.   (b) STL decomposition.

Fig. 1. An example of periodic and stochastic solar energy generation data and its STL decomposition result. The four signals in Fig. 1(b) represent original data, *seasonal* part, *trend* part, and *remainder* part, respectively.

problem and introduce in detail our proposed algorithms in Sects. 3, 4 and 5 respectively. Our experimental results are described in Sect. 6, followed by a brief conclusion in Sect. 7.

2   Related Work

Time series data compression is a widely researched topic, and various compression mechanisms have been proposed in recent decades. With respect to the application scenario, existing techniques can be divided mainly into two categories: local data compression and streaming data compression [4]. Local data compression aims to reduce the storage cost on local databases and enable faster processing of online queries with less I/O cost. During compression, new patterns of the raw data need to be recognised to help further data analysis. Streaming data compression, on the other hand, targets on data transmission. It learns and adjusts models for newly observed data, and has the superiority of extending sensors' battery life and making better use of limited communication bandwidth. Both methods play an important role in the field of time series research, especially in data management and data mining [6].

In terms of compression strategy, many techniques have been applied for generating an abstract representation of time series, including fourier transform [7], wavelet transform [8,9], symbolic representation [10–12] and piecewise regression [5]. These techniques seek to find an approximate representation that is smaller in size than the original time series, without losing much information contained in the original data. In particular, fourier transform and wavelet transform have been used to extract features from time series for compression and enable efficient subsequence matching. These techniques cannot provide an error guarantee for the compressed data which, however, is crucial for many real world data analysis tasks. Symbolic time series representation is based on data discretization and can achieve high compression ratio. Whereas, applications of the compressed data are limited to statistics and machine learning algorithms for some selected purposes. Piecewise regression is the most widely adopted compression technique which can approximate time series data with guaranteed quality at every data point. It divides a time series into fixed-length or variable-length intervals and represents each interval with different regression models. Currently, various models have been explored for piecewise regression, including constant models (e.g., PAA [13], PCA [14], APCA [15], PCH [16], etc.), linear models (e.g., PLA [17], PWLH [16], SWAB [18], SF [19], etc.), and nonlinear models (e.g., polynomial functions [20], CHEB [21], etc.), among which polynomial functions achieve the best balance between compression accuracy (under L_∞ norm) and efficiency.

As discussed in Sect. 1, existing compression techniques are largely inapplicable for time series data with both high periodicity and strong randomness. Therefore, we propose a decomposition-based compression algorithm with error guarantees under the L_∞ norm. The experimental results in Sect. 6 will demonstrate the limitations of existing compression methods and the superiority of our proposed algorithm in this case.

3   Problem Definition

In this work, we investigate the problem of time series data compression. We first present the formal definitions of time series and time series compression.

Definition 1 *(Time Series). A time series D is defined as a sequence of data points (t_i, v_i), i.e., $D = \{(t_i, v_i)\} = \{(t_1, v_1), (t_2, v_2), \ldots, (t_n, v_n)\}$, where t_i is the timestamp of v_i organised monotonically, i.e., $\forall j < k$, $t_j < t_k$. v_i represents the data value at time t_i, and it can be either one-dimensional or multi-dimensional.*

In practice, a time series dataset is sometimes extremely large from the following two perspectives: (1) there could be a huge amount of time series, and (2) each time series could be very long. In order to reduce its storage consumption and guarantee its usability in real world applications, a wisely designed compression strategy is indispensable. In particular, given a time series in the form of $\{(t_i, v_i)\}$, we focus on transforming it into another sequence of bits, such that the error for each v_i does not exceed a predefined threshold ϵ. The ultimate goal is to reduce the number of bits required for representing the original time series as much as possible. Note that we adopt the L_∞ metric for error constraint, which guarantees the accuracy of each data point (t_i, v_i). This is a widely used metric in recent time series compression techniques. Hence, the problem of time series compression with error bound is formally defined as follows.

Problem 1 *(Time Series Data Compression with Error Bound). Given a time series $D = \{(t_i, v_i^D)\}$ where $i \in [1, n]$, and an error threshold ϵ, the problem of time series compression is to find an approximated representation of D such that the reconstructed time series from the approximation, i.e., $D' = \{(t_i, v_i^{D'})\}$, satisfies the following condition: $\max(e_i) = \max(v_i^{D'} - v_i^D) \leq \epsilon$.*

Throughout the paper, we will use D' to denote both compressed time series and reconstructed time series whenever it is clear.

4   Decomposition-Based Compression Mechanism

We observe that most time series data in practice, such as illumination, traffic volume, solar energy generation, and energy consumption, etc., is a combination of necessity and contingency. In other words, they demonstrate both significant global periodicity and strong local randomness. Inspired by this unique property, we propose a decomposition-based compression technique that divides the original time series into several components corresponding to its periodicity and randomness respectively, and then compresses each component separately to guarantee the overall error bound and meanwhile achieve the largest compression ratio.

In particular, given the original time serious data $D = \{(t_i, v_i^D)\}$, we decompose D into three components: $S = \{(t_i, v_i^S)\}$, $T = \{(t_i, v_i^T)\}$, and $r = \{(t_i, v_i^r)\}$, where S, T and r represent the *seasonality*, *trend* and *remainder* parts of D respectively, such that:

$$D = S + T + r \quad i.e., \quad \forall i, v_i^D = v_i^S + v_i^T + v_i^r \tag{1}$$

Here, we divide the original time series with exiting STL (Seasonal Trend Decomposition using Loess) tools. Based on such a decomposition, we redefine the problem of time series compression as follows.

Problem 2 (Decomposition-Based Time Series Data Compression with Error Bound). *Given a time series $D = \{(t_i, v_i^D)\}$ that can be decomposed into $D = S + T + r$, and an error threshold ϵ, the problem of time series data compression is to find approximated representations of S, T and r such that the reconstructed time series D' and its components, i.e., $S' = \{(t_i, v_i^{S'})\}$, $T' = \{(t_i, v_i^{T'})\}$ and $r' = \{(t_i, v_i^{r'})\}$ respectively, satisfy the following conditions: $D' = \{(t_i, v_i^{D'})\} = S' + T' + r'$ and $\max(e_i) = \max(|v_i^{D'} - v_i^D|) \leq \epsilon$.*

4.1   Component Approximation

In this section, we will introduce our methods to compress each component, i.e., $S = \{(t_i, v_i^S)\}$, $T = \{(t_i, v_i^T)\}$ and $r = \{(t_i, v_i^r)\}$, of the original time series $D = (t_i, v_i^D)$ obtained by STL decomposition.

When conducting STL decomposition, the length of a single period needs to be determined. In practice, most time series have a natural period p (e.g., one day for illumination, solar energy generation, etc.) which, however, might be problematic if adopted directly for compression. The trend part for each natural period could be dramatically different, which will then lead to low compression ratio. But if a longer period is adopted, as in Fig. 1(b), the trend part will become smoother and easier to compress. Therefore, we define a manual period $d \cdot p$ where d is a positive integer, and use $d \cdot p$ to conduct STL decomposition. In this case, each natural period in a manual period follows similar patterns, whereas there are still slight differences (e.g., peak values) among these periods. To avoid the cost of storing the entire manual period, we only preserve its mean value as a reference. Specifically, we replace the periodic part S with its mean value $\bar{S} = \{(t_i, v_i^{\bar{S}})\}$, and add the difference between S and $\bar{S}$ to the remainder r such that the new remainder will be $r_p = r + S - \bar{S}$. More formally, let $\{S_1, S_2, \ldots, S_d\}$ be a manual period where S_j, $j \in [1, d]$ is the j-th natural period, namely $S_j = \{(t_{p \cdot (j-1)+1}, v_{p \cdot (j-1)+1}^S), \ldots, (t_{p \cdot j}, v_{p \cdot j}^S)\}$. We calculate the mean value of $\{S_1, S_2, \ldots, S_d\}$ based on Eq. 2, where $i\%p$ represents the modulo operation, i.e., $i\%p = i - \lfloor \frac{i}{p} \rfloor \cdot p$.

$$\forall i \in [1, d \cdot p], v_i^{\bar{S}} = \frac{\sum_{j=1}^{d} v_{p \cdot (j-1)+i\%p}^S}{d} \tag{2}$$

We then store only one period of $\bar{S}$ as the representative of the periodic part S. Since we have added the difference between S and $\bar{S}$ into the remainder part r_p, there is no information loss when compressing the periodic part $\bar{S}$ in this way. We will show the impact of d on the compression performance in Sect. 6.

We adopt piecewise linear approximation to compress the trend part T because it demonstrates an obvious linear property, as in Fig. 1(b). The information we need to store is just the coefficients of line segmentations in T'. For the remainder part r_p, we divide its value range $[\min(v^{r_p}), \max(v^{r_p})]$ into M intervals and construct a histogram on these M intervals. We then approximate each interval as well as it constituting values $v_i^{r_p}$ with the median of that interval. Finally, we adopt *Huffman coding* to encode these median values and transform the original remainder part r_p into a sequence of Huffman codes r_p'. *Run-length coding* can also be applied to further reduce the size of r_p'. The whole DBA approximation algorithm is presented in Algorithm 1.

4.2   Error Analysis

The error between D' and D equals to the sum of errors when approximating S, T and r respectively. Since we transfer the difference between S' and S into r_p, the error under L_∞ norm can be calculated as below.

$$e_i = |v_i^{D'} - v_i^{D}| = |(v_i^{T'} - v_i^{T}) + (v_i^{r_p'} - v_i^{r_p})| \tag{3}$$

We can approximate T using any piecewise linear approximation algorithm, such as PLA, PWLH, SWAB, SF, etc., that guarantees L_∞ norm accuracy, namely

$$\max(e_i^T) = \max(|v_i^{T'} - v_i^{T}|) \leq \epsilon_1 \tag{4}$$

For the remainder part r_p, since we divide its value range into M intervals and use the median values for Huffman coding, the maximum error of r_p is

$$\max(e_i^{r_p}) = \max(|v_i^{r_p'} - v_i^{r_p}|) \leq \frac{\max(v^{r_p}) - \min(v^{r_p})}{2M} \tag{5}$$

We can choose a proper M such that $M \geq \frac{\max(v^{r_p}) - \min(v^{r_p})}{2\epsilon_2}$ to guarantee $\max(e_i^{r_p}) \leq \epsilon_2$. Combining Inequations 4 and 5, we can conclude that

$$\max(e_i) \leq \max(e_i^T) + \max(e_i^{r_p}) \leq \epsilon_1 + \epsilon_2 \tag{6}$$

By simply setting $\epsilon_1 + \epsilon_2 \leq \epsilon$, the error bound $\max(e_i) \leq \epsilon$ is satisfied.

Algorithm 1. Decomposition-Based Approximation (DBA)

Input:
    D: original time series
    ϵ_1, ϵ_2: error bounds for trend and remainder
    p, d: natural and manual periods
Output:
    D': compression result of D
1: Divide D into S, T and r using STL decomposition
2: $S' \leftarrow \bar{S}$ calculated by Equation 2
3: $T' \leftarrow$ LINEARAPPRO(T, ϵ_1)
4: $r_p \leftarrow r + S - S'$
5: $M \leftarrow \frac{\max(v^{r_p}) - \min(v^{r_p})}{2\epsilon_2}$
6: Divide the value range of v^{r_p} into M intervals and encode the median m_k of each interval using Huffman coding, $k \in [1, M]$
7: $v^{r'_p} \leftarrow m_k$ of corresponding interval
8: $D' \leftarrow S' + T' + r'_p$

---

5   Online Compression Mechanism

In order to conduct STL decomposition, the original time series needs to be provided beforehand. Therefore, it is impossible to purely decompose a single data point (t_i, v_i) into the form $v_i = v_i^S + v_i^T + v_i^r$. In other words, the compression technique described in Sect. 4 can only be applied for offline cases. In this section, we introduce our optimisation to make Algorithm 1 capable of compressing stream data. We observe that the periodic part of original time series roughly follows certain distributions. Hence we can divide a time series $D = \{t_i, v_i^D\}$ into two components N_s and r_s, where $N_s = \{(t_i, v_i^{N_s})\}$ fits a pre-learned distribution and $r_s = \{(t_i, v_i^{r_s})\}$ is the remainder, such that:

$$D = N_s + r_s \quad i.e., \quad \forall i, v_i^D = v_i^{N_s} + v_i^{r_s} \tag{7}$$

In this way, $v_i^{r_s}$ can be calculated at real time and the algorithm can be applied for online stream data compression. Based on such a decomposition, we define the problem of stream data compression as follows.

Problem 3 (Decomposition-Based Stream Data Compression with Error Bound). *Given a time series $D = \{(t_i, v_i^D)\}$ that can be decomposed into $D = N_s + r_s$, and an error threshold ϵ, the problem of stream data compression is to find an approximated representation $r'_s = \{(t_i, v_i^{r'_s})\}$, such that $D' = \{(t_i, v_i^{D'})\} = N_s + r'_s$ and $\max(e_i) = \max(|v_i^{D'} - v_i^D|) \leq \epsilon$.*

5.1   Component Approximation

Recall that we divide D in a simpler form $D = N_s + r_s$ rather than STL decomposition. We first learn the distribution of the periodic part from historical data.

Algorithm 2. Decomposition-Based Approximation for Stream (DBAfS)

Input:

　　D: original time series

　　ϵ: error bound p, d: natural and manual periods

Output:

　　D': compression result of D

1: Learn the proper N_s from $d \cdot p$ period of D

2: $r_s \leftarrow D - N_s$

3: $M \leftarrow \frac{\max(v^{r_s}) - \min(v^{r_s})}{2\epsilon}$

4: Divide the value range of v^{r_s} into M intervals and encode the median m_k of each interval using Huffman coding, $k \in [1, M]$

5: $v^{r_s} \leftarrow m_k$ of corresponding interval

6: $D' \leftarrow N_s + r'_s$

In the case of solar energy generation, a Gaussian distribution $N_s = (\mu_s, \sigma_s)$ can be observed. But the problem is how many historical data should be utilised to learn the parameters of the distribution. On one hand, the distribution should be generalised enough to cover as much data as possible. On the other hand, it needs to be locally typical so that not too much difference is transferred into the remainder part. To this end, we adopt data within a manual period $d \cdot p$ to learn the distribution. Then we approximate r_s in the same way as r_p described in Sect. 4. Since $v_i^{r_s}$ can be calculated directly by $v_i^{r_s} = v_i^D - v_i^{N_s}$ at every timestamp t_i, this technique can be executed every time when a new data point arrives in the stream. The whole approximation algorithm is presented in Algorithm 2. After compression, we only need to store the coefficients of N_s and the Huffman coding results of r'_s. As before, run-length coding can be applied to further reduce the size of r'_s.

5.2 Error Analysis

The error between D' and D originates only from the error of approximating r_s. Hence, the error under L_∞ norm can be calculated as follows.

$$e_i = |v_i^{D'} - v_i^D| = |v_i^{r'_s} - v_i^{r_s}| \tag{8}$$

As discussed in Sect. 4, the maximum error of r_s is

$$\max(e_i) = \max(|v_i^{r'_s} - v_i^{r_s}|) \leq \frac{\max(v^{r_s}) - \min(v^{r_s})}{2M}. \tag{9}$$

We can also choose a proper M such that $M \geq \frac{\max(v^{r_s}) - \min(v^{r_s})}{2\epsilon}$ to make sure the error bound $max(e_i) \leq \epsilon$ is satisfied.

6　Experiments and Results

To evaluate the performance of our proposals, we conducted extensive experiments on a real world data set from the smart grid. The data set consists of information about solar energy generation for one year with each day containing 5,040 data collected at a 10-second resolution. All the experiments were conducted on a PC with Intel Core i7 processor 3.4 GHz and 16 GB main memory.

We compared DBA and DBAfS algorithms with current state-of-the-art lossy compression methods, and examined the impact of parameter setting on the compression performance which were measured in terms of both Compression Ratio (CR, Eq. 10) and accuracy.

$$CR = \frac{|D|}{|D'|} \tag{10}$$

$$RMSE = \sqrt{\frac{1}{n}\sum_{i=1}^{n}(v_i^D - v_i^{D'})^2} \tag{11}$$

In Eq. 10, $|D|$ and $|D'|$ represents the storage cost of original data D and compressed data D', respectively. We calculated accuracy with the Root Mean Square Error (RMSE, Eq. 11), a widely adopted metric for evaluating the quality of approximation. We noticed that the range of original data might vary a lot throughout the entire time series. Therefore, we used relative error (in percentage) in the experiments to represent RMSE and error bound ϵ. For the remaining part, we still use ϵ to denote relative error bound whenever it is clear.

6.1 Effect of Error Bound ϵ

As discussed in Sect. 2, piecewise regression is currently the most prevalent lossy compression technique. Hence, we selected several representative piece wise regression methods (i.e., PCA and APCA for constant models, PWLH and SF for linear models, and CHEB for nonlinear models) and compared with our DBA and DBAfS algorithms. Figures 2(a) and (b) demonstrate their respective compression ratio and RMSE, when error bound ϵ ranges from 1% to 10%. We can see that both compression ratio and RMSE of all methods rise with the increasing of ϵ. This is natural and also consistent with the results obtained from [5]. Moreover, DBA and DBAfS achieve the largest compression ratio at the cost of nearly the same RMSE with most counterparts, and such an improvement further enlarges when ϵ increases. In particular, when ϵ is set as 10%, the compression ratio of DBAfS is at least 10 times larger than all the existing piecewise regression methods. Comparing DBA and DBAfS, we observe that DBA is not as powerful as DBAfS with respect to compression ratio, especially when ϵ is large. This is because except for periodic part, DBA needs to compress both trend and remainder parts while DBAfS only handles remainder part. When ϵ is small, the storage cost of the compressed data is dominated by the remainder part, which results in similar compression ratio between DBA and DBAfS. However, when ϵ grows, remainder becomes less dominant and thus the difference between DBA and DBAfS gradually emerges.

To further explore the insights of DBA algorithm, we evaluated the impact of ϵ_1 and ϵ_2 on its compression performance. In this case, we set $\epsilon = 1\%$ and vary

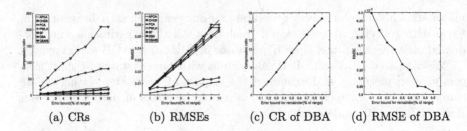

(a) CRs	(b) RMSEs	(c) CR of DBA	(d) RMSE of DBA

Fig. 2. Effect of error bound ϵ. Figures 2(a) and (b) compare the performance of our techniques with other state-of-the-art compression methods. Figures 2(c) and (d) show the impact of how we distribute ϵ_1 and ϵ on the performance of DBA.

ϵ_2 (i.e., the error bound for remainder part) from 0.1% to 0.9% (resp. ϵ_1 ranges from 0.9% to 0.1%). Figures 2(c) and (d) present the results. It can be observed that larger ϵ_2 leads to higher compression ratio and smaller RMSE. This also verifies the remainder part dominates the size and quality of compressed data in the solar generation data when total error bound ϵ is small, i.e. 1%.

6.2   Effect of Period Length d

As mentioned in Sect. 4, most periodic time series have a natural period p, and we can define a manual period $d \cdot p$, which is d times larger than the natural period, to balance between the compression of trend part and remainder part. In the case of solar energy generation, its natural period is one day. We evaluated the impact of d on the compression performance of DBA algorithm. Figures 3(a) and (b) present DBA's compression ratio and RMSE respectively when ϵ is set as 1%. We can see that both compression ratio and RMSE decrease with the increasing of d. It has been verified in previous experiments that in solar energy data, the storage size of compressed data is dominated by remainder part when error bound ϵ is small. The increase of d means more difference is transferred from periodic part to remainder part, which leads to larger deviation between $max(v^{r_p})$ and $min(v^{r_p})$, and hence larger M according to Eq. 5. Consequently, the compression of remainder part takes longer codes and thus more storage cost. This in turn results in smaller compression ratio when d enlarges.

For DBAfS, parameter d is also adopted to trade-off the generalisability and typicality of learned standard distribution of the original data. Therefore, we studied the impact of d on the compression performance of DBAfS, as depicted in Figs. 3(c) and (d) where $\epsilon = 1\%$. When d increases from 0 to 50 days, the learned distribution can capture more periods of original data. However, when d further enlarges, the learned distribution is too generalised and no longer typical for a particular part of the data. This means too much difference is transferred into the remainder part, which will lead to the decrease of compression ratio, as analysed above. Hence, the best setting of parameter d in the DBAfS algorithm for our solar generation data is selected as 50 days.

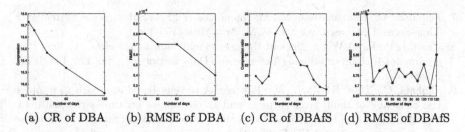

(a) CR of DBA   (b) RMSE of DBA   (c) CR of DBAfS   (d) RMSE of DBAfS

Fig. 3. Effect of period length d. Figures 3(a) and (b) demonstrate the impact of d on the compression performance of DBA. Figures 3(c) and (d) show the impact of d on the compression performance of DBAfS.

7   Conclusion

In this work, we observe a common characteristic in certain time series data, i.e., significant periodicity and strong randomness, which makes existing compression techniques, either lossless or lossy, largely inefficient in terms of compression ratio. Therefore, we propose two novel decomposition-based compression algorithms, namely DBA and DBAfS, to approximate such time series with max-error guarantees. Our experimental results on a real world data set demonstrate the superiority of our proposals, compared with current state-of-the-art lossy compression methods. As future work, we plan to apply DBA and DBAfS algorithms to more time series data, such as illumination, traffic volume, etc., to further examine their applicability and compression performance.

Acknowledgment. This research is partially supported by the Australian Research Council (Grant No.DP170101172) and the Queensland Government (Grant No.AQRF12516).

References

1. Ratanamahatana, C.A., Lin, J., Gunopulos, D., Keogh, E., Vlachos, M., Das, G.: Mining time series data. In: Maimon, O., Rokach, L. (eds.) Data Mining and Knowledge Discovery Handbook, pp. 1049–1077. Springer, Boston (2009)
2. Eichinger, F., Efros, P., Karnouskos, S., Böhm, K.: A time-series compression technique and its application to the smart grid. VLDB J. **24**(2), 193–218 (2015)
3. Tak-chung, F.: A review on time series data mining. Eng. Appl. Artif. Intell. **24**(1), 164–181 (2011)
4. Esling, P., Agon, C.: Time-series data mining. ACM Comput. Surveys (CSUR) **45**(1), 12 (2012)
5. Hung, N.Q.V., Jeung, H., Aberer, K.: An evaluation of model-based approaches to sensor data compression. IEEE Trans. Knowl. Data Eng. **25**(11), 2434–2447 (2013)
6. Keogh, E., Kasetty, S.: On the need for time series data mining benchmarks: a survey and empirical demonstration. Data Mining Knowl. Discov. **7**(4), 349–371 (2003)

7. Faloutsos, C., Ranganathan, M., Manolopoulos, Y.: Fast Subsequence Matching in Time-Series Databases, vol. 23. ACM, New York (1994)
8. Chan, K.P., Fu, A.W.C.: Efficient time series matching by wavelets. In: 1999 Proceedings of 15th International Conference on Data Engineering, pp. 126–133. IEEE (1999)
9. Shahabi, C., Tian, X., Zhao, W.: Tsa-tree: A wavelet-based approach to improve the efficiency of multi-level surprise and trend queries on time-series data. In: Proceedings of 12th International Conference on Scientific and Statistical Database Management, pp. 55–68. IEEE (2000)
10. Lin, J., Keogh, E., Li, W., Lonardi, S.: Experiencing SAX: a novel symbolic representation of time series. Data Mining Knowl. Discov. **15**(2), 107 (2007)
11. Shieh, J., Keogh, E.: i sax: indexing and mining terabyte sized time series. In: Proceedings of the 14th ACM SIGKDD International Conference on Knowledge Discovery and Data Mining, pp. 623–631. ACM (2008)
12. Yi, B.K., Faloutsos, C.: Fast time sequence indexing for arbitrary lp norms. In: VLDB (2000)
13. Keogh, E., Chakrabarti, K., Pazzani, M., Mehrotra, S.: Dimensionality reduction for fast similarity search in large time series databases. Knowl. Inf. Syst. **3**(3), 263–286 (2001)
14. Lazaridis, I., Mehrotra, S.: Capturing sensor-generated time series with quality guarantees. In: Proceedings 19th International Conference on Data Engineering (Cat. No.03CH37405), pp. 429–440 (2003)
15. Keogh, E., Chakrabarti, K., Pazzani, M., Mehrotra, S.: Locally adaptive dimensionality reduction for indexing large time series databases. In: Proceedings of the 2001 ACM SIGMOD International Conference on Management of Data, SIGMOD 2001, pp. 151–162. ACM, New York, NY, USA (2001)
16. Buragohain, C., Shrivastava, N., Suri, S.: Space efficient streaming algorithms for the maximum error histogram. In: 2007 IEEE 23rd International Conference on Data Engineering, pp. 1026–1035 (2007)
17. Chen, Q., Chen, L., Lian, X., Liu, Y., Yu., J.X.: Indexable pla for efficient similarity search. In: Proceedings of the 33rd International Conference on Very Large Data Bases, VLDB 2007, pp. 435–446. VLDB Endowment (2007)
18. Keogh, E., Chu, S., Hart, D., Pazzani, M.: An online algorithm for segmenting time series. In: Proceedings 2001 IEEE International Conference on Data Mining, pp. 289–296 (2001)
19. Elmeleegy, H., Elmagarmid, A.K., Cecchet, E., Aref, W.G., Zwaenepoel, W.: Online piece-wise linear approximation of numerical streams with precision guarantees. Proc. VLDB Endow. **2**(1), 145–156 (2009)
20. Ni, J., Ravishankar, C.V.: Indexing spatio-temporal trajectories with efficient polynomial approximations. IEEE Trans. Knowl. Data Eng. **19**(5), 663–678 (2007)
21. Cai, Y., Ng, R.: Indexing spatio-temporal trajectories with chebyshev polynomials. In: Proceedings of the 2004 ACM SIGMOD International Conference on Management of Data, SIGMOD 2004, pp. 599–610. ACM New York, USA (2004)

Similarity Search

Searching k-Nearest Neighbor Trajectories on Road Networks

Pengcheng Yuan[1], Qinpei Zhao[1(✉)], Weixiong Rao[1], Mingxuan Yuan[2], and Jia Zeng[2]

[1] School of Software Engineering, Tongji University, Shanghai, China
{qinpeizhao,wxrao}@tongji.edu.cn
[2] Huawei Noahs Ark Lab, Shatin, Hong Kong
{yuan.mingxuan,zeng.jia}@huawei.com

Abstract. With the proliferation of mobile devices, massive trajectory data has been generated. Searching trajectories by locations is one of fundamental tasks. Previous work such as [3,6,9] has been proposed to answer the search. Such work typically measures the distance between trajectories and queries by the distance between query points and GPS points of trajectories. Such measurement could be inaccurate because those GPS points generated by some sampling rate are essentially discrete. To overcome this issue, we treat a trajectory as a sequence of line segments and compute the distance between a query point and a trajectory by the one between the query point and line segments. Next, we index the line segments by R-tree and match each trajectory to the associated line segments by inverted lists. After that, we propose a k-nearest neighbor (KNN) search algorithm on the indexing structure. Moreover, we propose to cluster line segments and merge redundant trajectory IDs for higher efficiency. Experimental results validate that the proposed method significantly outperforms existing approaches in terms of saving storage cost of data and the query performance.

Keywords: KNN · Trajectory · Road network · Clustering · Index

1 Introduction

With the proliferation of mobile devices, massive trajectory data are generated. Many applications, such as urban planning and intelligent transportation, have used the trajectory data to understand people's mobility patterns. Among such applications, one fundamental task is to effectively and efficiently index and search trajectories. For example, given a few selected locations as an input query, a k-nearest neighbor (KNN) search demands to find the top-k trajectories that are closest to the selected locations.

In literature, various work [3,6,9] has been proposed to answer the KNN trajectory search. To compute the similarity between trajectories and queries, such work typically measures the distance between GPS points in trajectories and those selected locations in queries; in terms of query processing efficiency,

© Springer International Publishing AG 2017
Z. Huang et al. (Eds.): ADC 2017, LNCS 10538, pp. 85–97, 2017.
DOI: 10.1007/978-3-319-68155-9_7

these work focuses on precisely selecting a few GPS points (and thus a small amount of associated trajectories) in order to faster satisfy the KNN stopping condition by tighter lower and upper bounds of similarities between trajectories and queries.

Unfortunately, the GPS points in trajectories are discrete and regularly recorded by a fixed sampling rate (e.g., a GPS point per 300 s). Such discrete GPS points could lead to the *semantic issue* in terms of the similarity between trajectories and queries. For example, when mobile devices are continuously moving, the truly nearest position of a mobile device to a query point may not be recorded by GPS sensors. It is particularly true when a low sampling rate leads to very spare GPS points.

To overcome this issue above, in this paper, we propose a hybrid KNN algorithm to find k-Nearest Neighbor Trajectories on Road-network (KNNTR). In this algorithm, we treat a trajectory as continuous line segments and thus infinite GPS points (instead of discrete points), and then the distance between a query point and the trajectory by the one between the query point and the line segments is computed. Such computation can guarantee the correctness of the distance, no matter GPS sensors have sampled the point which leads to the correct distance or not. To enable line segments, we adopt a frequently used map-matching technique, given the map of road networks, to recover full trajectories from (sparse) GPS points. In this way, we then represent each trajectory by (partial) road segments.

Next, we index the line segments by a spatial structure such as R-tree [4], and each leaf node in R-Tree refers to a list of trajectories with line segments. Thus, each trajectory is matched to the associated line segments. With the indexing structure, we could follow the classic KNN query processing framework to find more candidates until the KNN stopping condition is satisfied. However, the challenge is that R-Tree itself cannot be directly used to compute the distance between query points and line segments, because the line segments are indexed in the form of rectangles in R-tree. To optimize the overhead to retrieve indexed line segments and distance computation, we approximate the distance lower/upper bounds simply using the internal nodes (such as Minimal Bounding Box: MBR) in R-Tree. Furthermore, due to distance locality, a trajectory could move on multiple line segments. It indicates that the trajectory could redundantly appear in multiple ID lists referred by those leaf nodes containing such line segments. Such redundance not only incurs larger space cost but also higher KNN query overhead. To this end, we propose a simple and efficient clustering approach to merge the redundant trajectory IDs. In this way we have chance to optimize the space cost and KNN query efficiency.

As a summary, we make the following contribution in this paper.

- We propose the line segments-based distance computation for more meaningful semantics, regardless of GPS points in the trajectories.
- We propose a fast hybrid KNN query algorithm based on a new distance measurement and the spatial indexing.
- We propose a simple clustering approach to merge the redundant data and therefore, optimize the space cost and the KNN query efficiency.

2 Problem Definition

We consider a query $Q = \{q_1...q_m\}$, represented by a set of m locations q_i with $1 \leq i \leq m$. A trajectory database $\mathbb{T}$ consists of a large amount of raw trajectories. Each trajectory $T \in \mathbb{T}$ contains a series of n GSP positions p_j with $1 \leq j \leq n$. Each position p_j is with the coordinations (*latitude, longitude*).

Given a query Q, we want to effectively and efficiently find the top-k nearest trajectories from a trajectory database $\mathbb{T}$. Firstly, we need to define a distance or similarity function between a trajectory $T \in \mathbb{T}$ and a query Q.

Definition 1. *After adopting a map-matching technique to map each trajectory onto a road network map, we represent the trajectory by the associated line segments S_T, and compute $Dist(T, q_i)$ by the distance between q_i and q_i's nearest point $p' \in S_T$, e.g., using Euclidean distance, i.e., $Dist(T, q_i) = Dist(S_T, q_i) = Dist(p', q_i)$.*

Next for a query Q, we define a monotonic aggregation function to measure the similarity between query Q and trajectory T.

$$Dist(T, Q) = \sum_{i=1}^{m} Dist(T, q_i) \tag{1}$$

In the equation above, a larger distance $Dist(T, q_i)$ leads to a larger $Dist(T, Q)$. Note that the technique proposed in this paper can be generally applied to other monotonic aggregation functions such as [12].

Problem 1. When Definition 1 and Eq. 1 are used to measure the distance $Dist(T, Q)$, we want to efficiently search a trajectory database $\mathbb{T}$ to find the k-NN trajectories for a query Q.

3 Solution Design

3.1 Overall Framework

We first briefly introduce the data structure to index trajectories. On the overall, in Fig. 1, the indexing structure consists of two components: a spatial structure and inverted lists. Specifically, when each trajectory is represented by a sequence of line segments (See Sect. 2), we use a spatial data structure, e.g., R-Tree, to index such line segments. Each leaf Minimal Bounding Rectangle (MBR) in the R-Tree maintains at least one line segment. After that, each leaf MBR refers to a list of trajectory IDs. Such trajectories are those containing the line segments appearing in the leaf MBR. Leaf MBRs and lists of trajectory IDs can be intuitively treated as inverted lists.

We take Fig. 1(b) for illustration. We have three example trajectories T_1, T_2, T_3 and eight line segments. A leaf MBR in the R-Tree contains four example segments $S_1...S_4$. For segment S_1, it refers to a list of 4 example trajectory IDs T_1, T_3, T_6, T_7; and S_2 refers to the list of T_1 and T_2.

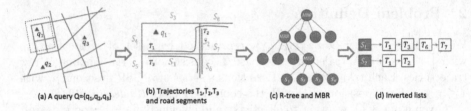

(a) A query Q={q₁,q₂,q₃}   (b) Trajectories T₁,T₂,T₃ and road segments   (c) R-tree and MBR   (d) Inverted lists

Fig. 1. The framework of query

Algorithm 1. KNN Search (Query Q, Index I, Number k)

1 Initiate $(m+2)$sets: PC, FC, C_i for each q_i $(i = 1...m)$, $\lambda \leftarrow k$;
2 **while** $|FC| < k$ **do**
3   **for** $i = 1...m$ **do** $C_i \leftarrow$ GetCandidate (q_i, λ, I);
4      $FC \leftarrow C_1 \cap C_2 \cap .. \cap C_m$, $PC \leftarrow C_1 \cup C_2 \cup .. \cup C_m$;
5      $\lambda \leftarrow \lambda + \Delta$;
6 **return** Refine(FC, PC, k);

Given the indexing structure above, Algorithm 1 gives the overall KNN search framework.

The Index I here refers to R-tree and Inverted Lists. Since we have m query points, we then have m associated candidate sets C_i. As shown in line 4 in Algorithm 1, we denote FC to the set of fully searched candidate set, and PC to be the set of partially searched candidate set. By performing multiple rounds of GetCandidate (see Algorithm 2). We search more candidate trajectories to meet the stopping condition in line 2: the number of trajectories in FC is no less than the number k, i.e., $|FC| \geq k$. After that, we stop the search and refine the candidates in PC to find the final k nearest trajectories (i.e., the function Refine in line 6). Initially, we set λ to k and the Δ is set to k. Our experiment in Sect. 6 will empirically evaluate the effect of different Δ.

Lemma 1. *For a KNN query, if $|FC| \geq k$ holds, then the set PC must contain the k nearest trajectories to query Q.*

The proof of Lemma 1 can be seen in [9].

4   Finding Candidate Trajectories and Refinement

In this section, we will present the algorithms GetCandidate to find candidate sets (Sects. 4.1 and 4.2) and Refine to find the final KNN result (Sect. 4.3).

4.1   Finding the Nearest Line Segments

On the overall, GetCandidate consists of two parts. Firstly, it needs to find at most λ nearest line segments that are not searched yet. Secondly, among those

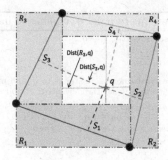

Fig. 2. Illustration of $Dist(S, q) \neq Dist(R, q)$.

trajectories associated with the found λ nearest line segments, the algorithm searches λ newly founded trajectories which do not appear inside C_i.

The key of GetCandidate is its first part: how to efficiently find the λ nearest line segments.

Example 1. We take Fig. 2 for illustration. It is not hard to find that the distance between a MBR and query point q differs from the one between a line segment and q, e.g., $Dist(R_3, q) \neq Dist(S_3, q)$ as shown in Fig. 2. If we find the nearest line segments simply by using the distance between MBRs and query points, since $R_1 \rightarrow R_2 \rightarrow R_3 \rightarrow R_4$, we will falsely get a sorted lists of four nearest line segments: $S_1 \rightarrow S_2 \rightarrow S_3 \rightarrow S_4$. Instead, by the real distance between line segments and query points, we have the correct sorted list $S_2 \rightarrow S_1 \rightarrow S_4 \rightarrow S_3$. Thus, we cannot simply use the distance between MBRs and query points to find the nearest line segments.

We highlight the idea of GetCandidate to find the λ nearest line segments. Though $Dist(S, q) \neq Dist(R, q)$, we can use the distance $Dist(R, q)$ to be the *lower bound* of $Dist(S, q)$; otherwise, the line segment S must be out of the bounding rectangle R. We then use $Dist(R, q)$ as the lower bound of $Dist(S, q)$ to find the λ nearest line segments as follows.

Among the sorted list of founded MBRs R by ascending order of $Dist(R, q)$, we process each of the MBRs R one by one by retrieving the line segments S inside R. Suppose that we have retrieved at least k line segments. Denote that LB_R to be the largest distance to q among all processed MBRs R, and UB_S to the kth smallest distance to q among all retrieved line segments S. If $LB_R > UB_S$ holds, we determine that the k nearest line segments needed by GetCandidate must be inside those retrieved ones. We then stop the retrieval of line segments and return the k nearest line segments from the retrieved ones.

4.2 Finding the Nearest Trajectories

Algorithm 2 gives the pseudocode of GetCandidate. Firstly, based on the idea of Sect. 4.1, lines 3–6 are to find the nearest line segments. After that, lines 7–10 are

---

Algorithm 2. GetCandidate(query point q, number λ, index I)

---

1 Initiate an empty candidate set $, $Cnt_{traj} \leftarrow 0$;
2 **while** $Cnt_{traj} < \lambda$ **do**
3     $LB_R \leftarrow 0, UB_S \leftarrow +\infty$;
4   **while** $LB_R < UB_S$ **do**
5   find a new MBR R with the nearest distance to q, and update LB_R;
6   retrieve new line segments S from the new MBR R and update UB_S;
7   **for** *each found line segment S* **do**
8   **for** *each trajectory t moving on S* **do**
9   **if** *t does not inside* $ *and* $Cnt_{traj} < \lambda$ **then**
10   add t to $, Cnt_{traj}++;

11 **return** set $;

---

to find the λ nearest trajectories which are moving on the found line segments. Such trajectories are found by using the inverted list structure. In this way, we guarantee that the found λ trajectories by Algorithm 2 must have smaller distance to q than other remaining trajectories.

4.3 Candidate Refinement

Finally, the refinement is given by Algorithm 3. First we create a maximal heap H by adding each trajectory $T_j \in FC$. The items in heap H is sorted by descending order of $Dist(T_j, Q)$. The head item $T_j \in H$, i.e., the largest $Dist(T_j, Q)$, becomes the threshold for the trajectories in PC to update the items in H.

As shown in the **for** loop (lines 3–6), we process the remaining trajectories $T_\ell \in PC$ one by one except those in FC. For each T_ℓ, if the lower bound of $Dist(T_\ell, Q)$, denoted by $LB(T_\ell, Q)$, is larger than the threshold τ, we do not necessarily perform line 6; otherwise, we need to compute $Dist(T_\ell, Q)$, refresh the heap H and update the threshold τ. The lower bound here is similar with that in [9].

Algorithm 3. Refine(sets FC and PC, Number k)

---

1 Create a maximal heap H and add all trajectories in FC to H;
2 threshold $\tau \leftarrow$ the kth smallest $Dist(T_j, Q)$ for $T_j \in H$; $PC \leftarrow PC - FC$;
3 **for** $\ell = 1 \dots |PC|$ **do**
4     $T_\ell \leftarrow$ the ℓ-th trajectory in PC;
5   **if** $LB(T_\ell, Q) \geq \tau$ **then** continue;
6   **else if** $Dist(T_\ell, Q) < \tau$ **then** { add T_ℓ to H; $H.pop()$; update τ};

7 **return** H;

---

5 Redundancy Reduction

In this section, in order to reduce redundant trajectory IDs in inverted lists, we propose a set of techniques to optimize the space cost of inverted lists as well as KNN query efficiency.

5.1 Basic Idea

Our basic idea is to first find similar line segments, and then merge such segments in order to reduce the redundant trajectory IDs. We take Fig. 3 for illustration.

Fig. 3. Reduction of redundant trajectory IDs

Example 2. In Fig. 3 (a), we have six segments $S_1 \ldots S_6$, each of which refers to an inverted list of trajectory IDs. The three inverted lists of $S_1 \ldots S_3$ all contain the three trajectories $T_1...T_3$. Thus, we consider that the three segments $S_1 \ldots S_3$ are similar. Next, we merge the three redundant trajectories $T_1...T_3$ by creating a new inverted list referred by a virtual segment 1^*. After that, the redundant trajectories $T_1...T_3$ in the inverted lists of $S_1 \ldots S_4$ are replaced by the segment 1^*, as shown in Fig. 3(b). Similarly, the three segments $S_4 \ldots S_6$ all contain two redundant trajectories T_3, T_7. We again create a new inverted list referred by the virtual segment 2^* and replace the redundant trajectories in three associated inverted lists as before.

Given the basic idea above, the key problem is how to efficiently and effectively find similar line segments. We might simply follow traditional clustering algorithms, such as k-means, to find similar line segments. Unfortunately, when given a large amount of line segments and trajectories, traditional algorithms suffer from inefficiency issues. Instead, we adopt the technique of Locality Sensitive Hash (LSH) [5] to efficiently find approximately similar line segments. Nevertheless, LSH gives only an approximation approach, and we thus design an algorithm to select truly similar ones among such approximately similar line segments. After the similar segments are found, we then merge redundant trajectories as above.

5.2 Locality Sensitive Hashing

To enable LSH, we need to define the line segment similarity and hash functions as follows.

Definition 2. *Consider two line segments S_i and S_j which are referred by two inverted lists of trajectories: $S_i = \{ T_i^1 \ldots T_i^{|S_i|} \}$ and $S_j = \{ T_j^1 \ldots T_j^{|S_j|} \}$. We define a Jaccard similarity $sim(S_i, S_j) = \frac{|S_i \cap S_j|}{|S_i \cup S_j|} = \frac{|\{T_i^1 \ldots T_i^{|S_i|}\} \cap \{T_j^1 \ldots T_j^{|S_j|}\}|}{|\{T_i^1 \ldots T_i^{|S_i|}\} \cup \{T_j^1 \ldots T_j^{|S_j|}\}|}$.*

Definition 3. *Hash Functions: The random permutations $h_1, h_2, \ldots, h_n$ is a set of hash functions where h_i is a permutation of values $T_1, T_2, \ldots, T_n$. The values $T_1, T_2, \ldots, T_n$ are the trajectories.*

Given the definition above, we create the clusters of line segments using LSH as follows. First of all, we need to transform each inverted list of a line segment into a boolean vector (with size equal to the number of trajectories). In the vector, if a trajectory is moving on the line segment, then the associated vector element is one and otherwise zero. After that, given m hash functions, we generate a minhash signature with respect to the line segment. Suppose that we have n line segments. We then have n associated minhash signature, and apply LSH to find the similar line segments.

5.3   Cluster Reorganization

Since LSH gives an approximation algorithm, we have to re-organize the clusters of line segments. To this end, for a cluster $\mathcal{C}$ of $|\mathcal{C}|$ line segments $S_i \in \mathcal{C}$ with $1 \leq i \leq |\mathcal{C}|$, we define the following three metrics to measure the clustering quality.

$$Sim_{avg}(\mathcal{C}) = \frac{\sum_{i=1}^{|\mathcal{C}|} \sum_{j=i+1}^{|\mathcal{C}|} Sim(S_i, S_j)}{|\mathcal{C}| \times (|\mathcal{C}| - 1)/2}$$

$$Sim_{cum}(S_i, \mathcal{C}) = \sum_{j=1, j \neq i}^{|\mathcal{C}|} Sim(S_i, S_j)$$

$$Space(\mathcal{C}) = |\mathcal{C}| + \sum_{i=1}^{|\mathcal{C}|} |S_i|$$

In the three equations above, $Sim_{avg}(\mathcal{C})$ measures the average similarity of pairwise line segments in the cluster $\mathcal{C}$, and a higher $Sim_{avg}(\mathcal{C})$ indicates more dense line segments and better clustering quality; $Sim_{cum}(S_i)$ cumulates the similarity between a given segment S_i and all other segments $S_j \in \mathcal{C}$, and $Space(\mathcal{C})$ instead computes the space cost used to maintain the inverted lists of trajectories for all line segments $S_i \in \mathcal{C}$. Note that we use the count $|\mathcal{C}|$ of line segments and the count $|S_i|$ of trajectories in segment S_i together to measure $Space(\mathcal{C})$. A smaller $Space(\mathcal{C})$ means less space overhead used for the cluster $\mathcal{C}$.

Given the metrics above, we show the idea to optimize the clusters as follows. If the line segments in a cluster $\mathcal{C}$ is not well clustered, we split the cluster $\mathcal{C}$, such that the split sub-clusters can achieve better clustering quality and less space cost measured by the metrics above. The Split function is shown in Algorithm 4.

Algorithm 4. Split(Cluster $\mathcal{C}$)

---

1 Initiate an empty set M;
2 **for** *each line segment $S_i \in \mathcal{C}$* **do**
3   ⌊ **if** $Sim_{cum}(S_i) < Sim_{avg}(\mathcal{C}) * (|\mathcal{C}| - 1)$ **then** add S_i to M;
4 **if** $Space(M) + Space(\mathcal{C} - M) < Space(\mathcal{C})$ **then**
5   ⌊ create two subclusters for M and $\mathcal{C} - M$, respectively;
6 **return** M;

First, we find those line segments which have the accumulated similarity smaller than the average one (line 3) and add them to a set M. After that, we then have two sets M and $\mathcal{C} - M$. If the space cost of such two sets is smaller than the one before the split, we then perform the Split operation and create two sub-clusters M and $\mathcal{C} - M$. Note that we can recursively perform the Split operation on the sub-clusters until no chance to optimize smaller space cost.

Take Fig. 3 as an example. LSH might first cluster the three segments S_4, S_5, S_6 together into the same group, as shown in Fig. 3(b). After the reorganization, we split the cluster and have two subsets $\{S_4, S_5\}$ and $\{S_6\}$. The redundant trajectory IDs in subset $\{S_4, S_5\}$ are merged as shown in Fig. 3(c) for smaller space cost.

Finally note that the merging operation has changed the structure of inverted lists above. We thus have an updated GetCandidate Algorithm. In the new algorithm, an item in the inverted list could be either a merged set of trajectory IDs or an individual trajectory ID. Thus, besides trajectory IDs, we will also confirm whether the merged set is searched. If not, the trajectory IDs of the set will be added into the searched set. We use a single merged item here to replace a set of redundant trajectory IDs, and thus avoids redundant processing efforts.

6 Experiments

To evaluate our work, we use two real world data sets, Shanghai with 25000 trajectories and Porto [1] with 100000 trajectories. The road network data of Shanghai and Porto have 46556 road segments and 7647 road segments respectively. We perform a map-matching algorithm [8] on the trajectories to the road network of both datasets. After that, we implement the proposed algorithm, namely KNNTR, in Java, and evaluate the performance on a Ubuntu Linux platform with Xeon 2.40 GHz cpu (6 cores), 64 GB of RAM, and 4 standard 512 GB SSD drive. The space cost of raw data was 100.1M, while it costs only 52.5M when using our algorithm. In the following, we will mainly focus on the running time of algorithm in performance analysis.

For comparison, we choose the state of art algorithms including the Incremental k-NN based Algorithm (IKNN) [3], Global Heap-based algorithm (GH) [9] and Spatial Range-Based Approach (SRA) [6] as three competitors. In the experiments, we study the effect of the following parameters:

- The number m of query points: following the previous work [3,6,9], we do not choose a large number for m, and instead vary its value from 2 to 10.

- The number k of trajectories for the KNN query result: since a very large number k is meaningless, we require that the number k is smaller than 32.
- By default, we conduct all experiments on the whole dataset. Each of the experiments are the average running time of 100 pairs of different query points.

6.1 Baseline Study

In this section, we mainly compare the performance of proposed KNNTR with the three competitors (IKNN, GH and SRA). In this experiment, we fix $k = 10$ and vary m from 2 to 10. When the number m is given, we then randomly choose the query points inside corresponding maps.

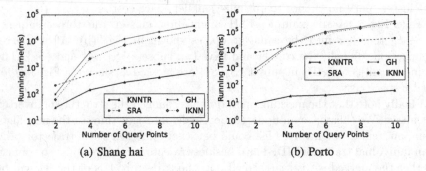

(a) Shang hai   (b) Porto

Fig. 4. Running Time of four algorithms in two datasets

As shown in Fig. 4, KNNTR outperforms the other three competitors no matter the number m in both datasets. Both figures show the same trend with the number of query points increases. IKNN and GH have a good performance when the number m of query points is very small, such as 2. SRA shows a better performance when the number of query points increases, for it compute the distance between the newly searched trajectory points and query points, which helps it get a tighter bound. AS shown in the Fig. 4(a), the time cost of KNNTR is three times shorter than SRA while KNNTR outperforms SRA in Fig. 4(b) by at least two order of magnitude. We then study the effect of the parameters on the algorithms in the following section.

6.2 Performance of Optimization

The KNNTR algorithm outperforms the other three algorithms dramatically on its second efficiency. In the KNNTR algorithm, the LSH technique has been employed to reduce running time and space cost. In the following No-LSH refers to the baseline, LSH refers to the baseline with LSH and LSH-S refers to the optimization with split algorithm.

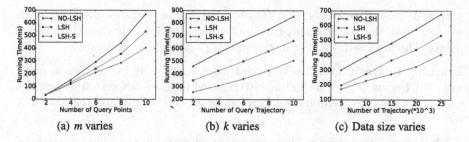

Fig. 5. Comparisons when parameters varies for Shanghai data

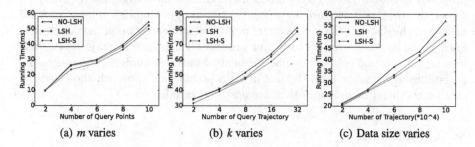

Fig. 6. Comparisons when parameters varies for Porto data

Effect of Query Points' Number. First, we evaluate the algorithms' performances by tuning different number of query points. Query points are all randomly generated from the whole scope of the corresponding city. As shown in the Figs. 5(a) and 6(a), both figures show the same trend. The LSH-S increases in the slowest rate.

Effect of Query Trajectories' Number. The query trajectories' number k also affects the query time a lot. As shown in Figs. 5(b) and 6(b), with the increasing of k, LSH-S is always the best because it reduce the time that many adjacent road segments need to be added into the candidate set successively.

Scalability of Algorithms. Here we design the experiments to show the scalability of three algorithms. As shown in Figs. 5(c) and 6(c), LSH-S does not outperforms LSH notably when the number of trajectories is not large because the adjacent road segments are not similar to each other. With the increasing of the trajectory number, we can notice that the LSH-S shows a robust performance for the optimization part which can get better clusters.

7   Related Work

Distance Measure: To answer the KNN trajectory search, the similarity and/or distance measure is the first key point, and many previous work has proposed various metrics [12]. [2,10] proposed various distance measures while we note that the trajectories of moving objects record only discrete location points. The proposed distance metric in this paper is more accurate.

Learning to Hash: Learning to Hash is an efficient work in NN search, many work [7,11] show a good performance compared to exact NN search while these work is not appropriate in our application for we need to perform the exact NN search.

KNN Query Processing: There existing some works for searching trajectories by locations these years. [3] uses an incremental k-NN based algorithm and [9] proposed a Global Heap (GH) based method. Spatial Range-based (SRA) [6] shows the best performance by using a tighter upper bound. These work all measured the distance between a query point and its nearest point in trajectories. Such previous work are to index and process GPS points in trajectories, and instead our solution is based on line segments. Since the number of line segments is significantly smaller than the one of GPS points, our approach thus leads to smaller overhead including both space cost and running time.

8   Conclusion

In this paper, we propose to compute the distance between the query point and road segments for more meaningful semantics of KNN trajectory search. For efficiency search, we design an indexing structure consisting of both R-tree and inverted lists. Moreover, we propose a simple and efficient clustering approach to merge the redundant trajectory IDs. With help of the indexing structure and clustering approach, our approach can answer the KNN trajectory search by using faster running time and smaller space.

Acknowledgment. This work is partially sponsored by National Natural Science Foundation of China (Grant No. 61572365, 61503286), Science and Technology Commission of Shanghai Municipality (Grant No. 14DZ1118700, 15ZR1443000, 15YF1412600) and Huawei Innovation Research Program (HIRP).

References

1. https://www.kaggle.com/c/pkdd-15-predict-taxi-service-trajectory-i
2. Agrawal, R., Faloutsos, C., Swami, A.: Efficient similarity search in sequence databases. In: Lomet, D.B. (ed.) FODO 1993. LNCS, vol. 730, pp. 69–84. Springer, Heidelberg (1993). doi:10.1007/3-540-57301-1_5
3. Chen, Z., Shen, H.T., Zhou, X., Zheng, Y., Xie, X.: Searching trajectories by locations: an efficiency study. In: Proceedings of the 2010 ACM SIGMOD International Conference on Management of Data, SIGMOD 2010, pp. 255–266. ACM, New York (2010)
4. Guttman, A.: R-trees: a dynamic index structure for spatial searching. In: Proceedings of the 1984 ACM SIGMOD International Conference on Management of Data, SIGMOD 1984, pp. 47–57. ACM, New York (1984)

5. Indyk, P., Motwani, R.: Approximate nearest neighbors: towards removing the curse of dimensionality. In: Proceedings of the Thirtieth Annual ACM Symposium on Theory of Computing, STOC 1998, pp. 604–613. ACM, New York (1998)
6. Qi, S., Bouros, P., Sacharidis, D., Mamoulis, N.: Efficient point-based trajectory search. In: Claramunt, C., Schneider, M., Wong, R.C.-W., Xiong, L., Loh, W.-K., Shahabi, C., Li, K.-J. (eds.) SSTD 2015. LNCS, vol. 9239, pp. 179–196. Springer, Cham (2015). doi:10.1007/978-3-319-22363-6_10
7. Song, J., Gao, L., Liu, L., Zhu, X., Sebe, N.: Quantization-based hashing: a general framework for scalable image and video retrieval. Pattern Recognit. (2017)
8. Song, R., Sun, W., Zheng, B., Zheng, Y.: PRESS: a novel framework of trajectory compression in road networks. Proc. VLDB Endow. 7(9), 661–672 (2014)
9. Tang, L.-A., Zheng, Y., Xie, X., Yuan, J., Yu, X., Han, J.: Retrieving k-nearest neighboring trajectories by a set of point locations. In: Pfoser, D., Tao, Y., Mouratidis, K., Nascimento, M.A., Mokbel, M., Shekhar, S., Huang, Y. (eds.) SSTD 2011. LNCS, vol. 6849, pp. 223–241. Springer, Heidelberg (2011). doi:10.1007/978-3-642-22922-0_14
10. Vlachos, M., Kollios, G., Gunopulos, D.: Discovering similar multidimensional trajectories. In: Proceedings 18th International Conference on Data Engineering, pp. 673–684 (2002)
11. Wang, J., Zhang, T., Sebe, N., Shen, H.T.: A survey on learning to hash. IEEE Trans. Pattern Anal. Mach. Intell. PP(99), 1 (2017)
12. Zheng, Y.: Trajectory data mining: an overview. ACM Trans. Intell. Syst. Technol. 6(3), 29:1–29:41 (2015)

Efficient Supervised Hashing via Exploring Local and Inner Data Structure

Shiyuan He[1], Guo Ye[1], Mengqiu Hu[1], Yang Yang[1(✉)], Fumin Shen[1],
Heng Tao Shen[1], and Xuelong Li[2]

[1] School of Computer Science and Engineering, Center for Future Media,
University of Electronic Science and Technology of China, Chengdu, China
shiyuanhe.david@gmail.com, michaelyeah7@gmail.com, hmq.uestc@gmail.com,
dlyyang@gmail.com, fumin.shen@gmail.com, shenhengtao@hotmail.com
[2] State Key Laboratory of Transient Optics and Photonics,
Center for OPTical IMagery Analysis and Learning (OPTIMAL),
Xi'an Institute of Optics and Precision Mechanics,
Chinese Academy of Sciences, Beijing, China
xuelong_li@opt.ac.cn

Abstract. Recent years have witnessed the promising capacity of hashing techniques in tackling nearest neighbor search because of the high efficiency in storage and retrieval. Data-independent approaches (e.g., Locality Sensitive Hashing) normally construct hash functions using random projections, which neglect intrinsic data properties. To compensate this drawback, learning-based approaches propose to explore local data structure and/or supervised information for boosting hashing performance. However, due to the construction of Laplacian matrix, existing methods usually suffer from the unaffordable training cost. In this paper, we propose a novel supervised hashing scheme, which has the merits of (1) exploring the inherent neighborhoods of samples; (2) significantly saving training cost confronted with massive training data by employing approximate anchor graph; as well as (3) preserving semantic similarity by leveraging pair-wise supervised knowledge. Besides, we integrate discrete constraint to significantly eliminate accumulated errors in learning reliable hash codes and hash functions. We devise an alternative algorithm to efficiently solve the optimization problem. Extensive experiments on two image datasets demonstrate that our proposed method is superior to the state-of-the-arts.

Keywords: Supervised hashing · Approximate anchor graph · Inherent neighborhood

1 Introduction

Hashing has become a popular method in computer vision and machine learning. Especially in recent years, the demand for retrieving relevant content among massive images is stronger than ever under such a big-data era. Hashing methods

---

S. He and G. Ye contributed equally to this work.

© Springer International Publishing AG 2017
Z. Huang et al. (Eds.): ADC 2017, LNCS 10538, pp. 98–109, 2017.
DOI: 10.1007/978-3-319-68155-9_8

map high-dimensional images into short binary codes. In this way, searching similar images converts to finding neighbor hashing codes in Hamming space. This technique leads to significant efficiency in retrieval [8,12,15,21,23] and pattern matching [5,16,22].

At the early time, many tree-based indexing approaches have been developed for approximate nearest neighbor (ANN) search like KD tree [1]. However, as the dimension increases, KD tree needs enough space to store data which costs a lot and the performance degrades quickly. In consideration of the inefficiency of tree-based indexing methods, hashing approaches have been proposed to map entire dataset into discrete codes and the similarity can be measured by Hamming distance which cost little time to calculate. One of the most popular hashing methods is Locality-Sensitive Hashing (LSH) [3] that has been widely used to handling massive data. LSH uses a hash function that randomly projects or permutates nearby data points into same binary codes. However, LSH needs long binary codes to achieve promising retrieval performance which increases the storage space and computation costs. Moreover, LSH ignores the underlying distributions and manifold structure of the data on account of random-projection.

Realizing this deficiency, Weiss et al. proposed Spectral Hashing (SH) [19] utilizing the subset of thresholded eigenvectors of the graph Laplacian by relaxing the original problem which improves the retrieval accuracy to some extent yet charges more time to build a neighborhood graph. Liu et al. delivered some improvements to SH and proposed the Anchor Graph Hashing (AGH) [11] that uses anchor graphs to obtain low-rank adjacency matrices. Formulation of AGH costs constant time by extrapolating graph Laplacian eigenvectors to eigenfunctions.

However, hashing methods mentioned above can not achieve high retrieval performance with a simple approximate affinity matrix [9]. Due to the semantic gap, returning the nearest neighbors in metric space can not guarantee search quality [18]. To solve this problem, images that are artificially labeled as similar or dissimilar are used by supervised hashing methods like KSH [10] and SDH [14] to satisfy the semantic similarity. By leveraging pairwise labeled information, the performance has been remarkably improved.

In this paper, we aim to design a supervised hash method which can efficiently generate high-quality compact codes. We utilize the anchor graph which is built based on the pairwise similarity to exploit the inner structure of the original data, in the process of learning hash function, we also take the supervision information to preserve the pairwise similarity to improve the accuracy of retrieval. To avoid accumulated errors caused by continuous relaxation, we choose to directly optimize the binary codes. With the discrete constraints added to objective function, we propose a novel hashing framework, termed Local and Inner Data Structure Supervised Hashing (LISH) which is able to efficiently generate codes and satisfy the semantic similarity at the same time. Our main contributions are summarized as follows:

- Our method uses graph laplacian to captures the local neighborhoods to enhance hashing codes' quality. And semantic gap can be properly solved by utilizing labeled information. By this way, both metric and semantic similarity are preserved by our method which contribute a lot to improve the performance significantly.

- Most existing hash method solve the problem with the relaxation of the discrete constraints, since we directly optimize our method and each bit can be sequentially learned by the algorithm, our method outperforms in an alternative and efficient manner.
- We evaluate our method on two popular large-scale image datasets and obtain superior accuracy than state-of-the-arts.

In the remainder of this paper, we present the detailed formulation of proposed LISH method in Sect. 2. Section 3 shows experimental results and conclusions are given in Sect. 4.

2   Local and Inner Data Structure Supervised Hashing

In this section, we mainly introduce the algorithm of our method in detail. We propose an alternative optimization model and sequentially learn each bit. The hash functions are learned during the optimization process simultaneously.

Suppose we have n samples $\mathbf{x}_i \in \mathbb{R}^d$, $i = 1, \cdots, n$, deposited in matrix $\mathbf{X} = [\mathbf{x}_1, \mathbf{x}_2, \cdots, \mathbf{x}_n]^T \in \mathbb{R}^{n \times d}$, where d is the dimensionality of the feature space. Denote the label matrix $\mathbf{Y} = [\mathbf{y}_1, \mathbf{y}_2, \cdots, \mathbf{y}_n] \in \{0, 1\}^{n \times c}$ where c is the number of classes of labels. When $\mathbf{x}_i$ belongs to class j, $\mathbf{y}_{ij}$ equals 1, otherwise equals 0. By means of finding the mapping relation between the original feature space and hamming space, hashing generates binary codes to represent the features.

2.1   Anchor Graphs

Since conventional graph structure is inefficient in large scale, we use Anchor Graphs [11] to build the graph affinity matrix $\mathbf{A}$. As in [11], by defining a subset $\mathbf{U} = \{\mathbf{u}_j \in \mathbb{R}^d\}_{j=1}^m$ where $\mathbf{u}_j$ represents an anchor to approximate the neighborhood structure of the training dataset $\mathbf{X}$. The regression matrix $\mathbf{Z}$ that measures the underlying relationship between $\mathbf{X}$ and $\mathbf{U}$ can be calculated as follows:

$$Z_{ij} = \begin{cases} \frac{\exp\left(-\mathcal{D}^2(\mathbf{x}_i, \mathbf{u}_j)/t\right)}{\sum_{j' \in \langle i \rangle} exp(-\mathcal{D}^2(\mathbf{x}_i, \mathbf{u}_{j'})/t)}, & \forall j \in \langle i \rangle, \\ 0, & otherwise. \end{cases} \tag{1}$$

The distance function $\mathcal{D}(\cdot)$ used here is ℓ_2 distance and t is the bandwidth parameter. The anchor graph gives the diagonal matrix $\mathbf{\Lambda} = diag(\mathbf{Z}^T \mathbf{1}) \in \mathbb{R}^{m \times m}$. Since approximate affinity matrix $\mathbf{A} = \mathbf{Z}\mathbf{\Lambda}^{-1}\mathbf{Z}^T$ is positive semidefinite (PSD) and has unit row and column sums, $\mathbf{A}$ can be calculated with high efficiency. Each data point map to an r-bit binary code $\mathbf{b}_i$. The code matrix $\mathbf{B} = [\mathbf{b}_1, \mathbf{b}_2, \cdots, \mathbf{b}_n]^T \in \mathbb{R}^{n \times r}$. In order to assure that the similar inputs have the minimal Hamming distance of the hashing codes, the following objective formulation is proposed:

$$\min_B \frac{1}{2} \sum_{i,j=1}^{n} \|\mathbf{b}_i - \mathbf{b}_j\|^2 \mathbf{A}_{ij} = tr(\mathbf{B}^T \mathbf{L} \mathbf{B}),$$

$$s.t. \begin{cases} \mathbf{B} \in \{\pm 1\}^{n \times r}, \\ \mathbf{1}^T \mathbf{B} = 0, \mathbf{B}^T \mathbf{B} = n\mathbf{I}_r. \end{cases} \tag{2}$$

Constraint $\mathbf{1}^T \mathbf{B} = 0$ ensures each bit to be balanced, and $\mathbf{B}^T \mathbf{B} = n\mathbf{I}_r$ enforces hashing codes to be uncorrelated to minimize the redundancy among these bits. However, problem (2) is still NP-hard. By using the anchor graph Laplacian $\mathbf{L} = \mathbf{I}_n - \mathbf{A}$ and defining a set $\Omega = \{\mathbf{V} \in \mathbb{R}^{n \times r} \mid \mathbf{1}^T \mathbf{V} = 0, \mathbf{V}^T \mathbf{V} = n\mathbf{I}_r\}$ [11], problem (2) can be simplified as:

$$\max_B \ tr(\mathbf{B}^T \mathbf{A} \mathbf{B}) - \frac{\rho}{2} dist^2(\mathbf{B}, \Omega),$$

$$s.t. \ \mathbf{B} \in \{\pm 1\}^{n \times r}, \tag{3}$$

where ρ is a tuning parameter. Since $tr(\mathbf{B}^T \mathbf{B}) = tr(\mathbf{V}^T \mathbf{V}) = nr$, problem (3) equals to the following problem:

$$\max_{\mathbf{B}, \mathbf{Y}} \ tr(\mathbf{B}^T \mathbf{A} \mathbf{B}) + \rho tr(\mathbf{B}^T \mathbf{V}),$$

$$s.t. \begin{cases} \mathbf{B} \in \{\pm 1\}^{n \times r}, \mathbf{V} \in \mathbb{R}^{n \times r}, \\ \mathbf{1}^T \mathbf{V} = 0, \mathbf{V}^T \mathbf{V} = n\mathbf{I}_r. \end{cases} \tag{4}$$

2.2 Proposed Model

For most data-independent hashing methods like LSH, they use linear random projections which lack good discrimination over data. So we extract the underlying structure of the data leveraging the RBF kernel mapping [7,10]. The nonlinear embedding algorithm we choose is formulated as:

$$F(\mathbf{x}) = \phi(\mathbf{x})\mathbf{P}, \tag{5}$$

where $\phi(\mathbf{x}) = \left[exp(\|\mathbf{x} - \mathbf{a}_1\|^2 / \sigma), \cdots, exp(\|\mathbf{x} - \mathbf{a}_m\|^2 / \sigma) \right]$, $\{\mathbf{a}_j\}_{j=1}^{m}$ are the randomly chosen anchors. To learn a discrete matrix, combining the relaxed empirical fitness term from problem (4) and the relaxed regularization term, we propose a novel optimization model as:

$$\min_{\mathbf{B}, \mathbf{W}, \mathbf{U}, F} \ tr(\mathbf{B}^T \mathbf{L} \mathbf{B}) + \omega \|\mathbf{Y} - \mathbf{B} \mathbf{W}\|^2 + \lambda \|\mathbf{W}\|^2$$

$$+ \nu \|\mathbf{B} - F(\mathbf{X})\|^2 + \eta \|\mathbf{P}\|^2 - \rho tr(\mathbf{B}^T \mathbf{V}),$$

$$s.t. \begin{cases} \mathbf{B} \in \{\pm 1\}^{n \times r}, \mathbf{V} \in \mathbb{R}^{n \times r}, \\ \mathbf{1}^T \mathbf{V} = 0, \mathbf{V}^T \mathbf{V} = n\mathbf{I}_r. \end{cases} \tag{6}$$

The first term approximates the underlying structure of the data and the second term is the loss function measuring the approximation error between the prediction results and labels. There are several loss function like logistic loss, least square loss and hinge loss function. The least loss function is popular in quantization and classification problems. The norms here refer to the matrix norms induced by vector norms. Therefore, here we choose the least square loss function to evaluate the variance. By minimizing the object function (6), we can get the discriminative hash matrix $\mathbf{B}$, each row of it represents the corresponding binary code.

2.3 Alternating Manipulation

Our hashing problem is a nonlinear mixed-integer program involving a discrete variable $\mathbf{B}$, a continuous variable $\mathbf{V}$ and two regular variables $\mathbf{W}$ and $\mathbf{P}$. In this case, problem (6) is a NP-hard problem and also difficult to find a approximate solution. As we can see, only when $\omega, \lambda, \nu, \eta, \rho = 0$ and $r = 1$, problem (6) is a Max-Cut problem [9], there exists no polynomial-time algorithm or approximate solution which can achieve the global optimum unless P = NP [4]. For this purpose, an effective solution to the problem (6) is using alternating manipulation algorithm. In this way, our hashing problem can be decomposed into four subproblems: $\mathbf{B}$-subproblem, $\mathbf{V}$-subproblem, $\mathbf{W}$-subproblem and $\mathbf{P}$-subproblem.

B-Subproblem. Notice that term $\lambda\|\mathbf{W}\|^2$ and $\eta\|\mathbf{P}\|^2$ are constant for $\mathbf{B}$-subproblem, we simply cast out these two terms. Acknowledging the linear algebra theorem that $\|\cdot\|^2 = tr((\cdot)^T(\cdot))$, the rest regularization terms can be changed into $tr(\cdot)$ form for succeeding manipulation. And we have the anchor graph Laplacian $\mathbf{L} = \mathbf{I}_n - \mathbf{A}$. By these three steps, the optimization model can be simplified as:

$$\max_{\mathbf{B}} \; tr(\mathbf{B}^T \mathbf{A} \mathbf{B}) - \omega tr(\mathbf{Y} - \mathbf{B}\mathbf{W})(\mathbf{Y}^T - \mathbf{W}^T \mathbf{B}^T)$$
$$- \nu tr\Big((\mathbf{B} - F(\mathbf{X}))(\mathbf{B}^T - F^T(\mathbf{X}))\Big) + \rho tr(\mathbf{B}^T \mathbf{V}), \tag{7}$$
$$s.t. \begin{cases} \mathbf{B} \in \{\pm 1\}^{n \times r}, \mathbf{V} \in \mathbb{R}^{n \times r}, \\ \mathbf{1}^T \mathbf{V} = 0, \mathbf{V}^T \mathbf{V} = n\mathbf{I}_r. \end{cases}$$

In order to directly optimize problem (7), we can further simplify it into

$$\max_{\mathbf{B} \in \{\pm 1\}^{n \times r}} \; tr(\mathbf{B}\mathbf{B}^T \mathbf{A} - \omega \mathbf{B}\mathbf{W}\mathbf{W}^T \mathbf{B}^T + \mathbf{B}C_1), \tag{8}$$

where $C_1 = 2\omega \mathbf{W}\mathbf{Y}^T + 2F^T(\mathbf{X}) + \rho \mathbf{V}^T$ is the constant term that can be obtained easily. However, it is NP hard to achieve $\mathbf{B}$. To solve this problem, we use DCC algorithm [14] to learn $\mathbf{B}$ bit by bit. In the iteration, we draw out the lth, $l = 1, \cdots, r$ column of $\mathbf{B}$ as $\mathbf{b}_l$ and the $\mathbf{B}'$ is the matrix $\mathbf{B}$ which excludes $\mathbf{b}_l$. In the same way, $\mathbf{w}_l$ and $\mathbf{c}_l$ respectively are the lth column of $\mathbf{W}$

Algorithm 1. B-subproblem of LISH.

    Input : $\mathbf{B}^0 \in \{-1,1\}^{n \times r}, \mathbf{V} \in \mathbb{R}^{n \times r}, \mathbf{P} \in \mathbb{R}^{m \times r}, \mathbf{W} \in \mathbb{R}^{r \times c}$ and $\mathbf{Y} \in \mathbb{R}^{n \times c}$;
    Output: $\mathbf{B} = [b_1, b_2, \cdots b_l, \cdots b_r]$;
1   $l := 1$;
2 **repeat**
3     $k := 1$;
4   **repeat**
5        $C_1 = 2\omega \mathbf{W}\mathbf{Y}^T + 2F^T(\mathbf{X}) + \rho \mathbf{V}^T$;
6        $C_2 = \left(c_l^T - 2\omega w_l^T \mathbf{W}'^T_l \mathbf{B}'^T_l \right)$;
7        $\nabla f(b_l^k) = 2\mathbf{A}b_l^k + C_2^T$;
8        $\mathbf{b}_l^{k+1} = sgn(2\mathbf{A}b_l^k + C_2^T)$;
9        $k := k + 1$;
10   **until** $\mathbf{b}_l$ *converges*;
11     $l := l + 1$;
12 **until** $l = r$;
13 **return B**

---

and C, $\mathbf{W}'$ and C' are the the matrix $\mathbf{W}$ and C which exclude $\mathbf{w}_l$ and $\mathbf{c}_l$. Then we reformulate the problem (8) as:

$$\max_{\mathbf{b}_l \in \{\pm 1\}^{n \times 1}} tr(\mathbf{b}_l \mathbf{b}_l^T \mathbf{A} - 2\omega \mathbf{b}_l \mathbf{w}_l^T \mathbf{W}_l'^T \mathbf{B}_l'^T + \mathbf{b}_l c^T), \qquad (9)$$

which is equivalent to

$$\max_{\mathbf{b}_l \in \{\pm 1\}^{n \times 1}} f(\mathbf{b}_l) = tr(\mathbf{b}_l^T \mathbf{A} \mathbf{b}_l + C_2 \mathbf{b}_l), \qquad (10)$$

where $C_2 = \left(c_l^T - 2\omega w_l^T \mathbf{W}'^T_l \mathbf{B}'^T_l \right)$.

In the k-th iteration, a proxy function $\hat{f}_k(\mathbf{b}_l)$ is defined to linearize $f(\mathbf{b}_l)$ at the point $\mathbf{b}_l^k$. This majorization method was first introduced by [2]. Since $\mathbf{A}$ is positive semidefinite, f is a convex function then we have $\hat{f}_k(\mathbf{b}_l) \leq f(\mathbf{b}_l)$, the fact $f(\mathbf{b}_l^{k+1}) \geq \hat{f}_k(\mathbf{b}_l^{k+1}) \geq \hat{f}_k(\mathbf{b}_l^k) \equiv f(\mathbf{b}_l^k)$ guarantees that $f(\mathbf{b}_l^k)$ and $\mathbf{b}_l^k$ could converge [9]. The next discrete point $\mathbf{b}_l^{k+1}$ is

$$\mathbf{b}_l^{k+1} \in arg \max_{\mathbf{b}_l \in \{\pm 1\}^{n \times 1}} \hat{f}_k(\mathbf{b}_l) = f(\mathbf{b}_l^k) + \left\langle \nabla f(\mathbf{b}_l^k), (\mathbf{b}_l - \mathbf{b}_l^k) \right\rangle, \qquad (11)$$

where $\nabla f(\mathbf{b}_l^k) = 2\mathbf{A}b_l^k + C_2^T$, thus we have $\hat{f}_k(\mathbf{b}_l) = f(\mathbf{b}_l^k) - (2\mathbf{A}b_l^k + C_2^T)\mathbf{b}_l^k + (2\mathbf{A}b_l^k + C_2^T)\mathbf{b}_l$, we can clearly see that the first and second term are constant. To maximize the problem (11), the following function is proposed:

$$\mathbf{b}_l^{k+1} = sgn(2\mathbf{A}b_l^k + C_2^T). \qquad (12)$$

The procedure to solve the **B**-subproblem is described in Algorithm 1. Figure 1 shows that our algorithm converges rapidly to reach the optimal solution.

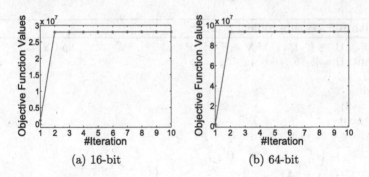

Fig. 1. Convergence curves of Eq. (11) in 16 and 64-bit.

V-Subproblem. When $\mathbf{B}$, $\mathbf{P}$ and $\mathbf{W}$ are fixed in the problem (6), the objective function of the $\mathbf{V}$-subproblem is

$$\max_{\mathbf{V}\in\mathbb{R}^{n\times r}} tr(\mathbf{B}^T\mathbf{V}),$$
$$s.t. \ \mathbf{1}^T\mathbf{V} = 0, \mathbf{V}^T\mathbf{V} = n\mathbf{I}_r. \tag{13}$$

We solve the problem with the aid of singular value decomposition (SVD). $\mathbf{B}^*$ denotes a zero-mean matrix with row-wise, where $\mathbf{B}^* = \mathbf{B}(\mathbf{I}^n - \frac{1}{n}\mathbf{1}\mathbf{1}^T)$. We write the $\mathbf{B}^*$ as $\mathbf{B}^* = \mathbf{J}\mathbf{\Sigma}\mathbf{K}^T$ by using the SVD, where $\mathbf{J} \in \mathbb{R}^{n\times r'}$ and $\mathbf{K} \in \mathbb{R}^{r\times r'}$ are left and right matrices of singular vectors correspondingly. Then we apply eigendecomposition for the small $r \times r$ matrix $\mathbf{B}^{*T}\mathbf{B}^* = [\mathbf{K} \ \bar{\mathbf{K}}]\begin{bmatrix}\Sigma_b^2 & 0 \\ 0 & 0\end{bmatrix}[\mathbf{K} \ \bar{\mathbf{K}}]^T$

after introducing a matrix $\bar{\mathbf{K}} \in \mathbb{R}^{r\times(r-r')}$ by employing the Gram-Schmidt orthogonalization to the zero eigenvalues. Under the restriction of $\mathbf{J}^T\mathbf{J} = \mathbf{I}_{r-r'}$, $[\mathbf{J} \ \mathbf{1}]^T\bar{\mathbf{J}} = 0$, we naturally add a matrix $\bar{\mathbf{J}} \in \mathbb{R}^{n\times(r-r')}$ to satisfy the constraint $\mathbf{1}^T\mathbf{V} = 0$. Then we have $\mathbf{J} = \mathbf{B}^*\mathbf{K}\mathbf{\Sigma}^{-1}$. By the adoption of relevant theorem in DGH [9] we can simply have

$$\mathbf{V}^* = \sqrt{n}\,[\mathbf{J} \ \bar{\mathbf{J}}][\mathbf{K} \ \bar{\mathbf{K}}], \tag{14}$$

which is a high degree approximate solution to the $\mathbf{V}$-subproblem.

P-subproblem and W-subproblem. When $\mathbf{B}$, $\mathbf{V}$ and $\mathbf{W}$ are fixed in the problem (6). Taking the advantage of Eq. (5), the objective function of the $\mathbf{P}$-subproblem is

$$\min_{\mathbf{P}} \ \|\mathbf{B} - \mathbf{P}^T\phi(\mathbf{x})\|^2 + \frac{\eta}{\nu}\|\mathbf{P}\|^2. \tag{15}$$

When $\mathbf{B}$, $\mathbf{V}$ and $\mathbf{P}$ are fixed in the problem (6), the objective function of the $\mathbf{W}$-subproblem is

$$\min_{\mathbf{P}} \ \|\mathbf{Y} - \mathbf{B}\mathbf{W}\|^2 + \frac{\lambda}{\omega}\|\mathbf{W}\|^2. \tag{16}$$

Algorithm 2. Local and Inner Data Structure Supervised Hashing (LISH)

Input : $X \in \mathbb{R}^{n \times d}$: training data,
        r: hash code length,
        m: number of anchor points,
        ω, λ, ν, η and ρ: initial parameters;
Output: $\mathbf{B} \in \{-1, 1\}^{n \times r}$: binary codes,
        $P \in \mathbb{R}^{m \times r}$: projection matrix;

1 **Initialization**: randomly initialize $\mathbf{B}$, $\mathbf{P}$, $\mathbf{W}$ and $\mathbf{V}$;
2 **repeat**
3   **B-subproblem:** Update $\mathbf{B}$ according to **Algorithm 1**;
4   **P-subproblem:** Update $\mathbf{P}$ according to **Eq. (17)**;
5   **W-subproblem:** Update $\mathbf{W}$ according to **Eq. (18)**;
6   **V-subproblem:** Update $\mathbf{V}$ according to **Eq. (14)**;
7 **until** *there is no change to* $\mathbf{B}, \mathbf{P}, \mathbf{W}, \mathbf{V}$;
8 **return** $\mathbf{B}$, $\mathbf{P}$, $\mathbf{W}$ *and* $\mathbf{V}$

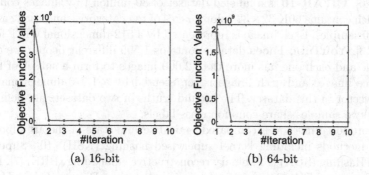

(a) 16-bit   (b) 64-bit

Fig. 2. Convergence curves of Eq. (6) in 16 and 64-bit.

For these two subproblems, we can get the solution by the regularized least squares problem. Then we have the closed-form solution:

$$\mathbf{P} = \left(\phi(\mathbf{X})^T \phi(\mathbf{X}) + \frac{\eta}{\nu} \mathbf{I}_r \right)^{-1} \phi(\mathbf{X})^T \mathbf{B}, \qquad (17)$$

$$\mathbf{W} = \left(\mathbf{B}^T \mathbf{B} + \frac{\lambda}{\omega} \mathbf{I}_r \right)^{-1} \mathbf{B}^T \mathbf{Y}. \qquad (18)$$

We implement our idea by Algorithm 2. Figure 2 shows the objective function values and the number of iterations where we can see the rapid convergence of our algorithm.

3 Experiments

We conduct extensive experiments to evaluate our proposed method on three publicly available large-scale image datasets: **CIFAR-10** [6] and **YouTube**

Table 1. Results with performance (MAP) and training time (seconds) of different comparing algorithms on **CIFAR-10** dataset.

Method	$n = 1000$				$n = 2000$			
	MAP			Time	MAP			Time
	8-bit	16-bit	32-bit	32-bit	8-bit	16-bit	32-bit	32-bit
BRE	0.1403	0.1439	0.1618	79.01	0.1339	0.1507	0.1612	402.39
LFH	0.1146	0.1796	0.1979	0.53	0.1402	0.1539	0.2000	0.80
KSH	**0.2182**	0.2446	0.2618	117.48	0.2369	0.2696	0.2975	412.02
SDH	0.1512	0.2491	0.2751	0.22	0.1913	0.2744	0.3170	0.90
LISH	0.2122	**0.2524**	**0.2834**	6.93	**0.2420**	**0.2862**	**0.3173**	29.68
DGH	0.1163	0.1142	0.1135	0.76	0.1315	0.1346	0.1332	0.97

Faces [20]. **CIFAR-10** is a labeled dataset of 80-million tiny images collection [17], which contains 60K 32×32 color images of ten categories and each category has 6,000 samples. Each image is represented by a 512-dimensional GIST feature vector [13]. **YouTube Faces** dataset contains 1,595 different people, we choose 38 people and each one has more than 2,000 images to form a subset of totally 100K face images and each image is represented by a 1,770-dimensional LBP feature vector in this dataset. The ground truths of two datasets are defined by whether two samples share common class labels.

The proposed method is compared against several state-of-the-art supervised hashing methods including kernel supervised hashing (KSH) [10], Supervised Discrete Hashing (SDH) [14], binary reconstructive embedding (BRE) [7], Latent Factor Hashing (LFH) [24] and an unsupervised method Discrete Graph Hashing (DGH) [9]. We use the public codes and suggested parameters of these methods from authors. In consideration of the fair comparison, we choose 1000 and 2000 samples for learning. For KSH, DGH and SDH, we randomly select 300 anchors when using 1000 samples and 500 anchors when using 2000 samples.

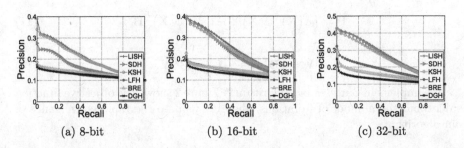

(a) 8-bit   (b) 16-bit   (c) 32-bit

Fig. 3. Different hashing methods' Precision-recall curves using 8, 16 and 32-bit codes on the **CIFAR-10** dataset with 2000 training samples.

In the experimental part, we evaluate the above hashing methods in term of two standard criterions: mean average precision (MAP) and precision-recall (PR) curve. The performance curves are shown from Fig. 3 to Fig. 4. The comparison results of MAP and the training time are presented from Table 1 to Table 2.

3.1 CIFAR-10

The comparative results in MAP are shown in Table 1 and the precision-recall curves using 8, 16, 32 bits are shown in Fig. 3. We partition the entire dataset into two parts: a training subset of 59,000 images and a test subset of 1,000 images. We also randomly select 300 anchors when we use 1000 train data and 500 anchors for 2000 train data. We generate 8, 16 and 32-bit binary codes by LFH, SDH, BRE, KSH, DGH and our proposed LISH.

As shown in Table 1 and Fig. 3, we can see the performance of all methods become higher as the training samples increase. It is understandable that the unsupervised hash method DGH do not perform well as others. Nevertheless, LFH have a lower accuracy at 8-bit than DGH when $n = 1000$. KSH's performance is slight higher than our method at 8-bit, although LISH outperform in all other situations. Moreover, KSH costs much longer time than ours. LISH achieves the highest retrieval accuracy (precision and MAP) with almost all code lengths and significantly outperforms the compared hashing methods at 16-bit. It is clear that LISH is very effective with different code lengths for **CIFAR-10**.

3.2 YouTube Faces

In **YouTube Faces** dataset, it contains 98,000 images of 1595 individuals, 38 people are chosen and each one has more than 2,000 images to form a subset of totally 100K faces and the test set is constituted of 3.8K images which are sampled from the 38 classes. Evaluation results in term of Precision and MAP are shown in the Table 2 and Fig. 4 respectively, we can see from Fig. 4, LFH

Table 2. Results with performance (MAP) and training time (seconds) of different comparing algorithms on **YouTube Faces** dataset.

Method	$n = 1000$				$n = 2000$			
	MAP			Time	MAP			Time
	16-bit	32-bit	64-bit	64-bit	16-bit	32-bit	64-bit	64-bit
BRE	0.4982	0.5909	0.6437	605.09	0.5384	0.5932	0.6479	2170.45
LFH	0.3731	0.7011	0.8442	3.42	0.4230	0.7242	0.8532	4.30
KSH	**0.8454**	0.9003	0.9250	214.67	0.8969	0.9414	0.9532	870.09
SDH	0.8367	0.7539	0.9278	28.60	0.9030	0.8191	0.9597	70.02
LISH	0.8408	**0.9071**	**0.9326**	19.91	**0.9049**	**0.9770**	**0.9630**	60.05
DGH	0.3088	0.4437	0.5286	0.55	0.1876	0.2920	0.2952	2.37

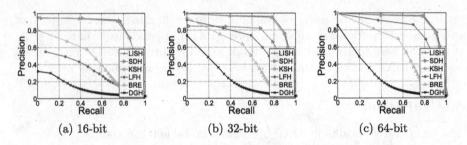

(a) 16-bit   (b) 32-bit   (c) 64-bit

Fig. 4. Different hashing methods' Precision-recall curves using 16, 32 and 64-bit codes on the **YouTube Faces** dataset with 2000 training samples.

perform badly in **YouTube Faces**, the reason may be it overfit the **cifar-10**. As the Table 2 shows, KSH have higher performance at short bits. Since **YouTube Faces** dataset is made up of people's faces whose characteristics are obvious to recognize, the results of BRE, KSH, SDH and our method increase rapidly. Our method integrates the merits of those methods to have superior performance both in time cost and retrieval accuracy. Especially when $n = 1000$ at 32-bit, its MAP is 16.89% higher than SDH.

4   Conclusions

In this paper, we exploited underlying manifold structure of samples by graph Laplacian. The approximate anchor graph was used to save training cost. To capture and preserve the semantic label information in the Hamming space, we explicitly formulated the tractable optimization function integrated with ℓ_2 loss and decomposed it into several sub-problems which could be iteratively solved by our algorithm. We proposed a discrete supervised paradigm to directly generate hash codes without continuous relaxation, by working in the discrete code space, the retrieval accuracy of the short binary codes can be boosted. Empirical evaluations in retrieving semantically similar neighbors on three benchmark databases showed that our method has superior performance over state-of-the-arts.

Acknowledgments. This work was supported in part by the National Natural Science Foundation of China under Project 61572108, Project 61632007 and Project 61502081, the National Thousand-Young-Talents Program of China, and the Fundamental Research Funds for the Central Universities under Project ZYGX2014Z007 and Project ZYGX2015J055.

References

1. Bentley, J.L.: Multidimensional binary search trees used for associative searching. Commun. ACM **18**(9), 509–517 (1975)
2. De Leeuw, J.: Applications of convex analysis to multidimensional scaling. Department of Statistics, UCLA (2005)

3. Gionis, A., Indyk, P., Motwani, R., et al.: Similarity search in high dimensions via hashing. In: VLDB, vol. 99, pp. 518–529 (1999)
4. Håstad, J.: Some optimal inapproximability results. J. ACM **48**(4), 798–859 (2001)
5. Hu, M., Yang, Y., Shen, F., Zhang, L., Shen, H.T., Li, X.: Robust web image annotation via exploring multi-facet and structural knowledge. IEEE TIP **26**(10), 4871–4884 (2017)
6. Krizhevsky, A., Hinton, G.: Learning multiple layers of features from tiny images (2009)
7. Kulis, B., Darrell, T.: Learning to hash with binary reconstructive embeddings. In: Proceedings of NIPS, pp. 1042–1050 (2009)
8. Kulis, B., Jain, P., Grauman, K.: Fast similarity search for learned metrics. IEEE TPAMI **31**(12), 2143–2157 (2009)
9. Liu, W., Mu, C., Kumar, S., Chang, S-F.: Discrete graph hashing. In: Proceedings of NIPS, pp. 3419–3427 (2014)
10. Liu, W., Wang, J., Ji, R., Jiang, Y-G., Chang, S-F.: Supervised hashing with kernels. In: Proceedings of CVPR, pp. 2074–2081 (2012)
11. Liu, W., Wang, J., Kumar, S., Chang, S-F.: Hashing with graphs. In: Proceedings of ICML, pp. 1–8 (2011)
12. Luo, Y., Yang, Y., Shen, F., Huang, Z., Zhou, P., Shen, H.T.: Robust discrete code modeling for supervised hashing. Pattern Recognit. (2017). doi:10.1016/j.patcog.2017.02.034
13. Oliva, A., Torralba, A.: Modeling the shape of the scene: a holistic representation of the spatial envelope. IJCV **42**(3), 145–175 (2001)
14. Shen, F., Shen, C., Liu, W., Shen, H.T.: Supervised discrete hashing. In: Proceedings of CVPR, pp. 37–45 (2015)
15. Shen, F., Yang, Y., Liu, L., Liu, W., Tao, D., Shen, H.T.: Asymmetric binary coding for image search. IEEE Trans. Multimedia **19**(9), 2022–2032 (2017). doi:10.1109/TMM.2017.2699863
16. Strecha, C., Bronstein, A.M., Bronstein, M.M., Fua, P.: LDAHash: improved matching with smaller descriptors. IEEE TPAMI **34**(1), 66–78 (2012)
17. Torralba, A., Fergus, R., Freeman, W.T.: 80 million tiny images: a large data set for nonparametric object and scene recognition. IEEE TPAMI **30**(11), 1958–1970 (2008)
18. Wang, J., Kumar, S., Chang, S.-F.: Semi-supervised hashing for large-scale search. IEEE TPAMI **34**(12), 2393–2406 (2012)
19. Weiss, Y., Torralba, A., Fergus, R.: Spectral hashing. In: Proceedings of NIPS, pp. 1753–1760 (2008)
20. Wolf, L., Hassner, T., Maoz, I.: Face recognition in unconstrained videos with matched background similarity. In: Proceedings of CVPR, pp. 529–534. IEEE (2011)
21. Yang, Y., Luo, Y., Chen, W., Shen, F., Shao, J., Shen, H.T.: Zero-shot hashing via transferring supervised knowledge. In: Proceedings of ACM MM, pp. 1286–1295 (2016)
22. Yang, Y., Shen, F., Huang, Z., Shen, H.T., Li, X.: Discrete nonnegative spectral clustering. In: IEEE TKDE (2017)
23. Yang, Y., Shen, F., Shen, H.T., Li, H., Li, X.: Robust discrete spectral hashing for large-scale image semantic indexing. IEEE TBD **1**(4), 162–171 (2015)
24. Zhang, P., Zhang, W., Li, W.-J., Guo, M.: Supervised hashing with latent factor models. In: Proceedings of SIGIR, pp. 173–182 (2014)

Learning Robust Graph Hashing for Efficient Similarity Search

Luyao Liu[1(✉)], Lei Zhu[1(✉)], and Zhihui Li[2(✉)]

[1] School of Information Technology and Electrical Engineering,
The University of Queensland, Brisbane, Australia
luyao.liu@uq.edu.au, leizhu0608@gmail.com
[2] Beijing Etrol Technologies Co., Ltd, Beijing, China
zhihuilics@gmail.com

Abstract. Unsupervised hashing has recently drawn much attention in efficient similarity search for its desirable advantages of low storage cost, fast search speed, semantic label independence. Among the existing solutions, graph hashing makes a significant contribution as it could effectively preserve the neighbourhood data similarities into binary codes via spectral analysis. However, existing graph hashing methods separate graph construction and hashing learning into two independent processes. This two-step design may lead to sub-optimal results. Furthermore, features of data samples may unfortunately contain noises that will make the built graph less reliable. In this paper, we propose a Robust Graph Hashing (RGH) to address these problems. RGH automatically learns robust graph based on self-representation of samples to alleviate the noises. Moreover, it seamlessly integrates graph construction and hashing learning into a unified learning framework. The learning process ensures the optimal graph to be constructed for subsequent hashing learning, and simultaneously the hashing codes can well preserve similarities of data samples. An effective optimization method is devised to iteratively solve the formulated problem. Experimental results on publicly available image datasets validate the superior performance of RGH compared with several state-of-the-art hashing methods.

1 Introduction

Similarity search plays an important role in machine learning [5], data mining [10], and database [1]. However, for large-scale high-dimensional data, similarity search becomes dramatically time and memory consuming. Therefore, it is important to accelerate the similarity search while satisfying memory savings.

Hashing [2, 29–31, 34–36] is an effective indexing technique that can achieve the above objective. With binary embedding of hashing, the original time-consuming similarity computation is transformed to efficient bit operations. The similarity search process could be greatly accelerated with constant or linear time complexity [33]. And moreover, binary representation could significantly shrink the memory cost of data samples, and thus accommodate large-scale similarity

© Springer International Publishing AG 2017
Z. Huang et al. (Eds.): ADC 2017, LNCS 10538, pp. 110–122, 2017.
DOI: 10.1007/978-3-319-68155-9_9

search with only limited memory. Due to these desirable advantages, hashing has been widely explored for efficient large-scale similarity search [3,19,25,27,32].

Locality Sensitive Hashing (LSH) [6,11,14] is a typical data-independent hashing technique. It simply exploits random mapping to project the data samples into binary Hamming space. In practice, for its ignorance on underlying data characteristics, LSH generally requires more hashing tables and longer codes to achieve satisfactory performance [19,23]. This requirement will unfortunately bring in more storage cost. To solve the limitation, learning-based hashing methods [12,16–20,22,23,28,33] are proposed to generate the projected hashing codes with various advanced machine learning models. Their basic learning principal is that if two data samples are similar in the original space, they should have a small Hamming distance between their corresponding binary codes.

According to the learning dependence on semantic labels, existing learning-based hashing methods can be categorized into two major families: unsupervised hashing [9,12,16,19,26,28] and supervised hashing [17,18,22,33]. Supervised hashing learns effective binary codes based on large amounts of semantic labels. It usually achieves better performance than unsupervised hashing methods. However, high-quality labels are hardly or expensive to obtain in many applications. Unsupervised hashing is developed without this limitation. Graph hashing [12,16,19,20,23,28] is one of the representative unsupervised hashing methods that achieve state-of-the-art performance. It generally operates with two subsequent steps: First, similarity graph is constructed by computing similarities of data samples. Second, hashing codes and functions are learned on the built graph with spectral analysis. Spectral Hashing (SPH) [28], as a pioneering graph hashing methods, extends spectral clustering to hashing. In SPH, hashing codes are computed by eigenvalue decomposition on Laplacian matrix computed from similarity graph. Hashing functions are constructed with an efficient Nystrom method. Anchor Graph Hashing (AGH) [20] utilizes low-rank matrix to approximate the similarity graph to reduce the complexity of SPH. In this method, hashing functions are learned by binarizing the eigen-functions. Inductive Manifolds Hashing (IMH) [23] considers the intrinsic manifold structure and generates non-linear hashing functions. Scalable Graph Hashing (SGH) [12] leverages feature transformation to approximate the similarity graph, and thus avoids explicit graph computing. In SGH, hashing functions are learned in a bit-wise manner with a sequential learning. Discrete Graph Hashing (DGH) [19] proposes a discrete optimization approach to learn the binary codes in binary Hamming space directly. Discrete Proximal Linearized Minimization (DPLM) [24] can also handle the discrete constraint imposed on graph hashing. In DPLM, hashing codes are solved by iterative procedures with each one admitting an analytical solution.

Although graph hashing can achieve promising results, there still exist several problems that impede its performance.

- In graph hashing methods, hashing learning fully relies on a previously constructed similarity graph. The separation of graph construction and hashing learning may lead to sub-optimal result, as the manually constructed graph may not be optimal for the subsequent hashing learning.

– Graph is generally built with specific distance measures. Extra parameters, such as the bandwidth parameter for similarity computation and the number of nearest neighbours for graph construction, will inevitably be brought into graph construction. These parameters are highly dependent on experimental experience. They may not achieve the satisfactory performance due to the high complexity of graph modelling.
– The similarity graph may be unreliable as the data samples from the real world always contain adverse noise. These noises will damage the local manifold structure and bring negative impact on the hashing performance.

In this paper, we propose a Robust Graph Hashing (RGH) to address the aforementioned problems in existing graph hashing methods. Specifically, RGH automatically learns a robust graph based on self-representation of data samples to avoid the adverse noises. Moreover, it seamlessly integrates graph construction and hashing learning into a unified learning framework where two learning processes can mutually reinforce each other. An effective optimization method is devised to iteratively solve the formulated learning problem. Experimental results on two publicly available image datasets demonstrate the superior performance of RGH compared with several state-of-the-art hashing methods.

It is worthwhile to highlight the key contributions of our proposed approach:

– We formulate a unified learning framework that integrates graph construction and hashing learning. The framework ensures that the optimal graph can be automatically constructed for hashing learning, and simultaneously the hashing codes can effectively preserve discriminative information of original data samples.
– We exploit self-representation of data samples to construct the similarity graph. It can avoid parametric distance measures. In addition, we introduce an error matrix that mitigates the negative interferences and makes the constructed graph more robust.

The rest of the paper is structured as follows. Section 2 gives the details of the proposed approach. In Sect. 3, we present experimental results and analysis. Section 4 concludes the paper.

2   Methodology

2.1   Overall Objective Formulation

As indicated by the principal of subspace clustering [7,8,15,21], a point set in a union of subspaces generally locates in an underlying low-dimension subspace instead of distributing equably in the entire space. In presentation, it means that a data sample could be expressed as a linear combination of other data samples in low-dimension subspace.

We construct the graph with the inspiration of above principal. For a given data set $X = \{x_1, x_2, \ldots, x_n\} \in \Re^{n \times d}$ (n is the number of data samples and d is the dimension of feature representation). Each sample can be represented with the representation of the remaining samples in data set. Formally, X can be represented as $X = XZ$, where $Z = \{z_1, z_2, \ldots, z_n\} \in \Re^{n \times n}$ is the coefficient matrix, z_i is the representation of the raw data x_i based on the subspaces. Z actually characterizes the correlations of data samples and it can be exploited for affinity matrix construction of graph. On the other hand, real world data samples inevitably contain noises that are adverse for subsequent learning. In this paper, we resort to a outlying entries matrix $E = \{e_1, e_2, \ldots, e_n\} \in \Re^{n \times d}$ to alleviate them. Then, $X = XZ$ is rewritten as $X = XZ + E$. In optimization, we try to minimize the difference between X and $XZ + E$ and automatically learn robust Z. This process can be represented as

$$\min_{Z,E} \|X - XZ - E\|_F^2 + \lambda \|E\|_1 \quad s.t. \ diag(Z) = 0, Z^T \mathbf{1} = 1 \tag{1}$$

where $\lambda > 0$ plays the trade-off between two terms, the constraint $diag(Z) = 0$ avoids the trivial case that a data point x_i is represented with a combination of itself. The constraint $Z^T \mathbf{1} = 1$ indicates that the data point locates in a union of affine subspaces. Data sample can be linearly represented with other samples. E is introduced to strengthen the robustness of the model to the corrupted noise involved in the real world data. The l_1 norm on E is to promote the sparseness of the noises.

The non-zero elements in z_i only correspond to the points from the same subspace. With Z, affinity matrix of graph can be defined as

$$W = \frac{|Z| + |Z^T|}{2} \tag{2}$$

As long as we have the affinity matrix, a conceptive graph hashing can be performed on such a self-representation based affinity matrix. The basic principal is that close data samples in original space should have small Hamming distance between their corresponding hashing codes. Formally, the hashing codes can be obtained by solving

$$\min_Y Tr(Y^T(D - W)Y) \quad s.t. \ Y^T Y = I, Y \in \{-1, 1\}^{n \times c} \tag{3}$$

where D is the diagonal matrix with diagonal elements and the i_{th} entry is defined as $d_i = \sum_j \frac{|z_{ij}| + |z_{ji}|}{2}$, $Y \in \{-1, 1\}^{n \times c}$ is the hashing codes of data samples. The constraint $Y^T Y = I$ ensures the different hashing bits to be independent with each other.

Conventional graph hashing methods separate graph construction and hashing learning in two subsequent process, which may lead to suboptimal results. Different from them, in this paper, we integrate them into a unified learning framework. We derive the overall objective formulation as

$$\min_{Z,E,Y} \|X - XZ - E\|_F^2 + \beta Tr(Y^T(D - W)Y) + \lambda \|E\|_1$$
$$s.t. \ diag(Z) = 0, Z^T \mathbf{1} = 1, Y^T Y = I, Y \in \{-1, 1\}^{n \times c} \tag{4}$$

where $\beta > 0$ balances robust graph construction and hashing learning.

2.2 Optimization

Directly solving hashing codes is NP-hard. In this paper, we first relax the discrete constraint to continues ones and then propose an iterative algorithm to compute the approximate results. Specifically, we iteratively optimize each variable by fixing the others.

Update Matrix Z. By fixing Y, E, the optimization for Z could be represented as the following problem:

$$\min_{Z} \|X - XZ - E\|_F^2 + \beta Tr(Y^T(D - \frac{|Z| + |Z^T|}{2})Y)$$

$$s.t.\ diag(Z) = 0, Z^T\mathbf{1} = 1. \tag{5}$$

As derived in [8], the solution of above problem could be rewritten as

$$z_k = \begin{cases} v_k - \frac{\beta p_k}{4}, & if\ v_k > \frac{\beta p_k}{4}, \\ v_k + \frac{\beta p_k}{4}, & if\ v_k < -\frac{\beta p_k}{4}, \\ 0, & otherwise. \end{cases} \tag{6}$$

where z^T presents the i-th row of Z, if $k = i, z_k = z_i = 0$ and $v = \frac{X - (XZ - xz^T) - E)^T x}{x^T x}$, p presents the i-th row of P and $P_{ij} = \|y^i - y^j\|_F^2$ where y^i presents the i-th row of matrix Y.

Update Matrix E. By fixing Z, Y, the optimization for E can be represented as

$$\min_{E} \|X - XZ - E\|_F^2 + \lambda\|E\|_1 \tag{7}$$

The solution of the above problem can be obtained as [8]:

$$E_{ij} = \begin{cases} (X - XZ)_{ij} - \frac{\lambda}{2}, & if\ (X - XZ)_{ij} > \frac{\lambda}{2}, \\ (X - XZ)_{ij} + \frac{\lambda}{2}, & if\ (X - XZ)_{ij} < -\frac{\lambda}{2}, \\ 0, & otherwise. \end{cases} \tag{8}$$

Update Matrix Y. By fixing Z, E, the optimization for Y can be formulated as

$$\min_{Y} Tr(Y^T(D - \frac{|Z| + |Z^T|}{2})Y)\quad s.t.\ Y^TY = I. \tag{9}$$

Let $A = D - \frac{|Z| + |Z^T|}{2}$, Eq. (9) becomes:

$$\min_{Y} Tr(Y^T AY),\ s.t. Y^TY = I, \tag{10}$$

The optimal solution of Y are the eigenvectors corresponding to the smallest k eigenvalue of the Laplacian matrix A.

We iteratively conduct the above procedures and obtain the optimal relaxed results of hashing codes.

Algorithm 1. Solving the relaxed hashing codes

Input: Training data $X = \{x_1, x_2, \ldots, x_n\} \in \Re^{n \times d}$, code length c, parameters λ, β.
Output: Relaxed hashing codes Y.

    Initialize Y as sparse matrix, $E = 0$, parameters λ, β.
    repeat
        Update Z by Eq. (6):
            For $k = i, z_k = 0$; For $k \neq i$,

$$z_k = \begin{cases} v_k - \frac{\beta p_k}{4}, & if \ v_k > \frac{\beta p_k}{4}, \\ v_k + \frac{\beta p_k}{4}, & if \ v_k < -\frac{\beta p_k}{4}, \\ 0, & otherwise. \end{cases}$$

        Update E by Eq. (8):

$$E_{ij} = \begin{cases} (X - XZ)_{ij} - \frac{\lambda}{2}, & if \ (X - XZ)_{ij} > \frac{\lambda}{2}, \\ (X - XZ)_{ij} + \frac{\lambda}{2}, & if \ (X - XZ)_{ij} < -\frac{\lambda}{2}, \\ 0, & otherwise. \end{cases}$$

        Update Y by Eq. (10):

$$\min_{Y^T Y = I} Tr(Y^T A Y),$$

        Solve it by eigenvalue decomposition of A.
    until Converges or N iteration steps

2.3 Iterative Rotation

Directly binarizing the relaxed hashing codes (continues values) will lead to quantization errors. In this paper, we apply a iterative rotation on the relaxed hashing codes Y to minimize the quantization loss. This process can be formulated as:

$$\min_{B,Q} \|B - YQ\|_F^2, \tag{11}$$

$$s.t. \ B \in \{-1, 1\}^{n \times c}, Q^T Q = I.$$

$Q \in \Re^{c \times c}$ is an arbitrary orthogonal matrix for rotation.

This optimization problem can be solved by applying iteratively alternative minimization as [9].

2.4 Hashing Function Learning

With the hashing codes, we construct the hashing functions. In this paper, we leverage linear projection to achieve the aim for its high efficiency. The objective is to minimize the loss between the hashing codes and the projected ones. The formulation is

$$\min_H \|Y - XH\|_F^2 + \eta \|H\|_F \tag{12}$$

where $\eta > 0$ has the same function with the aforementioned β and λ, $H \in \Re^{d \times c}$ denotes the projection matrix. The optimal H is calculated as

$$H = \left(X^T X + \eta I\right)^{-1} X^T Y \tag{13}$$

With H, hashing functions can be constructed as

$$F(x) = \frac{\text{sgn}(xHQ) + 1}{2} \tag{14}$$

2.5 Online Search

In online search, we first leverage hashing functions to generate hashing codes for queries. Then, the similarities between queries and database samples are calculated with Hamming distance computation. Finally, similarities are ranked according to the distance ascending or descending, and their corresponding data samples are returned.

3 Experiment

3.1 Experimental Dataset

In this paper, two widely used benchmark datasets, *CIFAR-10* [13] and *NUS-WIDE* [4], are applied to evaluate the performance of our method on similarity search.

- *CIFAR-10* has 60,000 32×32 color images which are separated into 10 classes. It has 6,000 images for per class and each images is represented by 512-dimensional GIST feature. For evaluation, we randomly select 1,000 images as our query set and 1,000 images as training images. The rest images are determined as the database images to be retrieved. In this dataset, images are considered to be relevant only if they belong to the same category.
- *NUS-WIDE* is a web image database which contains 269,648 images. It provides ground-truth of 81 concepts. In experiment, we prune the original *NUS-WIDE* to construct a new dataset consisting of 195,834 images by preserving the images that belong to one of the 21 most frequent concepts. Each image is described with 500-D bag-of-words[1]. The dataset partition for evaluation in *NUS-WIDE* is the same with *CIFAR-10*. In *NUS-WIDE*, as images are labelled into multiple concepts, they are determined as relevant if they share at least one concept.

---

[1] SIFT is employed as local feature.

3.2   Evaluation Metrics

In experiment, mean average precision (mAP) [9,12,16,19,26,28] is adopted as the evaluation metric for effectiveness. It is defined as the average precision (AP) of all queries. Larger mAP indicates the better retrieval performance. For a given query, AP is calculated as

$$AP = \frac{1}{NR} \sum_{r=1}^{R} pre(r)rel(r) \tag{15}$$

where R is the total number of retrieved images, NR is the number of relevant images in retrieved set, $pre(r)$ denotes the precision of top r retrieval images, which is defined as the ratio between the number of the relevant images and the number of retrieved images r, and $rel(r)$ is indicator function which equals to 1 if the r_{th} image is relevant to query, and 0 vice versa. In experiments, we set R as 100 to collect experimental results. Furthermore, *Precision-Scope* curve is also reported to illustrate the retrieval performance variations with respect to the number of retrieved images.

Table 1. mAP of all approaches on two datasets The best result in each column is marked with bold.

Methods	CIFAR-10				NUS-WIDE			
	16	32	64	128	16	32	64	128
SH	0.2434	0.2880	0.3201	0.3360	0.4425	0.4547	0.4551	0.4491
AGH	0.2755	0.2995	0.3039	0.3024	0.4321	0.4479	0.4493	0.4559
PCAH	0.2694	0.2989	0.3125	0.3048	0.4525	0.4581	0.4532	0.4420
SGH	0.2713	0.3186	0.3498	0.3838	0.4448	0.4596	0.4621	0.4655
DPLM	0.2603	0.2702	0.2992	0.3011	0.4173	0.4292	0.4331	0.4315
RGH	**0.2809**	**0.3244**	**0.3547**	**0.3871**	**0.4547**	**0.4712**	**0.4818**	**0.4965**

Table 2. Effects of robust graph construction and one-step hashing learning. RGH* is the variant method that removes E in Eq. (4). RGH† is the approach which separates graph construction and hashing learning into two steps.

Methods	NUS-WIDE			
	16	32	64	128
RGH*	0.4408	0.4631	0.4733	0.4857
RGH†	0.4500	0.4612	0.4767	0.4897
RGH	**0.4547**	**0.4712**	**0.4818**	**0.4965**

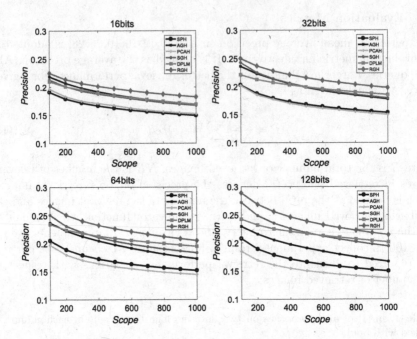

Fig. 1. *Precision-Scope* curves on *CIFAR-10* varying code length.

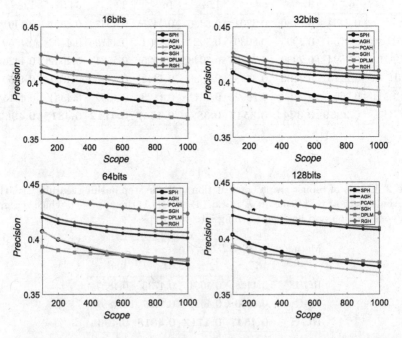

Fig. 2. *Precision-Scope* curves on *NUS-WIDE* varying code length.

3.3   Compared Approaches

We compare RGH with several state-of-the-art unsupervised hashing approaches. They include Spectral Hashing (SPH) [28], Anchor Graph Hashing (AGH) [20], PCA based Hashing (PCAH) [26], Scalable Graph Hashing (SGH) [12], and Discrete Proximal Linearized Minimization (DPLM) [24].

3.4   Implementation Setting

RGH has two parameters, λ and β, in hashing objective function Eq. (4). They are used to play the balance between the formulated regularization terms. In experiment, we set the value of parameters as $\{\lambda = 10^{-3}, \beta = 10^{-1}\}$ and $\{\lambda = 10^{-1}, \beta = 10^{-2}\}$ to achieve the best performance on *CIFAR-10* and *NUS-WIDE* respectively. The maximum number of iterations to solve Eq. (4) and ITQ is set to 20 and 50 respectively.

In experiments, hashing code length on all datasets is varied in the range of [16, 32, 64, 128] to observe the performance. Further, The retrieval scope on two datasets is set from 100 to 1000 with step size 100. In the first step of Algorithm 1, the initial Y is a sparse matrix, values of E are set to 0. When minimizing the quantization error, we initialize Q with an arbitrary orthogonal matrix.

3.5   Experimental Results

We report mAP results of all compared methods in Table 1. The *Precision-Scope* curves on *CIFAR-10* and *NUS-WIDE* are presented in Figs. 1 and 2 respectively. We can easily find that RGH outperforms the competitors on all cases. As shown in Table 1, the largest mAP gain of RGH over the second best approach is about 3.5% and 6.7%, on 16bits of *CIFAR-10* and 128 bits of *NUS-WIDE* respectively. Moreover, Fig. 2 shows that, on *NUS-WIDE* with 16, 64, 128 bits, the performance of RGH is much better than the second best approach with the increasing of code length.

We also compare RGH with two variants of our approach: RGH* and RGH†. RGH* learns hashing codes without any specific sample noise accommodation. In implementation, we remove the variable E in objective function Eq. (4) and conduct hashing learning. RGH† separates graph construction and hashing learning into two subsequent steps. The experimental results are presented in Table 2. From it, we observe that RGH improves the performance from 1% to 3% than other two variants. These results clearly validate the effects of robust graph construction and our one-step graph hashing learning.

4   Conclusion

In this paper, we propose a Robust Graph Hashing (RGH) to solve efficient similarity search. We formulate a unified learning framework that integrates graph construction and hashing learning. The framework guarantees that the optimal graph can be automatically constructed for hashing learning, and the hashing codes can well preserve the original data similarity. RGH automatically learns a robust graph based on self-representation to avoid the parameterized similarity graph construction. And an error matrix is introduced to mitigate the negative interferences of noises and enhance the robustness of the constructed graph. An effective optimization method is devised to iteratively solve the formulated problem. Experimental results demonstrate that the proposed method can achieve superior performance than several state-of-the-art methods.

References

1. Agrawal, R., Faloutsos, C., Swami, A.N.: Efficient similarity search in sequence databases. In: Lomet, D.B. (ed.) FODO 1993. LNCS, vol. 730, pp. 69–84. Springer, Heidelberg (1993). doi:10.1007/3-540-57301-1_5
2. Andoni, A., Indyk, P.: Near-optimal hashing algorithms for approximate nearest neighbor in high dimensions. Commun. ACM **51**(1), 117–122 (2008)
3. Andoni, A., Razenshteyn, I.: Optimal data-dependent hashing for approximate near neighbors. In: Proceedings of Annual ACM Symposium on Theory of Computing, STOC 2015, pp. 793–801. ACM (2015)
4. Chua, T.S., Tang, J., Hong, R., Li, H., Luo, Z., Zheng, Y.: NUS-WIDE: a real-world web image database from national university of singapore. In: Proceedings of ACM International Conference on Image and Video Retrieval, CIVR 2009, pp. 48: 1–48: 9. ACM (2009)
5. Cost, S., Salzberg, S.: A weighted nearest neighbor algorithm for learning with symbolic features. Mach. Learn. **10**(1), 57–78 (1993)
6. Datar, M., Immorlica, N., Indyk, P., Mirrokni, V.S.: Locality-sensitive hashing scheme based on p-stable distributions. In: Proceedings of Annual Symposium on Computational Geometry, SCG 2004, pp. 253–262. ACM (2004)
7. Elhamifar, E., Vidal, R.: Sparse subspace clustering. In: Proceedings of 2009 IEEE Conference on Computer Vision and Pattern Recognition, pp. 2790–2797 (2009)
8. Gao, H., Nie, F., Li, X., Huang, H.: Multi-view subspace clustering. In: Proceedings of 2015 IEEE International Conference on Computer Vision (ICCV), pp. 4238–4246 (2015)
9. Gong, Y., Lazebnik, S., Gordo, A., Perronnin, F.: Iterative quantization: a procrustean approach to learning binary codes for large-scale image retrieval. IEEE Trans. Pattern Anal. Mach. Intell. **35**(12), 2916–2929 (2013)
10. Hastie, T., Tibshirani, R.: Discriminant adaptive nearest neighbor classification. IEEE Trans. Pattern Anal. Mach. Intell. **18**(6), 607–616 (1996)
11. Indyk, P., Motwani, R.: Approximate nearest neighbors: towards removing the curse of dimensionality. In: Proceedings of Annual ACM Symposium on Theory of Computing, STOC 1998, pp. 604–613. ACM (1998)
12. Jiang, Q.Y., Li, W.J.: Scalable graph hashing with feature transformation. In: Proceedings of International Conference on Artificial Intelligence, IJCAI 2015, pp. 2248–2254. AAAI Press (2015)
13. Krizhevsky, A., Hinton, G.: Learning multiple layers of features from tiny images. Technical report, University of Toronto (2009)
14. Kulis, B., Grauman, K.: Kernelized locality-sensitive hashing. IEEE Trans. Pattern Anal. Mach. Intell. **34**(6), 1092–1104 (2012)

15. Li, C.G., Vidal, R.: Structured sparse subspace clustering: a unified optimization framework. In: Proceedings of 2015 IEEE Conference on Computer Vision and Pattern Recognition (CVPR), pp. 277–286 (2015)
16. Li, X., Hu, D., Nie, F.: Large graph hashing with spectral rotation (2017)
17. Liong, V.E., Lu, J., Wang, G., Moulin, P., Zhou, J.: Deep hashing for compact binary codes learning. In: Proceedings of 2015 IEEE Conference on Computer Vision and Pattern Recognition (CVPR), pp. 2475–2483 (2015)
18. Liu, W., Wang, J., Ji, R., Jiang, Y.G., Chang, S.F.: Supervised hashing with kernels. In: Proceedings of 2012 IEEE Conference on Computer Vision and Pattern Recognition, pp. 2074–2081 (2012)
19. Liu, W., Mu, C., Kumar, S., Chang, S.F.: Discrete graph hashing. In: Proceedings of International Conference on Neural Information Processing Systems, NIPS 2014, pp. 3419–3427. MIT Press (2014)
20. Liu, W., Wang, J., Kumar, S., Chang, S.F.: Hashing with graphs. In: Getoor, L., Scheffer, T. (eds.) Proceedings of International Conference on Machine Learning (ICML-2011), pp. 1–8. ACM (2011)
21. Luxburg, U.: A tutorial on spectral clustering. Stat. Comput. $17(4)$, 395–416 (2007)
22. Shen, F., Shen, C., Liu, W., Shen, H.T.: Supervised discrete hashing. In: Proceedings of 2015 IEEE Conference on Computer Vision and Pattern Recognition (CVPR), pp. 37–45 (2015)
23. Shen, F., Shen, C., Shi, Q., van den Hengel, A., Tang, Z.: Inductive hashing on manifolds. In: Proceedings of 2013 IEEE Conference on Computer Vision and Pattern Recognition, pp. 1562–1569 (2013)
24. Shen, F., Zhou, X., Yang, Y., Song, J., Shen, H.T., Tao, D.: A fast optimization method for general binary code learning. IEEE Trans. Image Process. $25(12)$, 5610–5621 (2016)
25. Wang, J., Kumar, S., Chang, S.F.: Semi-supervised hashing for scalable image retrieval. In: Proceedings of 2010 IEEE Computer Society Conference on Computer Vision and Pattern Recognition, pp. 3424–3431 (2010)
26. Wang, J., Kumar, S., Chang, S.F.: Semi-supervised hashing for large-scale search. IEEE Trans. Pattern Anal. Mach. Intell. $34(12)$, 2393–2406 (2012)
27. Wang, J., Xu, X.S., Guo, S., Cui, L., Wang, X.L.: Linear unsupervised hashing for ANN search in euclidean space. Neurocomputing 171, 283–292 (2016)
28. Weiss, Y., Torralba, A., Fergus, R.: Spectral hashing. In: Koller, D., Schuurmans, D., Bengio, Y., Bottou, L. (eds.) Proceedings of Advances in Neural Information Processing Systems, vol. 21, pp. 1753–1760. Curran Associates, Inc. (2009)
29. Xie, L., Shen, J., Zhu, L.: Online cross-modal hashing for web image retrieval. In: Proceedings of AAAI Conference on Artificial Intelligence (AAAI), pp. 294–300 (2016)
30. Xie, L., Zhu, L., Chen, G.: Unsupervised multi-graph cross-modal hashing for large-scale multimedia retrieval. Multimed. Tools Appl. $75(15)$, 9185–9204 (2016)
31. Xie, L., Zhu, L., Pan, P., Lu, Y.: Cross-modal self-taught hashing for large-scale image retrieval. Signal Process. 124, 81–92 (2016)
32. Xu, H., Wang, J., Li, Z., Zeng, G., Li, S., Yu, N.: Complementary hashing for approximate nearest neighbor search. In: Proceedings of 2011 International Conference on Computer Vision, pp. 1631–1638 (2011)
33. Zhang, P., Zhang, W., Li, W.J., Guo, M.: Supervised hashing with latent factor models. In: Proceedings of International ACM SIGIR Conference on Research 38; Development in Information Retrieval, SIGIR 2014, pp. 173–182. ACM (2014)
34. Zhu, L., Shen, J., Xie, L., Cheng, Z.: Unsupervised topic hypergraph hashing for efficient mobile image retrieval. IEEE Trans. Cybern. $PP(99)$, 1–14 (2016)

35. Zhu, L., She, J., Liu, X., Xie, L., Nie, L.: Learning compact visual representation with canonical views for robust mobile landmark search. In: Proceedings of the Twenty-Fifth International Joint Conference on Artificial Intelligence, pp. 3959–3965 (2016)
36. Zhu, L., Shen, J., Xie, L., Cheng, Z.: Unsupervised visual hashing with semantic assistant for content-based image retrieval. IEEE Trans. on Knowl. and Data Eng. **29**(2), 472–486 (2017)

Data Mining

Provenance-Based Rumor Detection

Chi Thang Duong[1], Quoc Viet Hung Nguyen[2]($\boxtimes$), Sen Wang[2],
and Bela Stantic[2]

[1] École Polytechnique Fédérale de Lausanne, Lausanne, Switzerland
[2] Griffith University, Gold Coast, Australia
quocviethung.nguyen@griffith.edu.au

Abstract. With the advance of social media networks, people are sharing contents in an unprecedented scale. This makes social networks such as microblogs an ideal place for spreading rumors. Although different types of information are available in a post on social media, traditional approaches in rumor detection leverage only the text of the post, which limits their accuracy in detection. In this paper, we propose a provenance-aware approach based on recurrent neural network to combine the provenance information and the text of the post itself to improve the accuracy of rumor detection. Experimental results on a real-world dataset show that our technique is able to outperform state-of-the-art approaches in rumor detection.

1 Introduction

With the advance of social media networks, people are sharing user-generated contents in an unprecedented scale. Due to its distributed and decentralized nature, social media provides a platform for information to propagate without any type of moderation. As a result, when an incorrect information propagates on social media networks, it may have a profound impact on real life. For instance, when a fake news claiming two explosions happened in the White House and Barrack Obama got injured was posted by a hacked Twitter account associated with a major newspaper, it caused panic in the society which incurred a loss of $136.5 billion in stock market. This incident shows how a rumor can have a severe impact on our life and it highlights the need for the detection of rumor among different events being discussed on social media networks.

In an attempt to combat fake news, several rumor debunking services such as snopes.com have been created to expose rumors and misinformation. These websites harness collaborative efforts from internet users to identify potential fake news and leverage experts to verify them. As they involve manual labor, the number of events that can be covered are limited and it would take a long time to fact-check an event.

In order to automate this process, several rumor detection models have been proposed. These techniques first design a wide range of features based on the content of the posts [6,9], their characteristics [1,16,20] or the network of users [8,19].

© Springer International Publishing AG 2017
Z. Huang et al. (Eds.): ADC 2017, LNCS 10538, pp. 125–137, 2017.
DOI: 10.1007/978-3-319-68155-9_10

However, feature engineering is a tedious and time-consuming process while the hand-crafted features may not be applicable to a new dataset.

Another important characteristics of rumors on social networks is their temporal nature. When some posts discuss an event, there would be several posts discussing the same event subsequently. These subsequent posts may just be reiteration of the original posts or they could add new information which sheds light on the event. Traditional approaches in rumor detection tend to ignore the temporal nature of the posts. The temporal dependency can also be indicator of rumors. For instance, an event that has many posts providing different views with subsequent posts arguing with previous ones tends to be a rumor as this event is controversial. To the best of our knowledge, there is only one work [10] that models the temporal dependency of the posts explicitly using recurrent neural network (RNN). However, in this work, although various information can be obtained from a post in social media, the users only leverage the textual contents of the posts to classify rumors. We posit that the provenance of the event plays a significant role in identifying rumors. The provenance of an event appears in a post in social media in the form of a link to an article discussing the event. Traditional approaches considered these links as part of the text, hence, provided no special treatment.

Based on this observation, in this work, we propose a provenance-based approach to rumor detection. Our approach considers the provenance of the events appearing as links in the posts as an important source of information. There are several challenges we need to solve in order to leverage the provenance information. First, as the provenance appears as links to some articles in the posts, we need to find a way to model the provenance information. Second, as both the provenance and textual information are present in a post, we also need a way to combine these information in a coherent manner. Third, there are cases that the provenance information is not available. For instance, a tweet may not refer to any article. In these cases, we need to handle the missing of the provenance information while making sure that the classification accuracy does not deteriorate. In order to handle these challenges, we propose a fusion approach that is based on the pooling operation to combine information from the provenance and the text itself. The pooling operation allows our approach to be robust with the missing of the provenance information. In addition, we also leverage RNN to capture the temporal dependency among the posts.

The contributions of this paper are as follows:

- We propose a provenance-based approach to classify events into rumors and non-rumors. Our model also leverages RNN to capture the temporal dependency between the posts.
- We have enriched a social media dataset by adding the provenance information. Our dataset can also be used by subsequent research in this direction.
- Our extensive experiments on a real-world dataset show that our approach is able to outperform state-of-the-art technique significantly.

The rest of the paper is organized as follows. Section 6 introduces related works on the field of rumor detection. Section 3 discusses our general framework to classify rumors. Section 2 explains in detail our provenance-based approach. Experimental evaluation and analysis are presented in Sect. 5 while Sect. 7 concludes the paper.

2   Recurrent Neural Network for Rumor Detection

2.1   Problem Statement

We consider a setting in which a set of users discuss n events which can be rumors or not. We denote a discussion for an event e_i as $d_{it} = \langle e_i, t \rangle$ where t is the time the discussion took place. In our setting, a discussion can be a tweet or a post by a user on a social network. It is worth noting that different events may have different number of discussions. In addition, each event is associated with a label indicating where it is a rumor or not.

The problem we want to solve is given a set of events together with their discussions, classify the events correctly. In particular, given a temporal sequence of discussions $D = \{e_i, t\}$, our goal is to assign a label $l_i \in \{0, 1\}$ for the event e_i where 1 denotes rumor and 0 otherwise. We achieve this by training a feedforward neural network which takes the discussions of an event as input and returns the label for the event. More precisely, given an event e_i and two classes $Y = \{0, 1\}$, we define a neural network that assigns probabilities to all $y \in Y$. The predicted class is then the one with the highest probability:

$$\hat{y} = arg \max_y P(Y = y | e) \tag{1}$$

Our network models the temporal characteristics of the dicussions using RNN and leverages the provenance of the tweets to achieve high accuracy.

2.2   Neural Network Model

A feedforward neural network estimates $P(Y = y | e)$ with a parametric function ϕ_θ (Eq. 1), where θ refers to all learnable parameters of the network. Given an event e, this function ϕ_θ applies a combination of functions such as

$$\phi_\theta(e) = \phi^L(\phi^{L-1}(\dots \phi^1(e) \dots)), \tag{2}$$

with L the total number of layers in the network.

We denote matrices as bold upper case letters ($\mathbf{X}, \mathbf{Y}, \mathbf{Z}$), and vectors as bold lower-case letters ($\mathbf{a}, \mathbf{b}, \mathbf{c}$). $\mathbf{A}_i$ represents the ith row of matrix $\mathbf{A}$ and $[\mathbf{a}]_i$ denotes the ith element of vector $\mathbf{a}$. Unless otherwise stated, vectors are assumed to be column vectors. We also denote $|\mathbf{a}|$ to be the dimensionality of the vector $\mathbf{a}$. We now introduce the layers when training linear classifiers with neural networks: the *recurrent neural network* layer, the *linear* layer and the *softmax* layer.

Recurrent Neural Network. Among different types of feed-forward neural networks, RNN is the one that can model the sequential characteristics of the input data such as time series or sentences. Given an input sequence $(x_1, ..., x_T)$, RNN processes each input sequentially (from x_1 to x_T), at each step, it updates its hidden state h_i and returns an output o_i. The hidden state vector h_i captures information of the elements that the RNN has seen. More precisely, at the time step i, the network does the following update operations [4]:

$$\mathbf{h_i} = tanh(\mathbf{Ux_i} + \mathbf{Wh_{i-1}} + \mathbf{b})$$
$$\mathbf{o_i} = \mathbf{Vh_i} + \mathbf{c}$$

where the matrices $\mathbf{U}, \mathbf{V}, \mathbf{W}$ are used, respectively, to convert input vector to hidden vector, hidden vector to output vector and hidden vector to hidden vector. Two vectors $\mathbf{b}, \mathbf{c}$ are the bias vectors and the function $tanh$ is a nonlinearity function. The matrices and the bias vectors are the trainable parameters of the RNN. In order to find these parameters, we compute the gradients of the network using back-propagation through time [18]. However, the RNN as discussed above suffers from the the vanishing or exploding gradients problem which makes it unable to learn from long sequences. A solution to this problem is to implement memory cells in the network to store information over time, which is the idea of Long Short-Term Memory (LSTM) [5,7] and Gated Recurrent Unit (GRU) [2]. In this work, we use LSTM instead of the vanilla RNN to capture long-term dependency in the inputs.

In our setting, we consider an RNN as the first layer in our network. It is modelled by the function $\phi_{\theta_1}^1(e)$ which takes as input the event e which contains $|e|$ discussions and returns the output in the last time step $o_{|e|}$. In particular, $o_{|e|} = \phi_{\theta_1}^1(e)$ and θ_1 is the parameter of the RNN that we need to find. There are two important hyperparameters of the RNN which is the size of the hidden state vector $\mathbf{h_i}$ and the output vector $\mathbf{o_i}$ (denoted as m).

Linear Layer. This layer applies a linear transformation to its inputs $\mathbf{x}$:

$$\phi^l(\mathbf{x}) = \mathbf{W}^l\mathbf{x} + \mathbf{b}^l \tag{3}$$

where $\mathbf{W}^l$ and $\mathbf{b}^l$ are the trainable parameters with $\mathbf{W}^l$ being the weight matrix, and $\mathbf{b}^l$ is the bias term.

In our model, we use two linear layers after the RNN layer to first convert the output of the RNN to a hidden vector space and then, convert from this hidden vector space to a score vector for the classes. In particular, the second layer of our network $\phi_{\theta_2}^2(\mathbf{o})$ takes as input the output of the first layer (the output vector $\mathbf{o}$) and returns a vector from a hidden space $\mathbf{p}$. More precisely, the layer ϕ^2 takes the vector $\mathbf{o} \in \mathbb{R}^m$ as input and uses the matrix $\mathbf{W}^2 \in \mathbb{R}^{p \times m}$ and $\mathbf{b}^2 \in \mathbb{R}^p$ to convert $\mathbf{o}$ to the hidden space vector $\mathbf{k} \in \mathbf{R}^p$. Similarly, the third layer of our network $\phi_{\theta_3}^3(\mathbf{k})$ takes as input the output of the second layer (the hidden space vector $\mathbf{k}$) and returns a score vector $\mathbf{s} \in \mathbb{R}^2$.

Softmax Layer. Given an input x, the penultimate layer outputs a score for each class $\mathbf{s} \in \mathbb{R}^2$. The probability distribution is obtained by applying the softmax activation function:

$$P(Y = y|e) \propto \phi_\theta(e, y) = \frac{\exp([\mathbf{s}]_y)}{\sum_{k=1}^{2} \exp([\mathbf{s}]_{y_k})} \tag{4}$$

2.3   Training

In summary, our network is modelled as a function ϕ_θ which is a combination of functions where each function represents a layer. The parameter θ, which combines all the trainable parameters in the network, is obtained by minimizing the negative log-likelihood using stochastic gradient descent (SGD):

$$L(\theta) = \sum_{(e,y)} -\log P(Y = y|e) \propto \sum_{(e,y)} -\log\left(\phi_\theta(e, y)\right). \tag{5}$$

3   Provenance-Based Approach

In this section, we discuss how to obtain the provenance of the events that we use in our models, and we present our technique for leveraging provenance to classify the events.

3.1   Provenance of an Event

When an event is discussed on social media, it usually has a source or several sources backing it up. When a post discusses an event, it tends to cite one of those sources. For instance, when a tweet mentions an event, it may include the link to the article. We consider these articles as the provenance of the event. As these articles contain detail information about the event, they provide several indicators of rumor. Based on this observation, we also include the provenance of the event into our model.

We consider a *discussion* (e.g. a tweet or a post) to be composed of a *text*, or both an *article* and a text. The article appears in a discussion in form a hyperlink. When both article and text are present, we assume that they are semantically related, e.g. the text is a summary of the article. Traditional techniques in rumor detection only leverage the textual information. As a result, they represent each discussion using only the text. In our setting, as we also consider the provenance of the event, we propose a technique to model a discussion using the text and article information. The text and article information are represented as feature vectors.

3.2   Feature Vectors

Before diving into the detail how to represent a discussion its text and article, we discuss the process to represent the article and text in a discussion as feature vectors $(\gamma(i), \psi(s))$ as they are the foundation to construct a post vector representing a discussion. A good article and text representation can affect the performance of our approach heavily. We model the article i in a discussion with an article feature vector $\gamma(i) \in \mathbb{R}^n$. Similarly, we represent the piece of text s with a textual feature vector $\psi(s) \in \mathbb{R}^m$.

Text Representation. In order to represent a text, we aim to convert it from its original format (i.e. words) to an n-dimensional vector. We first calculate the tf-idf values of all words in all the posts. The tf-idf value of a word reflects its importance based on its presence frequency in all the posts relative to the number of posts it appears in. We keep the top-K words with the highest tf-idf values as the vocabulary. Each post is then represented using the words in the vocabulary as a vector of length $|K|$. The value of the i-th element in the vector is 0 if the i-th word in the vocabulary does not appear in the text of the post.

Article Representation. An article contains different types of information such as its text, images. However, the images may not contain indicators of rumor. As a result, among different types of information in an article, we only consider its main text. We also follow the same approach from text representation by modeling each article by its tf-idf vector.

3.3   Joint Fusion

From the article and text vector of a discussion, there are many ways to construct a post vector representing the discussion. However, the technique to represent the discussion needs to take into account the missing of the article information. For instance, there are cases that a post may not always explicitly refer to an article and the technique must be robust to these absences. We propose joint fusion which is a technique to combine article and text vector which is able to handle the absence of the article.

Joint fusion takes the article vector and text vector $\gamma(i), \psi(s)$ as input and applies the pooling operation to obtain the post vector $\mathbf{x}$:

$$\mathbf{x} = \text{pooling}(\gamma(i), \psi(s)) \tag{6}$$

The pooling function can be either a *component-wise max* pooling, or an *average* pooling. In this work, we leverage the max pooling and we will extend this work with the average pooling in the future works.

It is worth noting that the pooling operation requires the vectors $\gamma(i) \in \mathbb{R}^n$ and $\psi(s) \in \mathbb{R}^m$ to have the same size. This can be done by adding an extra linear layer to γ (i.e. the network that extracts article feature vector). Assuming $n > m$, the linear layer is as follows:

$$\tilde{\gamma}(i) = \tilde{\mathbf{W}}\gamma(i) + \tilde{\mathbf{b}} \qquad (7)$$

where $\tilde{\gamma}(i) \in \mathbb{R}^m$, $\tilde{\mathbf{W}} \in \mathbb{R}^{n \times m}$ and $\tilde{\mathbf{b}} \in \mathbb{R}^m$. The input to joint fusion is then two vectors $\ddot{\gamma}(i)$ and $\psi(s)$. The trainable parameters of joint fusion are $\theta = \{\tilde{\mathbf{W}}, \tilde{\mathbf{b}}\}$.

4 Putting It All Together

Recall that our model takes an event as input and returns a prediction for the label of the event. As the event is composed of several discussions, and the discussion contains different types of information, we first model the discussion by combining information from the text and its provenance as discussed in Sect. 3. Then, we use the events with the post vectors as input to the network. The complete model of our approach is shown in Fig. 1. We train the model to find all the parameters in an end-to-end manner which means the parameters of the network and of the joint fusion are trained together. The training is similar to the one discussed in Sect. 2.3.

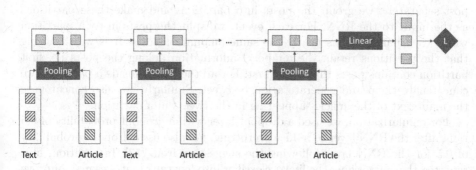

Fig. 1. Model of the network

5 Experiments

In this section, we evaluate our proposed approach on a real-world dataset.

5.1 Dataset

To the best of our knowledge, there is no large-scale dataset for rumor detection that contains both the texts and their provenance. For textual content only, there are the *twitter* and *weibo* dataset which were produced by Ma et al. [10]. This motivates us to construct a new dataset by adding the article information.

After inspecting the *weibo* dataset, we observe that there is no article information to be added. The reason is that each post from weibo contains only text without any link to an outside article. As a result, in this work, we only focus on the *twitter* dataset [10]. For this dataset, rumor and non-rumor events were identified using a real-time rumor debunking service[1]. The authors of the dataset then extracted keywords from Snopes and used the keywords to query tweets in real-time from Twitter. Due to legal restrictions, only the tweet IDs from the dataset are published instead of its content. Based on these tweet IDs, we crawled their contents including its texts. For the tweets that contain links to a article, we also followed the link and crawled the main text of the article. However, some of the tweets from the original dataset are missing as they were removed by the users or Twitter. As a result, we can only collected 586162 tweets. Over 60% of them contain a link to an article. The statistics of the dataset is shown in Table 1.

5.2 Experimental Settings

We compare our proposed approach (joint fusion) with a baseline that does not leverage the provenance information. This baseline is similar to the original approach to rumor detection by Ma et al. [10].

As some events has thousands of posts, it is inefficient to back propagate through time with such amount of posts. As a result, instead of considering each post separately, we group the posts into partitions and make these partitions as the input to the RNN. For each event, we split the posts in to N partition where each partition has nearly the same amount of posts. It is worth noting that the partitions retain the temporal information among the posts i.e. first partition contains posts that occur first. For all the tweets inside a partition, we concatenate them and generate a longer tweet. Similarly, we also concatenate the main text of the articles appearing in the tweets in a partition.

For regularization, we used a dropout layer with a dropout probability of 0.5 right after the RNN layer to reduce overfitting. We also use a dropout probability of 0.5 for the RNN layer following the suggestion from [17]. In addition, it is reported that factorizing the linear classifier into low rank matrices may improve the classification accuracy [13]. We also followed this approach by adding a linear layer right before the last layer to map the output vector from the RNN layer to a hidden vector space with a size of *nhid*. Regarding the hyperparameters, we tested different values of them on the validation set and select the ones that gave the best results. Table 2 describes other hyperparameters. Our models were trained with a learning rate set to 0.01.

The models were trained on a server equipped with a Tesla GPU. We use 10% events for testing and 10% for validation, the rest is used for training. We use the same splits for all the models in our experiments. All the source codes and the datasets will be released upon the publication of this work.

---

[1] snopes.com.

Table 1. Statistics of the dataset

Statistics	Twitter
Involved users	231535
Total posts	586162
Total events	992
Total rumors	498
Total non-rumors	494

Table 2. Model hyperparameters

Parameter	Value
Text vector size	$wvdim = 5000$
Article vector size	$avdim = 5000$
Hidden vector size of RNN	$nhid = 800$
Low rank output size	$lowrank = 400$
Classification main task weight	$\lambda = 3$
Fusion layer function	max

5.3   Effectiveness of the Provenance-Based Approach

Table 3 shows the experimental results of our approach in comparison with the baseline. The results show that our provenance-based approach is able to outperform the baseline significantly on three metrics: accuracy, recall and F-measure. For instance, our technique has an accuracy of 0.85, which is a 9% relative improvement in comparison with the baseline. The superior performance of our technique is also demonstrated by the recall metric. We are able to recall 92% of events while the baseline approach can only achieve 70%, which is 22% difference. These results show the effectiveness of our provenance-based approach as adding the provenance information allows us to improve the performance significantly. Although our approach has lower precision than the baseline, the difference is extremely small (2%).

Table 3. Performance of the provenance-based approach

Method	Accuracy	Precision	Recall	F-measure
LSTM [a]	0.78	0.83	0.70	0.76
ProvBased [b]	0.85	0.81	0.92	0.86

[a] $wvdim = 5000$, $nhid = 800$, $lowrank = 400$
[b] $wvdim = 5000$, $avdim = 5000$, $nhid = 800$, $lowrank = 400$

5.4   Effects of RNN Hidden Vector Size

In this experiment, we want to analyze the effects of the hidden vector size ($nhid$) to the performance of our approach and the baseline. In order to analyze the effect of $nhid$, we fix other parameters ($wvdim = 5000$ and $lowrank = 400$). The experimental results are shown in Fig. 2. It is clear that our approach has better accuracy, recall and F-measure over different values of $nhid$. For instance, the recall of our approach is always higher than 0.8 while the recall of the baseline is always lower.

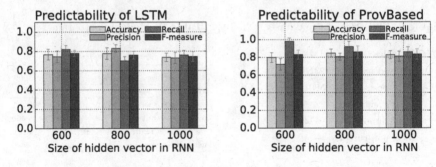

Fig. 2. Effects of hidden vector space on predictive power

5.5 Effects of Lowrank Vector Space

In this experiment, we want to analyze the effect of the lowrank vector space to the performance of two approaches. Similarly, in order to analyze this parameter, we fix $wvdim = 5000$ and $nhid = 800$. It is clear in this experiment that our approach outperforms the baseline significantly. For instance, when the size of the lowrank vector space is 200, our accuracy is 0.81 while the baseline's is only 0.72. We also observe the same pattern with the F-measure metric. When the lowrank vector space is 400, our approach achieves an F-measure of 0.83 while the baseline's F-measure is only 0.78. We are able to achieve the highest performance when the lowrank vector space is 400, which is the value we chose for our model (Fig. 3).

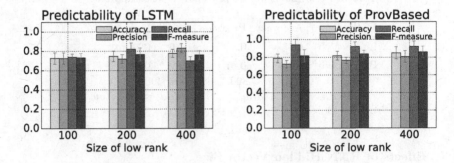

Fig. 3. Effects of low rank layer on predictive power

5.6 Effects of Word Vector Space

In this experiment, we want to analyze the effects of the input vector size to the two techniques. For the sake of simplicity, we set the article and text vector size to the same value. Similarly, we fix other parameters to $nhid = 800$ and $lowrank = 400$. Once again, our approach is able to outperform the baseline across different values of the word vector space. For instance, when the word

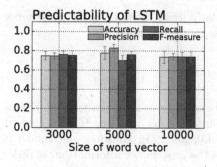 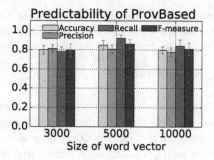

Fig. 4. Effects of word vector size on predictive power

vector space size is 5000, the baseline has only the precision value higher than 0.8. On the other hand, our approach has 3 metrics which have a higher value than 0.8. We also observe this phenomenon when the word vector space is 10000. Our approach consistently outperform the baseline across all metrics. These experiments confirm our observation that our approach is able to have higher performance due to the addition of the provenance information (Fig. 4).

6   Related Work

Rumor detection can be considered a binary classification task. Traditional works on automatic rumor detection aim to construct some classifiers based on hand-crafted features. Several works [3,6,12,15] leveraged linguistic features such as word usage or presence of conjunctions or pronouns. For instance, the authors of [6] found out that fake news usually contain swear words. Many works followed a different direction in which they do not take into account the content of the posts. For instance, some statistical features such as the number of retweets or replies are considered [1,11,20]. Similarly, some user-level features are also used such as the credibility or readability of the users [1,9,14]. A different approach is to examine network level features to detect rumors. For instance, by constructing a tree representing how the messages in an event are related, the authors of [19] is able to classify whether the root is a rumor or not. However, the problem with these hand-crafted features is that the feature engineering process is tedious and time-consuming. In addition, the selected features are usually data-specific and/or domain-specific, which hinders their generality.

Another problem with these approaches is that they do not consider the temporal information of the posts. As the posts are usually temporally related, ignoring this information has a negative effect on the accuracy of rumor detection. Recently, the authors of [10] has leveraged RNN to capture the temporal information of the posts while used deep learning to construct the features automatically. Although this work is the most similar to our work, there are some differences. The approach in [10] does not take into account the provenance of the events, which is an important information to detect rumors. As the tweets are short in nature and they may contain similar phrases, leveraging the provenance information allows us to improve the accuracy significantly.

7   Conclusions

In this paper, we propose a provenance-based approach to detect rumor. Our model is able to combine the provenance information with the textual content to improve the classification accuracy significantly. In addition, our model is robust as it is able to handle the missing of the provenance information. In order to showcase our model, we also enriched a real-world dataset with provenance information which allow us to test our approach in a real-world scenario. Future research directions will go towards adding different types of information such as network or user-level features. In addition, it is worth investigating whether the provenance information is reliable or not.

References

1. Castillo, C., Mendoza, M., Poblete, B.: Information credibility on Twitter. In: WWW, pp. 675–684 (2011)
2. Cho, K., Van Merriënboer, B., Bahdanau, D., Bengio, Y.: On the properties of neural machine translation: Encoder-decoder approaches. arXiv preprint arXiv:1409.1259 (2014)
3. Feng, V.W., Hirst, G.: Detecting deceptive opinions with profile compatibility. In: IJCNLP, pp. 338–346 (2013)
4. Goodfellow, I., Bengio, Y., Courville, A.: Deep Learning. MIT Press, Cambridge (2016). http://www.deeplearningbook.org
5. Graves, A.: Generating sequences with recurrent neural networks. arXiv preprint arXiv:1308.0850 (2013)
6. Gupta, A., Kumaraguru, P., Castillo, C., Meier, P.: TweetCred: real-time credibility assessment of content on Twitter. In: Aiello, L.M., McFarland, D. (eds.) SocInfo 2014. LNCS, vol. 8851, pp. 228–243. Springer, Cham (2014). doi:10.1007/978-3-319-13734-6_16
7. Hochreiter, S., Schmidhuber, J.: Long short-term memory. Neural Comput. 9(8), 1735–1780 (1997)
8. Hung, N.Q.V., Thang, D.C., Weidlich, M., Aberer, K.: Minimizing efforts in validating crowd answers. In: SIGMOD, pp. 999–1014 (2015)
9. Liu, X., Nourbakhsh, A., Li, Q., Fang, R., Shah, S.: Real-time rumor debunking on Twitter. In: Proceedings of the 24th ACM International on Conference on Information and Knowledge Management, pp. 1867–1870. ACM (2015)
10. Ma, J., Gao, W., Mitra, P., Kwon, S., Jansen, B.J., Wong, K.F., Cha, M.: Detecting rumors from microblogs with recurrent neural networks. In: IJCAI (2016)
11. Ma, J., Gao, W., Wei, Z., Lu, Y., Wong, K.F.: Detect rumors using time series of social context information on microblogging websites. In: Proceedings of the 24th ACM International on Conference on Information and Knowledge Management, pp. 1751–1754. ACM (2015)
12. Markowitz, D.M., Hancock, J.T.: Linguistic traces of a scientific fraud: the case of Diederik Stapel. PloS one 9(8), e105937 (2014)

13. Mikolov, T., Chen, K., Corrado, G., Dean, J.: Efficient estimation of word representations in vector space. arXiv preprint arXiv:1301.3781 (2013)
14. Nguyen, Q.V.H., Duong, C.T., Nguyen, T.T., Weidlich, M., Aberer, K., Yin, H., Zhou, X.: Argument discovery via crowdsourcing. VLDBJ 26(1), 511–535 (2017)
15. Nguyen, T.T., Duong, C.T., Weidlich, M., Yin, H., Nguyen, Q.V.H.: Retaining data from streams of social platforms with minimal regret. In: IJCAI (2017)
16. Nguyen, T.T., Nguyen, Q.V.H., Weidlich, M., Aberer, K.: Result selection and summarization for web table search. In: ICDE, pp. 231–242 (2015)
17. Pham, V., Bluche, T., Kermorvant, C., Louradour, J.: Dropout improves recurrent neural networks for handwriting recognition. In: 2014 14th International Conference on Frontiers in Handwriting Recognition (ICFHR), pp. 285–290. IEEE (2014)
18. Rumelhart, D.E., Hinton, G.E., Williams, R.J.: Learning representations by back-propagating errors. Cognit. Model. 5(3), 1 (1988)
19. Wu, K., Yang, S., Zhu, K.Q.: False rumors detection on Sina Weibo by propagation structures. In: 2015 IEEE 31st International Conference on Data Engineering (ICDE), pp. 651–662. IEEE (2015)
20. Yang, F., Liu, Y., Yu, X., Yang, M.: Automatic detection of rumor on Sina Weibo. In: Proceedings of the ACM SIGKDD Workshop on Mining Data Semantics, p. 13. ACM (2012)

An Embedded Feature Selection Framework for Hybrid Data

Forough Rezaei Boroujeni[✉], Bela Stantic, and Sen Wang

School of Information and Communication Technology,
Griffith University, Gold Coast, Australia
forough.rezaeiboroujeni@griffithuni.edu.au,
{b.stantic,sen.wang}@griffith.edu.au

Abstract. Feature selection in terms of inductive supervised learning is a process of selecting a subset of features relevant to the target concept and removing irrelevant and redundant features. The majority of feature selection methods, which have been developed in the last decades, can deal with only numerical or categorical features. An exception is the Recursive Feature Elimination under the clinical kernel function which is an embedded feature selection method. However, it suffers from low classification performance. In this work, we propose several embedded feature selection methods which are capable of dealing with hybrid balanced, and hybrid imbalanced data sets. In the experimental evaluation on five UCI Machine Learning Repository data sets, we demonstrate the dominance and effectiveness of the proposed methods in terms of dimensionality reduction and classification performance.

1 Introduction

Over the past few years, machine learning has been widely applied in various big data applications, such as social media mining [5–7] and medical data analysis [12,13]. Most of data used in these applications have different types of features with high dimensionality, making the learning tasks more complex and computationally demanding. To confront these problems, a reduced representation of a data set is obtained using data reduction strategies which can learn faster with higher accuracy, compared to the initial data set. One of the data reduction strategies is feature selection which is considered as an important pre-processing step in data mining. Feature selection removes irrelevant and redundant attributes in a data set leading to better classification performance and reducing the learning running time of the classifier. Feature selection methods are widely used in both supervised and semi-supervised learning tasks. For instance, Chang et al. [1,2] conducted feature selection in a semi-supervised manner for multi-label learning tasks.

In the area of data science, there are two main types of attributes: numerical and categorical. A numerical variable is measurable and takes on continuous values such as height. A categorical variable has a measurement scale consisting of a small number of discrete categories or classes such as favourite type of music

Z. Huang et al. (Eds.): ADC 2017, LNCS 10538, pp. 138–150, 2017.
DOI: 10.1007/978-3-319-68155-9_11

(classical, folk) or social class (upper, middle, lower). While many feature selection algorithms have been proposed in the last decades, the majority of them are designed to work with only numerical or categorical data and are not capable of dealing with both types of data. In real-world applications, however, data sets usually come with a mixture of both types of data which are called hybrid, mixed or heterogeneous data sets. Most medical data sets are examples of this type of data set. To utilise an algorithm which is only applicable to homogenous data, the data should be transformed to fully numerical or categorical variables. To obtain a categorical data set, numerical variables are discretised. This conversion makes the feature selection highly sensitive to the discretisation technique and may also lead to the loss of information. Alternatively, encoding categorical variables can be used to represent these variables in numerical space. This method may also be ineffective as it introduces an artificial order to the feature values. Therefore developing feature selection methods which can work with mixed data sets without changing the nature of features is required.

Feature selection methods are categorised in a number of different ways. For instance, Yu and Liu [14] classified feature selection methods according to their evaluation approach: individual evaluation (known as feature ranking) and subset evaluation. Feature selection methods can also be categorised based on their relationship with inductive learning methods into three separate categories: *filter*, *wrapper* and *embedded*.

Filter approaches rely on the intrinsic properties of the data to evaluate and select feature subsets without involving any inductive algorithms. These algorithms are fast and suitable for large data sets. However, Lamirel et al. [4] showed that many existing filter methods are not successful to deal with highly imbalanced, highly multidimensional and noisy data. A wrapper model requires one predetermined mining algorithm to select those features which improve mining performance. Although, these methods yield high fitting accuracy, they are often criticised for their high computational complexity and consequently they are not applicable for large data sets. Embedded methods, as the third type of feature selection algorithms, embed the feature selection procedure into a learning algorithm. While these methods depend directly on the nature of the classifier used, they are computationally less demanding than wrapper methods. Little attention has been paid in the literature to develop embedded methods for hybrid data set. Paul et al. [11] developed an embedded method based on Support Vector Machine (SVM) in order to deal with hybrid data sets. They recruited the clinical kernel and plugged this kernel into Recursive Feature Elimination (RFE-SVM). However, it suffers from low classification performance.

In this paper, we introduce five embedded feature selection methods based on SVM which are capable of working with hybrid data sets. These methods rely on plugging two types of kernel functions into embedded algorithms. The kernel functions are the clinical kernel, and the Gaussian kernel which is based on hybrid distances for numerical and categorical features. These kernel functions allow proposed methods to deal with hybrid data. In line with the literature for experimental evaluation, five data sets from UCI Machine Learning Repository [8] are employed. The results indicate better performance for the proposed methods in terms of dimensionality reduction, and classification performance on hybrid data sets when compared to the previous state-of-the-art method.

The remainder of the paper is organised as follows. In Sect. 2, the previous works on developing embedded methods based on SVM are elaborated. Then the foundation of proposed methods is described in Sect. 3. In Sect. 4, the experimental procedure for the validation of the proposed methods is provided, and in Sect. 5 the results and findings are presented. Section 6 concludes the paper and explains future works.

2   Related Works

Different embedded methods based on SVM, including kernel-penalised SVM (KP-SVM) and Holdout strategy for Backward Feature Elimination algorithm (HO-BFE), were proposed by Maldonado et al. [9,10]. KP-SVM attempts to find the best suitable RBF-type kernel function in order to eliminate features which have low relevance for the classifiers [9]. In the HO-BFE method, the unseen samples in the training data set are assessed and used to construct the loss function, instead of assembling the loss function with the same data set used for training [10]. HO-BFE employs two types of loss functions to work with balanced and imbalanced data sets. Imbalanced data refers to a data set which contains many more samples from one class than from the rest of the classes. In this case, learning algorithms often have high accuracy on a class with majority samples and poor accuracy in a class with minority samples. Imbalanced data sets can be found in many domains such as fraud detection and text categorization and several studies have been undertaken to deal with them [16]. Both KP-SVM and HO-BFE have been employed in numerical space only.

The RFE-SVM is another embedded method which has been employed for feature ranking in hybrid data sets [11]. In RFE-SVM method, those features which do not significantly decrease the margin are detected and removed as less important features [3]. The margin of the separating hyperplane is inversely proportional to the Euclidean norm of the weight vector, w:

$$W^2(\alpha) = \sum_{i=1}^{n}\sum_{j=1}^{n} \alpha_i\alpha_j y_i y_j k(x_i, x_j) \tag{1}$$

where α is the dual variable of a SVM obtained from training; x_i and x_j are the ith and j^{th} samples; y_i and y_j are the labels of x_i and x_j respectively; n is the number of samples; and k is a kernel function. A candidate feature f is detected as a less important feature if the following evaluation function does not change significantly after removing f from the training data set:

$$J_{SVM}(f) = |W^2(\alpha) - W^2_{-f}(\alpha)|, \ W^2_{-f}(\alpha) = \sum_{i=1}^{n}\sum_{j=1}^{n} \alpha_i\alpha_j y_i y_j k(x_i^{(-f)}, x_j^{(-f)}) \tag{2}$$

where $x_i^{(-f)}$ is the ith sample after removing feature f.

In order to deal with hybrid data sets, Paul et al. [11] employed the clinical kernel as the kernel function in a non-linear SVM structure. This kernel is represented as:

$$k(x_i, x_j) = \frac{1}{m} \sum_{f=1}^{m} k_f(x_{if}, x_{jf}), \ k_f(a,b) = \begin{cases} I(a=b) & f \text{ is categorical} \\ \frac{(max_f - min_f) - |a-b|}{max_f - min_f} & f \text{ is continuous} \end{cases}$$

(3)

where m is the total number of features; a and b are two scalar values; I is the indicator function; max_f and min_f are the maximum and minimum values of the feature f respectively; and x_{if} is the value of x_i for the feature f. Similar to the Gaussian kernel, the clinical subkernels lie between 0 and 1.

We have named the RFE-SVM under this kernel RFE-SVM$_C$ and have compared our proposed methods with this embedded method.

3   Materials and Methods

The components of the proposed methods are presented in this section. These components contain HO-BFE for balanced data sets proposed by Maldonado et al. [10], our proposed method for imbalanced data sets, and the kernel functions.

3.1   HO-BFE for Balanced Data

'Standard 0–1' lose function is the base of HO-BFE for balanced data. To construct this loss function, first, the data set $(\mathcal{T})$ is split into two groups of training data $(\mathcal{TR})$, and validation data $(\mathcal{V})$, using the holdout method. Afterward, SVM is trained using $\mathcal{TR}$, and the loss function is computed using $\mathcal{V}$. In each iteration, those features which lead to the minimum loss function, are removed from the data and ranked as less relevant. This loss function is calculated from the following equation:

$$\text{LOSS}_{0-1}((\alpha, b), F \setminus \{f\}, \mathcal{TR}, \mathcal{V})$$
$$= \sum_{l \in \mathcal{V}} \left| y_l^v - sgn \left(\sum_{i \in \mathcal{TR}} \alpha_i y_i K(x_i^{(-f)}, x_l^{v(-f)}) + b \right) \right|$$

(4)

where b is the bias obtained from training; F is the set of available features and $f \in F$; y_l^v is a class label of sample l; $x_i^{(-f)}$ is the training sample i with feature f removed; $x_l^{v(-f)}$ is the validation sample l with feature f removed; and sgn is the signum function. Algorithm 1 shows this feature selection method.

In Algorithm 1, $\mathcal{I}$ is a set of features to be eliminated as they are less important. Although for data sets with a moderate number of features, a single element of F can be removed in each iteration, this strategy is not efficient when data sets are high dimensional. However, removing a large portion of features at once may lead to discard relevant features [10].

Algorithm 1. HO-BFE for balanced data sets

Input: The original set of features, F; the training data set, T
Output: The ranked subset S
Initialisation: $S = \phi$
Repeat:
  $(TR, V) \leftarrow$ Splitting T using holdout technique
  Training SVM using TR and obtaining (α, b)
  $I \leftarrow argmin_I \sum_{f \in I} \text{LOSS}((\alpha, b), F \setminus \{f\}, TR, V)$
  $F \leftarrow F \setminus \{I\}$ and Append I to S
until: $F = \phi$

3.2 Proposed Embedded Feature Selection Method for Imbalanced Data (EFSI)

Our proposed method, EFSI, is based on balanced loss function [10]. The literature recommends the computation of this loss function based on positive and negative instances in each set of T, TR, and V i.e. T^+ and T^-, TR^+ and TR^-, V^+ and V^- respectively. However, in order to achieve a more reliable outcome, we introduce the following equation:

$$
\begin{aligned}
&\text{LOSS}_{\text{bl}}((\alpha, b), F \setminus \{f\}, TR, V) \\
&= \frac{\left(\sum_{l \in V^-} |y_l^v - sgn\left(\sum_{i \in TR} \alpha_i y_i K(x_i^{(-f)}, x_l^{\nu(-f)}) + b\right)|\right)}{|V^-|} \\
&+ \frac{\left(\sum_{l \in V^+} |y_l^v - sgn\left(\sum_{i \in TR} \alpha_i y_i K(x_i^{(-f)}, x_l^{\nu(-f)}) + b\right)|\right)}{|V^+|}.
\end{aligned}
\tag{5}
$$

In order to deal with imbalanced data, undersampling the majority class or oversampling the minority class are the two most common strategies which can be used. Maldonado et al. [10] employed the SMOTE method before hold-out splitting to oversample data sets. However, this strategy can result in overfitting problems and misleading results because there will be iterations where the training and validation sets contain the same samples. In order to solve this problem and to ensure the existence of sufficient samples from both classes in each iteration, Algorithm 2 is proposed. This algorithm introduces the novelty by splitting the data. First, T is divided into T^+ and T^-. Then T^+ (also T^-) is split, based on the holdout method, into TR^+ (also TR^-), and V^+ (also V^-). In this process TR is the union of TR^+ and TR^-, and V is the union of V^+ and V^-. In the next step, the minority class in TR will be oversampled using the SMOTE method.

3.3 Kernel Functions

In this work, two types of kernel functions are utilised in order to deal with hybrid data sets. The first is the 'clinical' kernel ((3)). This kernel is plugged

---

Algorithm 2. EFSI for imbalanced data sets

Input: The original set of features, F; the training data set, T
Output: The ranked subset S
Initialisation: $S = \phi$
Repeat:
  $(T^+, T^-) \leftarrow$ Splitting T
  $(TR^+, V^+) \leftarrow$ Splitting T^+ using holdout technique
  $(TR^-, V^-) \leftarrow$ Splitting T^- using holdout technique
  $TR \leftarrow TR^- \cup TR^+$, and $V \leftarrow V^- \cup V^+$
  $TR \leftarrow$ SMOTE(TR)
  Training SVM using TR and obtaining (α, b)
  $\mathcal{I} \leftarrow argmin_{\mathcal{I}} \sum_{f \in \mathcal{I}} \text{LOSS}((\alpha, b), F \setminus \{f\}, TR, V)$
  $F \leftarrow F \setminus \{\mathcal{I}\}$ and Append $\mathcal{I}$ to S
until: $F = \phi$

---

into HO-BFE and EFSI in order to build the new methods which we name HO-BFE$_C$ and EFSI$_C$ respectively.

The second kernel is the 'Gaussian' kernel function which Zeng et al. [15] developed based on different distance metrics for various types of attributes. They used this kernel function to build the fuzzy equivalence relation. We employ two distance metrics (6) to construct the Gaussian kernel in order to deal with hybrid data:

$$d_f(a, b) = \begin{cases} I(a \neq b) & \text{if } f \text{ is categorical} \\ \frac{|a-b|}{4\sigma_f} & \text{if } f \text{ is continuous} \end{cases} \tag{6}$$

$$k(x_i, x_j) = exp\left(\frac{D(x_i, x_j)^2}{2\sigma^2}\right), \quad D(x_i, x_j) = \sqrt{\sum_{f=1}^{m} d_f^2(x_{if}, x_{jf})} \tag{7}$$

where σ is a free parameter and σ_f is the standard deviation under the attribute f. We plugged this kernel into HO-BFE, EFSI, and RFE-SVM, and named them HO-BFE$_G$, EFSI$_G$ and RFE-SVM$_G$ respectively in the remainder of this work.

4 Evaluation Procedure

We compare the performance of our proposed methods with RFE-SVM$_C$ [11]. Comparison is undertaken in terms of dimensionality reduction, and classification performance. In this section, the details of data sets, classifiers, and the experimental protocol are explained.

4.1 Data Sets and Classifiers

To examine the performance of the proposed methods, in line with the literature, five binary classification data sets from UCI Machine Learning Repository [8] have been employed. Table 1 shows the main characteristics of these real data sets

Table 1. Summary of data sets

Data set	Discrete features	Continuous features	Class priors
Heart	7	6	164/139
Australian	8	6	307/383
Credit	9	6	307/383
Arrhythmia	51	191	44/245
Hepatitis	13	6	32/123

in terms of the number of features and class priors. The arrhythmia data set aims to classify the presence and absence of cardiac arrhythmia in one of the 16 groups. In this study, we only consider samples from two classes 01 (normal ECG), and 02 (one type of arrhythmia), and removed features with all values equal to zero. While the heart, the Australian and the credit data sets are balanced, the arrhythmia and the hepatitis data sets are imbalanced. Due to a small number of missing values in some data sets, mean imputation and mode imputation are considered for numerical and categorical features respectively.

Evaluating the performance of the feature selection method through a classification algorithm is necessary. It is of note that the classification performance depends on the classifier used. A common practice is to employ different learning algorithms in the validation procedure in order to obtain results which are as classifier-independent as possible. We use C4.5 and Naïve Bayes classifiers in the validation procedure.

4.2 Experimental Protocol and Parameter Settings

In their studies, Maldonado et al. [10] performed the m different holdout split and averaged the loss function before eliminating the features. They finally showed that HO-BFE is robust with respect to the variation of m. For this reason, we considered $m = 1$ in our experiments. However, in order to increase the statistical significance of the results, the proposed methods and RFE-SVM$_C$ were performed 5 times. After feature-ranking in each performance, the least important feature was removed from the ranked subset, and the predictive performance was measured by 10-fold cross validation classification. This process, which ensures the results are not biased towards the data sequence, was undertaken for all features. In this study, the geometric mean $(G - mean)$ is measured as the classification performance. $G - mean$ is a suitable measurement for balanced and imbalanced data sets, and is computed as the geometric mean of the true positive and true negative rates: $G - mean = \sqrt{\frac{TN}{N} * \frac{TP}{P}}$; where TN (also TP) is the number of true negative (also true positive) samples, and N (also P) is the number of negative (also positive) samples in a data set.

The next stage after feature ranking is selecting a small number of features. It should be noted that selecting a specific number of features from ranked features can be undertaken in favour of a lesser number of features (greater interpretative possibilities of the results) at the cost of some decrease in predictive performance, or vice versa. In this study, to maintain the trade-off between the size of data sets and predictive performance, we keep those features, the elimination of which would otherwise decrease the predictive performance obtained from the original data set by more than 2% for balanced data sets, and by more than 5% for imbalanced data sets.

After feature subset selection by each method, the number of selected features and predictive performances obtained from applying two classifiers to selected subsets are compared. To perform a fair comparison between proposed and previous methods, the constant box constraint parameter (C) and kernel parameter (σ) are selected for each data set. However, in order to choose the appropriate values for C and σ, a variety of experiments were performed. For selecting the best C, this parameter was altered within the range of $\{2^{-3}, \ldots, 2^5\}$ for constant value of σ. For each C, the highest and average $G - mean$ were calculated, the best C being selected. We obtained the same results as Maldonado et al. [10] achieved in their studies, which showed that the HO-BFE methods are robust to the changes of C. Due to this result, we considered C as 2^{-1} in our experiments. To select the σ for each data set, the value of this variable was altered within the range of $\{3^0, \ldots, 3^5\}$. The same procedure was followed to find the best σ.

5   Results and Analysis

The predictive performances of all methods across all feature sets are compared in this section. To do so, Figs. 1, 2, 3 and Tables 2, 3, 4, 5 have been provided. In several tables, two criteria are considered: the best performance of all the methods is shown in bold; and for methods with the same performance, a model which leads to fewer features in the final subset is shown in bold text.

5.1   Results for Balanced Datasets

Figure 1 demonstrates that the three proposed methods show similar behaviour for each balanced data set, whereas the behaviour associated with RFE-SVM$_C$ is different. Table 2 shows the $G - mean$ achieved by applying C4.5 and Naïve Bayes to the selected subsets. For each performance, the number of selected features is illustrated in parentheses.

This table shows that the three proposed methods identified the smaller number of features. For the heart data set, and based on our criteria for feature selection, the three proposed methods identified four out of thirteen features, whereas this value rose at six using RFE-SVM$_C$. In comparison, for the Australian and the credit data sets, one single feature was identified by employing the three proposed methods. By recruiting RFE-SVM$_C$, this number increased to four for the Australian data, and to three for the credit data.

In the heart data set, the $G - mean$ obtained from applying both classifiers to the selected subset from the RFE-SVM$_G$ is higher compared to $G - mean$

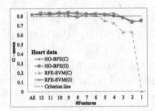 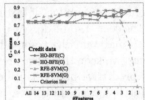 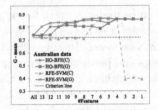

Fig. 1. $G-mean$ of Naïve Bayes after removing each feature in the balanced data sets.

Table 2. The $G-mean$ (%) of C4.5 and Naïve Bayes on selected features

Data set	Full set	RFE-SVM$_C$	RFE-SVM$_G$	HO-BFE$_C$	HO-BFE$_G$
C4.5					
Heart	77.66 (13)	79.57 (6)	**84.32 (4)**	83.48 (4)	83.48 (4)
Australia	85.93 (14)	85.97 (4)	**85.97 (1)**	**85.97 (1)**	**85.97 (1)**
Credit	85.67 (15)	72.65 (3)	**85.97 (1)**	**85.97 (1)**	**85.97 (1)**
Average	83.08 (14)	79.40 (4.3)	85.42 (2)	85.14 (2)	85.14 (2)
Naïve Bayes					
Heart	81.72 (13)	80.48 (6)	**83.02 (4)**	80.27 (4)	80.27 (4)
Australia	73.94 (14)	**86.26 (4)**	85.97 (1)	85.97 (1)	85.97 (1)
Credit	74.12 (15)	72.64 (3)	**85.96 (1)**	**85.96 (1)**	**85.96 (1)**
Average	76.59 (14)	79.80 (4.33)	84.32 (2)	84.07 (2)	84.07 (2)

obtained from the other methods, which shows that RFE-SVM$_G$ detected more effective features. For other data sets, the performances of all proposed methods are the same and almost higher than previous method. Generally speaking, the highest average predictive performances were achieved when RFE-SVM$_G$ was employed as the feature selection method.

5.2 Results for Imbalanced Data Sets

To compare the performance of two imbalanced data sets, the percentage of variables eliminated at every iteration was varied for each method. Figures 2 and 3 illustrate the $G-mean$ of Naïve Bayes after eliminating each percentage for the arrhythmia and hepatitis data sets respectively. For these data sets, Tables 3 and 4 show the maximum, average, and standard deviation of $G-mean$ along all subsets after each iteration.

These tables show that the predictive performance decreases when more features are eliminated at each iteration. However, for each method the results are stable with respect to the variation of the percentage based on an ANOVA test for both data sets (the p-value varies within $[0.26, 0.76]$ along the different percentages).

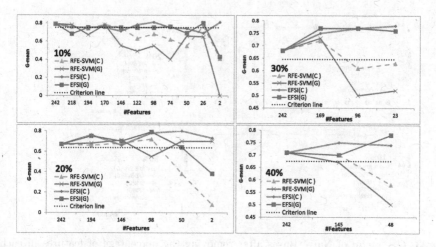

Fig. 2. $G - mean$ of Naïve Bayes after removing the percentage of features in each iteration for the arrhythmia data set.

After feature ranking, and based on the stability of the results along the variations of the percentage, the smallest subset leading to a reasonable predictive performance was chosen for each method (Table 5). Table 5 shows the predictive performance of two classifiers applied to each selected subset. The number of features and the percentage of eliminations for each performance are presented in parentheses. For arrhythmia data set, the EFSI$_C$ represents the highest $G-mean$ for both classifiers with the smallest number of features. For hepatitis data set, the number of selected features are almost similar for all methods. However, once again EFSI$_C$ shows the better performance. For both data sets, EFSI$_G$ stands in second place followed by RFE-SVM$_C$ and RFE-SVM$_G$ in third and fourth places respectively. These results show the dominance and effectiveness of EFSI$_C$ with respect to dimensionality reduction and classification performance for imbalanced data sets.

Table 3. Maximum, average, and standard deviation of Naïve Bayes performance along all ranked features. $G - mean$ analysis for the percentage of features in each iteration for the arrhythmia data set.

	RFE-SVM$_C$			RFE-SVM$_G$			EFSI$_C$			EFSI$_G$		
	Max	Mean	Std	Max	Mean	Std	Max	Mean	Std	Max	Mean	Std
10%	0,77	0.67	0.11	0.78	0.55	0.22	0.81	0.76	0.04	0.80	0.71	0.10
20%	0,72	0.5	0.24	0.71	0.67	0.06	0.80	0.75	0.05	0.79	0.65	0.14
30%	0,72	0.65	0.05	0.73	0.58	0.10	0.78	0.77	0.012	0.77	0.77	0.005
40%	0,70	0.64	0.06	0.67	0.59	0.09	0.75	0.75	0.01	0.78	0.74	0.04

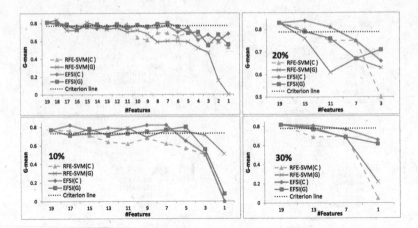

Fig. 3. $G - mean$ of Naïve Bayes after removing the percentage of features in each iteration for the hepatitis data set.

Table 4. Maximum, average, and standard deviation of Naïve Bayes performance along all ranked features. $G - mean$ analysis for the percentage of features in each iteration for the hepatitis data set.

	RFE-SVM$_C$			RFE-SVM$_G$			EFSI$_C$			EFSI$_G$		
	Max	Mean	Std	Max	Mean	Std	Max	Mean	Std	Max	Mean	Std
One-by-one	0.82	0.70	0.08	0.84	0.61	0.21	0.79	0.73	0.05	0.80	0.74	0.07
10%	0.76	0.57	0.21	0.79	0.71	0.08	0.82	0.66	0.25	0.80	0.65	0.21
20%	0.80	0.70	0.12	0.76	0.67	0.06	0.84	0.77	0.07	0.79	0.73	0.05
30%	0.70	0.48	0.30	0.79	0.56	0.25	0.81	0.75	0.06	0.77	0.69	0.06

Table 5. The $G - mean$ of both classifiers on selected features

Methods	Arrhythmia		Hepatitis	
	NB	C4.5	NB	C4.5
RFE-SVM$_C$	0.77 (26, 10%)	0.78 (26, 10%)	0.79 (11)	0.57 (11)
RFE-SVM$_G$	0.70 (2, 20%)	0.73 (2, 20%)	0.75 (5, 10%)	0.62 (5, 10%)
EFSI$_C$	**0.81** (2, 10%)	**0.81** (2, 10%)	**0.82** (7, 10%)	**0.75** (7, 10%)
EFSI$_G$	0.76 (23, 30%)	0.8 (23, 30%)	0.8 (5, 10%)	0.74 (5, 10%)

6   Conclusion

In this work, we address the problem of feature selection when dealing with hybrid data sets by proposing five embedded methods. To build these methods, two types of kernel functions (namely clinical and Gaussian) were plugged into the HO-BFE, RFE-SVM, and EFSI which is our proposed embedded method for imbalanced data. These methods were compared with one previous state-of-the-art embedded method. In line with the literature, we utilised five available hybrid balanced and imbalanced data sets from the UCI Machine Learning Repository. The results indicated that the RFE-SVM$_G$ for balanced data sets, and the EFSI$_C$ for imbalanced data sets, are effective in terms of data reduction and classification performance.

As for future research, we plan to apply the proposed methods to larger data sets in order to confirm the findings. Within our institution, a high dimensional data set is currently being collected with respect to obesity. The expected data will include almost 450,000 biomarkers per person. We believe that applying the proposed methods to this data set can provide a more precise evidence regarding their respective performances.

References

1. Chang, X., Nie, F., Yang, Y., Huang, H.: A convex formulation for semi-supervised multi-label feature selection. In: AAAI, pp. 1171–1177 (2014)
2. Chang, X., Yang, Y.: Semisupervised feature analysis by mining correlations among multiple tasks. In: IEEE TNNLS, pp. 1–12 (2016)
3. Guyon, I., Weston, J., Barnhill, S., Vapnik, V.: Gene selection for cancer classification using support vector machines. Mach. Learn. **46**(1–3), 389–422 (2002)
4. Lamirel, J.-C., Cuxac, P., Chivukula, A.S., Hajlaoui, K.: A new feature selection and feature contrasting approach based on quality metric: application to efficient classification of complex textual data. In: Li, J., Cao, L., Wang, C., Tan, K.C., Liu, B., Pei, J., Tseng, V.S. (eds.) PAKDD 2013. LNCS (LNAI), vol. 7867, pp. 367–378. Springer, Heidelberg (2013). doi:10.1007/978-3-642-40319-4_32
5. Li, J., Liu, C., Islam, M.S.: Keyword-based correlated network computation over large social media. In: ICDE, pp. 268–279. IEEE (2014)
6. Li, J., Liu, C., Zhou, R., Wang, W.: Top-k keyword search over probabilistic XML data. In: ICDE, pp. 673–684. IEEE (2011)
7. Li, J., Wang, X., Deng, K., Yang, X., Sellis, T., Yu, J.X.: Most influential community search over large social networks. In: ICDE, pp. 871–882. IEEE (2017)
8. Lichman, M.: UCI machine learning repository (2013). http://archive.ics.uci.edu/ml
9. Maldonado, S., Weber, R., Basak, J.: Simultaneous feature selection and classification using kernel-penalized support vector machines. Inf. Sci. **181**(1), 115–128 (2011)
10. Maldonado, S., Weber, R., Famili, F.: Feature selection for high-dimensional class-imbalanced data sets using support vector machines. Inf. Sci. **286**, 228–246 (2014)
11. Paul, J., Dupont, P., et al.: Kernel methods for heterogeneous feature selection. Neurocomputing **169**, 187–195 (2015)
12. Wang, S., Chang, X., Li, X., Long, G., Yao, L., Sheng, Q.Z.: Diagnosis code assignment using sparsity-based disease correlation embedding. IEEE TKDE **28**(12), 3191–3202 (2016)
13. Wang, S., Li, X., Yao, L., Sheng, Q.Z., Long, G., et al.: Learning multiple diagnosis codes for ICU patients with local disease correlation mining. ACM TKDD **11**(3), 31 (2017)
14. Yu, L., Liu, H.: Efficient feature selection via analysis of relevance and redundancy. J. Mach. Learn. Res. **5**, 1205–1224 (2004)

15. Zeng, A., Li, T., Liu, D., Zhang, J., Chen, H.: A fuzzy rough set approach for incremental feature selection on hybrid information systems. Fuzzy Sets Syst. **258**, 39–60 (2015)
16. Zheng, Z., Wu, X., Srihari, R.: Feature selection for text categorization on imbalanced data. ACM SIGKDD Explor. Newsl. **6**(1), 80–89 (2004)

A New Data Mining Scheme for Analysis of Big Brain Signal Data

Siuly Siuly[✉], Roozbeh Zarei, Hua Wang, and Yanchun Zhang

Centre for Applied Informatics, College of Engineering and Science,
Victoria University, Melbourne, Australia
{siuly.siuly,roozbeh.zarei,hua.wang,yanchun.zhang}@vu.edu.au

Abstract. Analysis and processing of brain signal data (e.g. Electroencephalogram (EEG) data) is a significant challenge in the medical data mining community due to its massive size and dynamic nature. The most crucial part of EEG data analysis is to discover hidden knowledge from a large volume of data through pattern mining for efficient analysis. This study focuses on discovering representative patterns from each channel data to recover useful information reducing the size of data. In this paper, a novel algorithm based on principal component analysis (PCA) technique is developed to accurately and efficiently extract a pattern from the vast amount of EEG signal data. This study considers PCA to explore the sequential pattern of each EEG channel data as PCA is a dominant tool for finding patterns in it. In order to represent the distribution of the pattern, the most significant satanical features (e.g. mean, standard deviation) are computed from the extracted pattern. Then aggregating all of the features extracted from each of the patterns in a subject, a feature vector set is created that is fed into random forest (RF) and random tree (RT) classification model, individually for classifying different categories of the signals. The proposed methodology is tested on two benchmark EEG datasets of BCI Competition III: dataset V and dataset IVa. In order to further evaluate performance, the proposed scheme is compared with some recently reported algorithms, where the same datasets were used. The experimental results demonstrate that the proposed methodology with RF achieves higher performance compared to the RT and also the recently reported methods. The present study suggests the merits and feasibility of applying proposed method in the medical data mining for efficient analysis of any biomedical signal data.

Keywords: Electroencephalogram · Big data · Data mining · Principal component analysis · Feature extraction · Classification

1 Background

Efficient and effective analysis of brain signal data such as electroencephalogram (EEG) signal data has attracted great importance in the medical field due to its significant applications in health and medicine. Analysing EEG signals provides

© Springer International Publishing AG 2017
Z. Huang et al. (Eds.): ADC 2017, LNCS 10538, pp. 151–164, 2017.
DOI: 10.1007/978-3-319-68155-9_12

essential information for brain diseases diagnosis and treatment [9,12,28,34], emotion/fatigue monitoring [17,23], brain computer interface [15,27,35,39] etc. The EEG recordings produce huge volume time-series data, which refers to the recording of the brain's spontaneous electrical activity. The patterns of EEG signals indicate the health states of the brain [28]. The data are aperiodic and non-stationary with dynamic behaviours. Today, One of the major challenges in the EEG data analysis is how to exploit these sheer volumes of data for effective and accurate analysis. Data mining is an interesting option in this respect. Data mining have a huge potential for analysing such large volumes of stored EEG signal data in order to discover hidden information from big data for accurately classifying data states, which can assist in automated decision-making [22].

Mostly, EEG data analysis process consists of two stages: *feature extraction* and *classification*. At the first stage, features are extracted from raw EEG data and then extracted features are fed into classifiers for decision-making. In the past a few years, a variety of methods have been reported for analysis and classification of big EEG signal data [9,12,15,17,22,23,27–29,34,35,39]. The traditional methods face lots of problems to extract meaningful and discriminative characteristics to represent individual EEG time series for classification when the data size is very massive (e.g. long-term data). Most of the methods extract spectrum information to characterize EEG time series, which is not suitable for handling big volume data. In spite of the usefulness in the analysis of short-term, most of them are insufficient to extract high-level structural information from long-term EEG signal data that are non-periodic [36]. The traditional EEG data analysis procedure does not consider a stage for pattern mining while it is very essential for managing a big volume of data with complex nature such as EEG data. Pattern mining is one of the most crucial part of EEG data analysis to discover representative pattern from each channel data for recovering useful information and also to reduce the size of data for efficient analysis. Patterns in the classification process provide essential knowledge to discover patterns hidden behind the data.

As per our knowledge, there is no efficient data mining based methodology available for analysis of big EEG data that can preserve all the important and significant characteristic properties of dynamic EEG signals reducing the size of data. Moreover, the adoption rate and research development in this space are still hindered by some fundamental problems inherent within the big data paradigm. Thus, the analysis of EEG time series data for knowledge discovery is far from straightforward and requires powerful technology that can extract useful information from a large volume of data through pattern mining for efficient analysis and classification.

The main motive of this study is to introduce a new data mining scheme with pattern extraction manner that is able to handle a large volume of EEG data reducing the dimensionality through minimizing the loss of original information for efficient and accurate analysis. In the proposed plan (see Fig. 1), firstly, we segment an EEG signal data into several groups to make it stationary considering a particular time-period. Secondly, we extract a representative pattern for that signal employing principal component analysis (PCA) technique. The reason for considering the PCA for pattern mining in EEG as PCA has the capability to discover representative pattern without loss of information removing redundant information because EEG contains a large amount of redundant information.

Thirdly, we compute some statistical features of that pattern and continue this process for all of the channel signals of a subject. Fourthly, we aggregate all of the features obtained from all of the channel signals of a subject as a feature vector set and finally use the feature set to a classification model to identify the various states of the signal. In this study, for classification, we employ two models: random forest (RF) and random tree (RT). The motivation of applying these methods to the classification of EEG signal data is that they are robust and faster classifier to achieve a high classification accuracy in high-dimensional data, which are non-stationary and very noisy.

The remainder of the paper is organized as follows: Sect. 2 presents a description of the proposed methodology. Section 3 provides a brief description of the datasets used in this study. The experimental results along with comparative analysis with existing state-of-art methods are discussed in Sect. 4. Concluding remarks are stated in Sect. 5.

2   Proposed Methodology

This research presents a new scheme of data mining for analysis of big EEG signal data. Figure 1 show a digram of the proposed plan that can be used for handling massive amount of brain signal data (e.g. EEG data). In this figure, we just provide an example how one channel EEG signal is processed for analysis in the proposed plan. The same procedure is continued for all of the channel signal data in a subject.The proposed methodology is divided into several parts such as data segmentation; pattern mining; feature computation and classification. The detail discussion of each part is provided in the following sections.

2.1   Data Segmentation

In this part, we segment an EEG signal data (came from a channel) into several time-window considering a particular time-period. As EEG signals are aperiodic and non-stationary time series data, it is important to having representative values of a specific time-period from every time window. The most important thing in the segmentation process is how to determine an appropriate size of time window. In this study, we choose the size of time window (e.g. 256) following some state-of-arts methods such as Stam et al. [30], Siuly et al. [14], Rankine et al. [21], Ines et al. [10], Siuly et al. [13].

2.2   Pattern Mining

The most crucial part of EEG data analysis is to discover patterns concealed in the data [28]. This section aims to extract a representative pattern that conveys

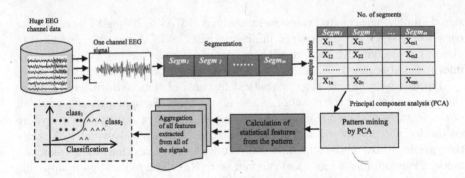

Fig. 1. Proposed plan for analysis of big brain signal data.

all meaningful knowledge from the shape of data. In this study, we apply PCA for mining pattern from each channel EEG signal. The reason for considering PCA is that the PCA is a powerful tool for summarizing high dimension data, which has the capability of identifying hidden patterns in data. In addition, the obtained pattern can compress high dimension data into a small dataset, reducing the number of dimensions, without loosing the most important information in the original data set. Another thing, as EEG signal data are highly correlated, and correlation indicates that there is redundancy in the data, in this circumstance, the PCA is a perfect choice to extract a representative pattern from the original signal. The detail description of PCA method is available in Ref [7, 26, 40].

As shown in Fig. 1, the segmentation process for one channel EEG signal data creates a matrix, $X_{m \times n}$ and we employ PCA method on that matrix to mine a representative pattern for that signal. In this study, we consider first principal component (FPC) as the representative pattern for each channel signal data as the FPC captures the largest variability of the data, while the next components (e.g. second, third, etc.) represent, respectively, less variation.

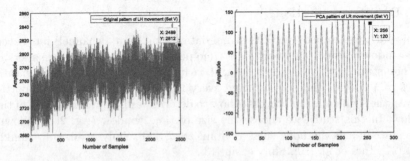

Fig. 2. An illustration of original signal vs PCA pattern in data set V, BCI Comp III.

As mentioned before, the proposed scheme is tested on two EEG data sets of BCI Comp III: V and IVa. The description of these two datasets is provided

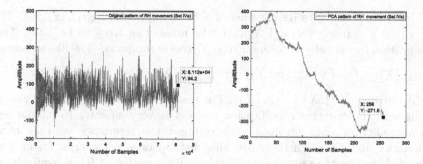

Fig. 3. An illustration of original signal vs PCA pattern in dataset IVa, BCI Comp III.

in Sect. 3. Figures 2 and 3 present two examples using two data sets, V and IVa, respectively, showing a difference between original signal and PCA pattern. These two figures also show how much data are reduced in the pattern signal with respect to the original signal. For example, in Fig. 2, 2489 samples are transformed into 256 data points in the pattern signal and in Fig. 3, 81120 samples in the original dataset are reduced to 256 in the pattern signal.

2.3 Feature Computation

Technically, a feature represents a distinguishing property, a recognizable measurement, and a functional component obtained from a section of a pattern. In order to characterize the distribution of a representative pattern, we calculates nine statistical features such as mean, standard deviation, median, first quartile (Q1), third quartile (Q3), inter-quartile range (IQR) (IQR = Q3-Q1), mode, minimum, maximum from each pattern of a signal in this study. After calculating features from the patterns of all of the signals in a subject, we aggregate all the features in a set, called feature vector set and this feature vector set is fed into a classification model discussed in the following section. The reasons for considering these nine statistical features are found in references [6, 24].

2.4 Classification

Random Forest (RF): In this study, we choose Random Forest (RF) model for classification where the obtained feature set is fed into as input. The decision making is performed based on the classification outcomes. RF is a powerful approach for classification, which consists of many individual classification trees [3]. Each tree is constructed using a tree classification by selecting a random subset of input features and a different bootstrap sample from the training data. The RF aggregates the results of all classification trees to classify new samples. Each tree casts a unit vote at the input data, and then the forest selects the class with the most votes for the input data. A majority vote among the trees provides the final result [4].

This study considers a RF classifier of N trees. Suppose, $(x_1, y_1), (x_2, y_2), \ldots,$ (x_n, y_n) is a training set and N_{tree} is the number of trees to be built. The classification by majority vote among the N tress is computed [2,4] like as below.

$$f_{avg}(X) := (p_1(X), \ldots, p_k(X)) := \frac{1}{N} \sum_{p_1}^{N} f_i(X) \text{ ; Here each b} = 1 \text{ to } N_{tree}$$

$f_{RF}(X) := argmax\{(X), \ldots, p_k(X)\}$. The reason for considering the RF model in this work is that this classification method is very effective to handle big dimensionality of data involved with complex characteristics such as non-stationarity and noisy. Furthermore, this method does not overfit, faster and little needs to fine-tune parameters. The detail discussion of RF is available in References [2–4].

Random Tree (RT): The random tree is a special case of RF model. The RT usually refers to randomly built trees. The RF model consists of many random trees. As in the RF, a random subset of candidate features is used, but instead of looking for the most discriminative thresholds, thresholds are drawn at random for each candidate feature, and the best of these randomly-generated thresholds is picked as the splitting rule. It is computational less expensive than a random forest.

3   Experimental Data

In order to test the effectiveness of the proposed method, we used two publicly available EEG datasets, V and IVa from BCI Competition III [1,5,20].

Dataset 1: Data set V is considered as dataset 1, which consists of mental imagery tasks EEG data recorded in brain-computer interface (BCI) competition III. This database contains EEG recordings from three normal subjects during three kinds of mental imagery tasks, which are imagination of repetitive self-paced left hand movements (class 1), imagination of repetitive self-paced right hand movements (class 2), and generation of different words beginning with the same random letter (class 3) [5,20]. The sampling rate of the raw EEG potential signals is 512 Hz. EEG recordings of 12 frequency components are obtained from each of the 8 channels, producing a 96-dimensional vector.

Dataset 2: In this study, dataset IVa [1,15,27] (of BCI competition III,) is named as dataset 2. This dataset contains motor imagery, EEG data recorded from five healthy subjects (labelled 'aa', 'al', 'av', 'aw', 'ay'), who performed right hand (denoted by 'RH') and right foot (denoted by 'RF') motor imagery. EEG was recorded using 118 electrodes and 280 trials were available for each subject, among which 168, 224, 84, 56 and 28 composed the training set for subject 'aa', 'al', 'av', 'aw', 'ay', respectively, the remaining trials composing the test set. In this study, we consider the training sets for all subjects as our experimental data as the algorithm requires a class label of each sample point. They provided a version of the data that was down-sampled at 100 Hz, which is used in this research.

4   Results and Discussions

This section presents the experimental results of the proposed method on two EEG datasets: Set V (mental imagery data set) and Set IVa (motor imagery data). The performance of the proposed scheme is evaluated on both mental imagery and motor imagery data in Sect. 4.1 and 4.2, respectively. A comparison between the proposed method and some existing methods is also provided for each dataset. All mathematical calculations are carried out in MATLAB R signal processing tool (version 7.11, R2010b). The classification executions for the RF and RT classifiers are executed in WEKA machine learning toolkit [8]. It is worth mentioning that the default parameter values for each classifier in WEKA is used as there are no specific guidelines for selecting these parameters.

4.1   Classification Results for the Mental Imagery EEG Data

For the mental imagery tasks EEG data, 96 vectors of 9 dimensions for each class of a subject are obtained using the proposed feature extraction method from the original data. In this dataset, we consider nine experiments (Exp 1–9) using different pairs of two-class EEG data. Each experiment uses 192 vectors of 9 dimensions for two-class data with 96 vectors from each class. The 10-fold cross-validation method is used to evaluate the performance of each classifier for each experiment. In this paper, each experiment is considered as an Exp. Exp 1–3 are created for Subject 1, Exp 4–6 are composed for Subject 2, and Exp 7–9 are compiled for Subject 3, which are given below:

For dataset V :
Exp 1: class 1 vs class 2; **Exp 2:** class 1 vs class 3; **Exp 3:** class 2 vs class 3
Exp 4: class 1 vs class 2; **Exp 5:** class 1 vs class 3; **Exp 6:** class 2 vs class 3
Exp 7: class 1 vs class 2; **Exp 8:** class 1 vs class 3; **Exp 9:** class 2 vs class 3

Table 1 presents the classification results of each classifier, the RF and the RT, for different pairs of two-class EEG signals from the mental imagery tasks EEG data. In Table 1, the results of each classifier are shown in terms of sensitivity, specificity, and accuracy. The average classification accuracy is calculated using all accuracy values for all Exps in each subject. For most of the Exps, the RF classifier achieves higher classification results in terms of sensitivity, specificity, and accuracy compared to the RT classifier. The RF classifier produces an average classification accuracy of 91.84% for subject 1, 75.18% for subject 2, and 82.81% for subject 3 while these values are 90.45%, 70.32%, and 76.04%, respectively, for the RT classifier. The average sensitivities of the RF classifier are 92.36%, 77.78%, and 80.21% for subject 1, subject 2, subject 3, respectively, while these values are 90.63%, 71.18%, 75.70%, respectively, for the RT classifier. The RF obtains 91.32%, 72.57%, and 85.41% classification specificity for subjects 1, 2, and 3, respectively, whilst those values are 90.28%, 69.45%, and 76.39%, respectively, for the RT classifier. Considering the results shown in Table 1, it is observed that the RF classifier is more capable of classifying the two-class EEG signals than the RT.

Table 1. Performance of the proposed scheme for data set V.

Subject	Different Experiments	Proposed algorithm with Random Forest			Proposed algorithm with random tree		
		Sensitivity (%)	Specificity (%)	Accuracy (%)	Sensitivity (%)	Specificity (%)	Accuracy (%)
1	Exp 1	90.63	92.71	91.67	93.75	95.83	94.79
	Exp 2	97.92	94.79	96.35	93.75	95.83	94.79
	Exp 3	88.54	86.46	87.50	84.38	79.17	81.77
	Average	**92.36**	**91.32**	**91.84**	**90.63**	**90.28**	**90.45**
2	Exp 4	68.75	65.63	67.17	57.29	65.63	61.49
	Exp 5	87.5	81.25	84.38	83.33	80.21	81.77
	Exp 6	77.08	70.83	73.96	72.92	62.50	67.71
	Average	**77.78**	**72.57**	**75.18**	**71.18**	**69.45**	**70.32**
3	Exp 7	85.42	91.66	88.54	85.42	88.54	86.98
	Exp 8	81.25	87.50	84.38	78.13	81.25	79.69
	Exp 9	73.96	77.08	75.52	63.54	59.38	61.46
	Average	**80.21**	**85.41**	**82.81**	**75.70**	**76.39**	**76.04**

Figure 4 presents the ROC areas for the RF and RT classifiers for proposed feature set, separately for each of the nine Exps in the mental imagery tasks EEG data. The area of the ROC curve is used as a measure for assessing the classifier performance (e.g., a higher value of the area indicates better performance of the classifier). As seen in Fig. 4, both classifiers obtain high ROC area for each Exp. From these experimental results, it is clear that the proposed feature extraction method is capable of extracting robust features from the mental imagery tasks EEG data and the RF classifier has the ability to accurately classify mental imagery tasks.

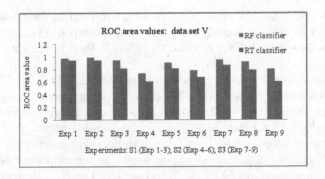

Fig. 4. ROC area for the RF and RT classifier in dataset V

Table 2 shows the performance comparison of the proposed method based on RF classifier with some reported methods in terms of subject-specific classification accuracy and average classification accuracy for the mental imagery tasks EEG data. From Table 2, it is observed that the proposed method obtains the highest performance in terms of average classification accuracy among the other methods. The average classification accuracy of the proposed method is 83.28% for the mental imagery tasks data while they are 68.35%, 61.69%, 56.66%,

Table 2. A comparative report with existing methods for Dataset V.

Authors	Methods	Classification accuracy (%)			
		S1	S2	S3	Average
Proposed method	**PCA based random forest model**	**91.84**	**75.18**	**82.81**	**83.28**
Lin et al. [16]	Neural networks based on improved particle swarm optimization(IPSO)	78.31	70.27	56.46	68.35
Siuly et al. [14]	Clustering technique-based LS-SVM (CT-LS-SVM)	68.19	64.77	52.12	64.69
Sun et al. [32]	Random electrode selection ensemble(RESE)	68.75	56.411	44.82	56.66
Sun et al. [33]	Ensemble methods	70.59	48.85	40.92	53.45
Sun et al. [31]	Adaptive common spatial patterns (ACSP)	67.70	68.10	59.55	65.12

Note: S1 = Subject 1; S2 = Subject 2; S3 = Subject 3.

53.45% and 65.12% for methods reported in [14, 16, 31–33], respectively. These results indicate the proposed method outperforms all six referenced methods and improves the average classification accuracy by at least 14.93%.

4.2 Classification Results for the Motor Imagery EEG Data

In this section, we provide the classification results of each classifier for the motor imagery EEG dataset. Using the proposed feature extraction technique, we obtain 118 feature vectors of 9 dimensions for one class from each subject from the motor imagery EEG dataset. As mentioned in Sect. 3 (data description section), this data set contains the EEG recorded data of five healthy subjects where each subject performed two tasks: the imagination of 'Right Hand' and 'Right Foot' movement. Each task is considered as a class of EEG data. In this dataset, we consider five experiments for five subjects and every experiment contains 236 feature vectors of 9 dimensions for two classes of a subject, with 118 vectors of the same dimension in each class. The 10-fold cross-validation method is used to evaluate the performance of each classifier for each experiment.

Table 3 shows the classification results of each classifier, the RF and the RT in terms of sensitivity, specificity and classification accuracy for each subject in the motor imagery EEG dataset. As shown in Table 3, the RF classifier achieves higher classification results for each subject compared to the RT classifier. The average the RT classifier. Considering the results shown in Table 3, it is observed that the RF classifier is more capable to classify the motor imagery tasks than the RT.

Figure 5 shows the ROC areas for the RF and RT classifiers for proposed feature set, separately for each of the five subjects in the motor imagery EEG dataset. As seen in Fig. 5, both classifiers achieve high ROC area close to 1 for

Table 3. Performance of the proposed scheme for data set IVa.

Subject	Proposed algorithm with Random Forest			Proposed algorithm with random tree		
	Sensitivity (%)	Specificity (%)	Accuracy (%)	Sensitivity (%)	Specificity (%)	Accuracy (%)
S1(aa)	98.30	99.15	98.73	98.30	98.30	98.31
S1(al)	91.53	89.83	90.68	86.44	87.29	86.86
S1(av)	98.31	100	99.15	98.31	100	99.15
S1(aw)	93.22	91.53	92.37	86.44	88.14	87.29
S1(ay)	99.15	99.15	99.15	99.15	99.15	99.15
Average	**96.10**	**95.93**	**96.02**	**93.73**	**94.58**	**94.15**

Note: S1 = Subject 1 (aa); S2 = Subject 2 (al); S3 = Subject 3 (av); S4 = Subject 4 (aw); S5 = Subject 5 (ay).

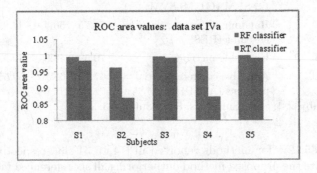

Fig. 5. ROC area for the RF and RT classifier in dataset IVa.

each subject. From these experimental results, the proposed method provides same evidence like as dataset V that the RF classifier has better ability to accurately classify motor imagery tasks compared to RT classifier.

Table 4 shows the comparison results between the proposed method based on RF classifier and the seven existing algorithms in terms of subject-specific classification accuracy and overall classification accuracy for the motor imagery EEG dataset. The overall performance of the proposed method is achieved 96.02% which is the highest performance among the other methods. Subject specific rates are 98.73% for subject 1, 90.68% for subject 2, 99.15% for subject 3, 92.37% for subject 4, and 99.15% for subject 5 achieved by our proposed approach. The average classification accuracies of sparse spatial filter optimization [40], R-CSP [39], composite CSP [38], SRCSP methods [37], CT-based LS-SVM [15], ISSPL [36], and CC-Based LS-SVM [35] for the motor imagery data are 73.50%, 74.20%, 76.22%, 78.62%, 88.32%, 94.21%, and 95.72%, respectively. The results show that our proposed method achieves by 0.3% to 22.52% improvements over all the seven existing algorithms in terms of average accuracy for the motor imagery EEG dataset.

Table 4. A comparative report with existing methods for Dataset IVa.

Authors	Method	Classification accuracy (%)					
		S1	S2	S3	S4	S5	Average
Proposed method	**PCA based random forest model**	**98.73**	**90.68**	**99.15**	**92.37**	**99.15**	**96.02**
Siuly and Li [25]	CC-Based LS-SVM	97.88	99.17	98.75	93.43	89.36	95.72
Wu et al. [37]	ISSPL	93.57	100	79.29	99.64	98.57	94.21
Siuly et al. [14]	CT-based LS-SVM	92.63	84.99	90.77	86.5	86.73	88.32
Lotte and Guan [18]	Spatially regularized common spatial pattern (SRCSP)	72.32	96.43	60.2	77.68	86.51	78.62
Kang et al. [11]	Composite common spatial pattern (compositeCSP) (method 1; n = 3)	67.66	97.22	65.48	78.18	72.57	76.22
Lu et al. [19]	Regulized common spatial pattern with generic learning (R-CSP)	69.6	83.9	64.3	70.5	82.5	74.20
Yong et al. [38]	Sparse spatial filter optimization	57.5	54.4	56.9	84.4	84.3	73.50

5 Conclusions

In this paper, we propose a new data mining scheme for Analysis huge volume of EEG data. In the proposed plan, the PCA is applied to extract a representative pattern from every channel EEG data for efficient analysis of the data. Then some important statistical characteristics are computed from this pattern to represent the distribution of the pattern. Combining all of the features extracted from every pattern, we generate a feature set for a subject, and finally, the feature set is used as input to the RF and RT classification model, separately to classy the different states of the data. This proposed scheme is evaluated on two EEG data sets BCI Competition III: V and IVa. Experimental results on two publicly available datasets demonstrate that the proposed data mining scheme is very effective for characterizing brain signals data such as EEG data. Furthermore, the PCA is not only insensitive to extract hidden information removing redundant information from the original huge volume data but also robust to noise. We compared the performance of the proposed method with several state-of-the-art approaches for both data sets in the literature. Experimental results show that the PCA based random forest algorithm achieves higher performances (e.g. classification accuracies) than those by the others.

References

1. BCI Competition III. http://www.bbci.de/competition/iii
2. Bentlemsan, M., Zemouri, E.T., Bouchaffra, D., Yahya-Zoubir, B., Ferroudji, K.: Random forest and filter bank common spatial patterns for eeg-based motor imagery classification. In: 5th International Conference on Intelligent Systems, Modelling and Simulation (ISMS), pp. 235–238. IEEE (2014)
3. Breiman, L.: Random forests. Mach. Learn. **45**(1), 5–32 (2001)
4. Chen, W., Wang, Y., Cao, G., Chen, G., Gu, Q.: A random forest model based classification scheme for neonatal amplitude-integrated eeg. Biomed. Eng. online **13**(2), S4 (2014)
5. Chiappa, S., Millán, J.: Data set v (mental imagery, multi-class). IDIAP Research Institute, Switzerland (2005). http://ida.first.fraunhofer.de/projects/bci/competition_iii/desc_V.htmli
6. De Veaux, R.D., Velleman, P.F., Bock, D.E.: Intro Stats, 3rd edn. Pearson/Addison Wesley, Boston (2008)
7. Duda, R.O., Hart, P.E., Stork, D.G.: Pattern Classification. John Wiley & Sons, New York (2012)
8. Frank, E., et al.: Weka-a machine learning workbench for data mining. In: Maimon, O., Rokach, L. (eds.) Data Mining and Knowledge Discovery Handbook. Springer, Boston (2009). doi:10.1007/978-0-387-09823-4_66
9. Hassan, A.R., Siuly, S., Zhang, Y.: Epileptic seizure detection in eeg signals using tunable-q factor wavelet transform and bootstrap aggregating. Comput. Methods Progr. Biomed. **137**, 247–259 (2016)
10. Ines, H., Slim, Y., Noureddine, E.: EEG classification using support vector machine. In: 10th International Multi-Conference on Systems, Signals & Devices (SSD), pp. 1–4. IEEE (2013)
11. Kang, H., Nam, Y., Choi, S.: Composite common spatial pattern for subject-to-subject transfer. IEEE Sig. Process. Lett. **16**(8), 683–686 (2009)
12. Kannathal, N., Choo, M.L., Acharya, U.R., Sadasivan, P.: Entropies for detection of epilepsy in eeg. Comput. Methods Progr. Biomed. **80**(3), 187–194 (2005)
13. Li, Y., Wen, P., et al.: Analysis and classification of eeg signals using a hybrid clustering technique. In: IEEE/ICME International Conference on Complex Medical Engineering (CME), pp. 34–39. IEEE (2010)
14. Li, Y., Wen, P.P., et al.: Clustering technique-based least square support vector machine for eeg signal classification. Comput. Methods Progr. Biomed. **104**(3), 358–372 (2011)
15. Li, Y., Wen, P.P., et al.: Modified cc-lr algorithm with three diverse feature sets for motor imagery tasks classification in eeg based brain-computer interface. Comput. Methods Progr. Biomed. **113**(3), 767–780 (2014)
16. Lin, C.J., Hsieh, M.H.: Classification of mental task from eeg data using neural networks based on particle swarm optimization. Neurocomputing **72**(4), 1121–1130 (2009)
17. Lin, Y.P., Wang, C.H., Jung, T.P., Wu, T.L., Jeng, S.K., Duann, J.R., Chen, J.H.: EEG-based emotion recognition in music listening. IEEE Trans. Biomed. Eng. **57**(7), 1798–1806 (2010)
18. Lotte, F., Guan, C.: Spatially regularized common spatial patterns for eeg classification. In: 2010 20th International Conference on Pattern Recognition (ICPR), pp. 3712–3715. IEEE (2010)

19. Lu, H., Plataniotis, K.N., Venetsanopoulos, A.N.: Regularized common spatial patterns with generic learning for eeg signal classification. In: Annual International Conference of the IEEE Engineering in Medicine and Biology Society, EMBC 2009, pp. 6599–6602. IEEE (2009)

20. Millan, J.R.: On the need for on-line learning in brain-computer interfaces. In: Proceedings of the 2004 IEEE International Joint Conference on Neural Networks, vol. 4, pp. 2877–2882. IEEE (2004)

21. Rankine, L., Stevenson, N., Mesbah, M., Boashash, B.: A nonstationary model of newborn eeg. IEEE Trans. Biomed. Eng. **54**(1), 19–28 (2007)

22. Shadabi, F., Sharma, D.: Artificial intelligence and data mining techniques in medicine-success stories. In: International Conference on BioMedical Engineering and Informatics, BMEI, vol. 1, pp. 235–239. IEEE (2008)

23. Shen, K.Q., Ong, C.J., Li, X.P., Hui, Z., Wilder-Smith, E.P.: A feature selection method for multilevel mental fatigue eeg classification. IEEE Trans. Biomed. Eng. **54**(7), 1231–1237 (2007)

24. Siuly, S., Li, Y., Wen, P.: EEG signal classification based on simple random sampling technique with least square support vector machine. Int. J. Biomed. Eng. Technol. **7**(4), 390–409 (2011)

25. Siuly, S., Li, Y.: Improving the separability of motor imagery eeg signals using a cross correlation-based least square support vector machine for brain-computer interface. IEEE Trans. Neural Syst. Rehabil. Eng. **20**(4), 526–538 (2012)

26. Siuly, S., Li, Y.: Designing a robust feature extraction method based on optimum allocation and principal component analysis for epileptic eeg signal classification. Comput. Methods Progr. Biomed. **119**(1), 29–42 (2015)

27. Siuly, S., Li, Y.: Discriminating the brain activities for brain-computer interface applications through the optimal allocation-based approach. Neural Comput. Appl. **26**(4), 799–811 (2015)

28. Siuly, S., Li, Y., Zhang, Y.: EEG Signal Analysis and Classification: Techniques and Applications. Springer, Cham (2017)

29. Siuly, S., Zhang, Y.: Medical big data: neurological diseases diagnosis through medical data analysis. Data Sci. Eng. **1**(2), 54–64 (2016)

30. Stam, C., Montez, T., Jones, B., Rombouts, S., Van Der Made, Y., Pijnenburg, Y., Scheltens, P.: Disturbed fluctuations of resting state eeg synchronization in alzheimer's disease. Clin. Neurophysiol. **116**(3), 708–715 (2005)

31. Sun, S., Zhang, C.: Adaptive feature extraction for eeg signal classification. Med. Biol. Eng. Compu. **44**(10), 931–935 (2006)

32. Sun, S., Zhang, C., Lu, Y.: The random electrode selection ensemble for eeg signal classification. Pattern Recogn. **41**(5), 1663–1675 (2008)

33. Sun, S., Zhang, C., Zhang, D.: An experimental evaluation of ensemble methods for eeg signal classification. Pattern Recogn. Lett. **28**(15), 2157–2163 (2007)

34. Supriya, S., Siuly, S., Wang, H., Cao, J., Zhang, Y.: Weighted visibility graph with complex network features in the detection of epilepsy. IEEE Access **4**, 6554–6566 (2016)

35. Wang, H., Xu, J.: Local discriminative spatial patterns for movement-related potentials-based eeg classification. Biomed. Sig. Process. Control **6**(4), 427–431 (2011)

36. Wang, J., Liu, P., She, M.F., Nahavandi, S., Kouzani, A.: Bag-of-words representation for biomedical time series classification. Biomed. Sig. Process. Control **8**(6), 634–644 (2013)

37. Wu, W., Gao, X., Hong, B., Gao, S.: Classifying single-trial eeg during motor imagery by iterative spatio-spectral patterns learning (isspl). IEEE Trans. Biomed. Eng. **55**(6), 1733–1743 (2008)

38. Yong, X., Ward, R.K., Birch, G.E.: Sparse spatial filter optimization for eeg channel reduction in brain-computer interface. In: IEEE International Conference on Acoustics, Speech and Signal Processing, ICASSP 2008. pp. 417–420. IEEE (2008)

39. Zarei, R., He, J., Siuly, S., Zhang, Y.: A PCA aided cross-covariance scheme for discriminative feature extraction from eeg signals. Comput. Methods Progr. Biomed. **146**, 47–57 (2017)

40. Zarei, R., He, J., Huang, G., Zhang, Y.: Effective and efficient detection of premature ventricular contractions based on variation of principal directions. Digit. Sig. Process. **50**, 93–102 (2016)

Extracting Keyphrases Using Heterogeneous Word Relations

Wei Shi[1](✉), Zheng Liu[2], Weiguo Zheng[1], and Jeffrey Xu Yu[1]

[1] The Chinese University of Hong Kong, Hong Kong, China
{shiw,wgzheng,yu}@se.cuhk.edu.hk
[2] Nanjing University of Posts and Telecommunications, Nanjing, China
zliu@njupt.edu.cn

Abstract. Extracting keyphrases from documents for providing a quick and insightful summarization is an interesting and important task, on which lots of research efforts have been laid. Most of the existing methods could be categorized as co-occurrence based, statistic-based, or semantics-based. The co-occurrence based methods do not take various word relations besides co-occurrence into full consideration. The statistic-based methods introduce more unrelated noises inevitably due to the inclusion of external text corpus, while the semantic-based methods heavily depend on the semantic meanings of words. In this paper, we propose a novel graph-based approach to extract keyphrases by considering heterogeneous latent word relations (the co-occurrence and the semantics). The underlying random walk model behind the proposed approach is made possible and reasonable by exploiting nearest neighbor documents. Extensive experiments over real data show that our method outperforms the state-of-art methods.

1 Introduction

Keyphrases are the topical phrases in a document, and keyphrase extraction is a process of automatic selection of important and topical phrases from the body of a document [15]. Such keyphrase extraction provides users with the knowledge of the topics of a document and can be effectively used to assist many other tasks such as text clustering [3], text classification [21], and document summarization [18].

In this paper, we focus on keyphrase extraction from a single document. There are two steps for keyphrase extraction. The first step is to extract a list of candidate phrases from the document. The second step is to select keyphrases from the candidate phrases. In the literature, the keyphrase extraction process could be fulfilled in a supervised manner or an unsupervised manner. The supervised methods deal with the keyphrase extraction as a binary classification task, and build up a classifier, for example, a Naive Bayes classifier, using a training dataset where phrases are labeled as keyphrase and non-keyphrase [6,7,16,19]. The supervised keyphrase extraction methods heavily rely on the

© Springer International Publishing AG 2017
Z. Huang et al. (Eds.): ADC 2017, LNCS 10538, pp. 165–177, 2017.
DOI: 10.1007/978-3-319-68155-9_13

features selected. The most frequent used features include statistical information (e.g., *tf-idf*), position information (e.g., phrase positions in the document), grammar information (e.g., part-of-speech tags) as well as the appearance in title and/or abstract. However, a binary classifier is unable to rank keyphrases, which is required in keyphrase extraction to select the most important keyphrases from a document. For example, to address the ranking issue, Jiang et al. [7] proposes a ranking method by applying ranking *SVM* over a training dataset where every data item is a pair of one keyphrase and one non-keyphrase. The difficult of the supervised methods is they need to use a large training dataset. In addition, the keyphrases of a single document may be unique in the sense that they are different from most keyphrases used in the training dataset. The unsupervised graph-based methods [2,4,5,8,11,17] deal with the keyphrase extraction by ranking nodes over a co-occurrence graph, constructed from the document. Here, a co-occurrence graph, is a weighted undirected graph, where a node represents a word, and an edge between two nodes indicates that the two words occur closely in a given window in the document D. There are two issues given such a co-öccurrence graph. The first issue is how to obtain the initial node/edge weights of the graph constructed. The second issue is how to rank the nodes, in order to select important words in the sense that a keyphrase is important if its words are important. The keyphrases are selected from the words that are ranked high and appear next to each other based on the graph after ranking. Experiments show that the unsupervised graph-based methods achieve comparable performance with supervised methods which need training datasets. On the other hand, the unsupervised graph-based methods heavily rely on the co-occurrence graph that mainly reflects how words appear in the document.

Our main contributions are summarized as follows: (1) We are the first to apply *Word2Vec* to discover semantic relationships among words and utilize them during the keyphrase extraction process. (2) We explore multiple heterogeneous relationships between words in constructing document graphs for keyphrase extraction in a unsupervised manner. (3) We carefully design a meaningful random walk model on a united graph with heterogeneous relationships and propose a biased ranking framework for calculating word importance score.

The rest of this paper is organized as follows. In Sect. 2, we discuss the existing unsupervised graph-based methods for keyphrase extraction. In Sect. 3, we introduce the preliminaries and give an overview of our solution framework. We describe the process of the united graph construction in Sect. 4, followed by the graph-based word ranking strategy in Sect. 5. We explain candidate phrase generation and selection in Sect. 6, followed by the experimental evaluation in Sect. 7. Finally we concludes this paper in Sect. 8.

2   Related Work

We discuss the existing unsupervised graph-based methods which are based on a co-occurrence graph G_C constructed for a document D, where a node represents a word and an edge between two nodes indicates that the two corresponding words appear together in a given window size w (≥ 2) in D.

The earliest work is *TextRank* [11], which assigns initial node-weights and edge-weights to G_C as follows. The initial node weight is $1/n$ for every node in G_C, where n is the total number of nodes in G_C, and the initial edge weights are selected in the range from 0 to 10 following the uniform distribution. Based on the weighted graph, *TextRank* selects the top-k nodes in G_C using *PageRank*, and extracts the keyphrases using the top-k selected nodes if they appear together in the document D. The importance of words is mainly determined by how they appear in D. *SingleRank* [17] takes the similar approach like *TextRank* to select keyphrases. Unlike *TextRank* that assigns an edge weight following uniform distribution, *SingleRank* uses the frequency of two words in the window of w size as the edge weight between the two corresponding nodes. Different from both *TextRank* and *SingleRank* that use *PageRank* to rank nodes, *BetweenRank* [5] utilizes the betweenness to rank nodes, where the betweenness is considered to reflect global properties: connectedness and compactness. In [18], in order to enhance the importance of words in D, it combines sentence extraction and keyphrase extraction together by building two additional graphs, namely, a sentence-sentence graph and a sentence-word graph, in addition to the co-occurrence (word-word) graph. The main idea is to enhance the importance of words using sentences in the sense that (i) a word is important if it appears in an important sentence and (ii) words in an importance sentence will be important. It applies eigenvector centrality to rank sentences and words. To investigate the effectivness of ranking for keyphrase extraction, [2] compares several graph centrality methods including degree, closeness, betweenness, eigenvector and *PageRank*, and shows that all perform in a similar manner, based on their performance studies conducted.

In addition to keyphrase extraction based on a single document, D, there are several attempts to extract keyphrases from a single document based on a collection of documents or an additional corpus. *ExpandRank* [17], selects a collection of similar documents, $D_1, D_2, \cdots, D_n$, to build a virtual document $\tilde{D}$ from which a graph $\tilde{G}$ is constructed. The node weight is $1/n$ assuming $\tilde{G}$ has n nodes. The edge weight is assigned as follows. (A) If the two corresponding words appear in D, the edge weight is the frequency of the two words appear in a window. (B) If the two corresponding words appear in D_i ($\neq D$), the edge weight is the frequency of the two words appear in a window in D_i multiplied by the document similarity between D and D_i. Then, *ExpandRank* ranks nodes in $\tilde{G}$ using *PageRank*. The node weights are enhanced for nodes representing words in G_C, which is a subgraph of $\tilde{G}$. Instead of using a collection of documents, *TPR* [8] utilizes a corpus to obtain topics for every word in the document D using Latent Dirichlet Allocation (*LDA*) [1]. Assume that there are t topics for the words in D or for the nodes in the corresponding G_C in total. *TPR* applies a topical *PageRank* to rank node per topic. At most, a node will have i ($\leq t$) topic weights if the word represented by the node has i topics. The final weight of a node is the sum of the topic weights of the node. [9] explored several popular similarity measure, e.g., cosine similarity, Point-wise Mutual Information and normalized Google similarity distance for clustering words using *Wikipedia* articles, in order to identify exemplar terms. *SemanticRank* [14], utilizes relations from *WordNet* and *Wikipedia* page linkages to build semantic graphs for keyphrase extraction.

3   A New Graph-Based Approach

Majority of the existing graph-based methods focus on a single co-occurrence graph for a given document D. Let the co-occurrence graph be a weighted undirected graph, denoted as $G_C = (V, E)$. Here, V is a set of nodes representing words, and E is a set of edges where an edge, (u, v), exists between two nodes $\text{dst}(u, v) \leq w$. $\text{dst}(u, v) \leq w$ means the words represented by u and v appear in a window size of w (≥ 2) in D. In the following, we denote the node set of a given graph G as $V(G)$ and the edge set as $E(G)$. Let $n = |V(G)|$ and $m = |V(E)|$, then n and m is the number of nodes and edges of G, respectively. It is worth noting that, for a given D and w, all the graph-based methods construct a graph G_C, which have the same topological structure with different node/edge weights. A question that arises is how much we can make use of the words appearance in a document to extract keyphrases. In this paper, in addition to the co-occurrence graph G_C, we introduce a semantic graph which is an undirected graph denoted as $G_S = (V, E)$. Here, V is a set of nodes representing words, and E is a set of edges where an edge, (u, v), exists between two nodes, u and v, if the corresponding words are closely relevant such that $\text{rel}(u, v) \geq \theta$. Let G_C and G_S be the co-occurrence graph and the semantic graph for a given document D, then we have $V(G_C) = V(G_S)$, and $E(G_C) \neq E(G_S)$ in general. On one hand, G_C maintains the local co-occurrence relations among words, which are specific for D. On the other hand, G_S maintains the global semantic relations among words, which are not specific to the document D.

In this paper, the co-occurrence graph and the semantic graph are united to formulate a united G_H, where $V(G_H) = V(G_C) = V(G_S)$ and $E(G_H) = E(G_C) \cup E(G_S)$. Then we carefully design a random walk model on the united graph G_H to evaluate the graph-based word importance scores. The word importance scores are further integrated with phrase features such as frequency and position for improving quality. Phrases with top-K largest scores are selected as the keyphrases. The overview framework of our solution are as follows. (1) We extract a list of candidate phrases using syntactic rules. (2) We build both the co-occurrence graph and the semantic graph for the input document and combine them into a united graph. (3) We calculate the graph-based word importance scores using the united graph and the phrase score by summing up the scores of words contained and integrating with frequency and position features. (4) We select the candidate phrases with top-K largest scores as the keyphrases of the input document.

4   Constructing the United Graph

In our keyphrase extraction solution, the united graph is formulated based on the co-occurrence graph and the semantic graph. Given a document D, let $G_C = (V, E_C)$ denote the co-occurrence graph of D and $G_S = (V, E_S)$ denote the semantic graph. It has been shown in the previous research such as *TextRank* [11] that not all words have the equal importance in representing the topics or ideas of a document. As a result, only nouns and adjectives are selected as nodes in graphs. In the following, we may use word and node to refer a node in a graph interlacedly without distinguishing their difference.

We preprocess the document text using *Stanford CoreNLP* [10], which is widely used in many related research works. *Stanford CoreNLP* includes many natural language processing (*NLP*) tools, to name a few, lemmatizer, part-of-speech (*POS*) tagger, the named entity recognizer (*NER*), and the parser. For each word in the document, we obtain its *POS*, lemma, and named entity label. Each word is labeled with a particular type of entity labels, for example, PERSON (the name label), NUMBER (the numerical label) and DATE (the temporal label). Words labelled as numerical and temporal, together with the stop words, are removed during the graph building, and only lemmas of nouns and adjectives are selected as nodes.

Building the Co-occurrence Graph

With the nodes of the co-occurrence graph, the second step is to add edges among these nodes. Given a window size W, if two words u and v co-occur within a window of W, then an edge $e_{u,v}$ is added between these two nodes. Each edge $e_{u,v} \in E_C$ is associated with an affinity weight $w_{u,v}$, where $w_{u,v} = count(u, v)$, the count of co-occurrence time between words u and v in the input document D. Formally, the adjacency matrix A_C of the co-occurrence graph G_C is defined as follows.

$$A_{Cuv} = \begin{cases} w_{u,v}, & if count(u, v) > 0; \\ 0, & otherwise. \end{cases} \tag{1}$$

Building the Semantic Graph

The co-occurrence relation only encodes the positional information of the input document, and it is not sufficient for extracting keyphrases. In this paper, we proposed to build a semantic graph G_S using *Word2Vec* as well, by taking into account the word semantic meanings. *Word2Vec* generates the vector representations of words through learning from their contexts, and words with similar contexts share similar representations. We use the *Google*'s pre-trained model[1] in this paper, which is based on a very large *Google News Corpus* and each word consists 300 features.

One can notice that semantic graph has the same node set as the co-occurrence graph. The edge set in the semantic similarity graph is constructed in the following way. We apply cosine similarity on each pair of words based on their vector representations. For vertex u and v, if their cosine similarity is larger than a pre-defined threshold θ, then an edge $e_{u,v}$ is added, whose weight is their cosine similarity. It is worth noting that other similarity metrics or weight assignments work just as well. Formally, the adjacency matrix A_S of the semantic graph G_S is defined as follows.

---

[1] https://code.google.com/archive/p/word2vec/.

$$A_{Suv} = \begin{cases} \cos(u,v), & if \cos(u,v) \geq \theta; \\ 0, & otherwise. \end{cases} \tag{2}$$

Uniting the Co-occurrence Graph and the Semantic Graph

With the co-occurrence graph G_C and semantic graph G_S for a given document D, recall that we have $V(G_C) = V(G_S)$, and $E(G_C) \neq E(G_S)$ in general, we could unite these two graphs to formulate a graph G, where $V(G) = V(G_C) = V(G_S)$ and $E(G) = E(G_C) \cup E(G_S)$. For a certain node pair, there might be two edges, one from the co-occurrence graph and the other from the semantic graph. Our following solution is compatible with any number of relations, so the united graph could contain more reasonable relations between words. For simplicity, we only discuss two relations in this paper, co-occurrence and *Word2Vec* semantic. The united graph G could be represented by a tensor $\mathcal{A} = (a_{i,j,k})$, where $i = 1, ..., m$, $j = 1, ..., m$ and $k = 1..l$. l is number of possible relations, and in this paper, $l = 2$. $\mathcal{A}$ is non-negative due to $a_{i,j,k} \geq 0$, and we have

$$a_{i,j,k} = \begin{cases} A_{Ci,j}, & k = 1; \\ A_{Si,j}, & k = 2. \end{cases} \tag{3}$$

5 Calculating Word Importance Scores

With the weighted united graph G, we study how to rank the words based on the graph structure from a global point of view, that is, by considering all possible relations. There are some graph-based ranking methods used for evaluating object importance, such as *PageRank* and *HITS*, by prorogating the initial important score inside the graph structure iteratively to obtain a stationary probability distribution as the score.

Traditional graph-based ranking methods focus on only one relation, and in this paper, we deal with the situation that there exists more than one relation. We carefully design a random surfing model on G in the next section, and similar to graph-based ranking methods on a single relation graph, we consider the stationary landing probabilities of random surfers on each node as its importance score.

5.1 A Random Walk Model on the United Graph

Let us consider a random walk model on the united graph G and focus on the stationary probability that a random surfer will arrive at a particular node using any possible relation. Suppose the random surfer is visiting node v_i at time $t - 1$, and will randomly visit a neighbor of v_i at time t. In the tradition random walk model on a single relation graph, the transitional probability of the random surfer is $p(X_t = v_i | X_{t-1} = v_j)$. But on the united graph G with multiple relations, transitional probability of the random surfer is $p(X_t = v_i | X_{t-1} = v_j, R_t = r_k)$, where r_k is the k-th relation in G. In the following of this paper, we use $p(v_i|v_j, r_k)$ to indicate this probability when there is no ambiguity. Similarly, $p(v_j, r_k)$ indicates $p(X_{t-1} = v_j, R_t = r_k)$.

Then the stationary landing probability of a random surfer for each node on the united graph G is

$$p(v_i) = \sum_{j=1}^{m} \sum_{k=1}^{l} p(v_i|v_j, r_k) \times p(v_j, r_k). \tag{4}$$

Recall that l is the total number of possible relations. Equation 4 shows that if we can estimate the transitional probability $p(v_i|v_j, r_k)$ and the joint probability $p(v_j, r_k)$, the problem of stationary landing probability on united graph G is solved.

Estimating the Intra-relation Transitional Probability

We call the transitional probability $p(v_i|v_j, r_k)$ the intra-relation transitional probability, which means the transitional probability of a random surfer moves from v_j to v_i using a particular relation r_k at time t. Like the transitional probability in a single relation graph, the intra-relation transitional probability can be defined using the following equation.

$$p(v_i|v_j, r_k) = \frac{a_{i,j,k}}{\sum_{i=1}^{m} a_{i,j,k}}. \tag{5}$$

If $a_{i,j,r} = 0$ for all $1 \le i \le m$, then this node is called dangling node [13], and its $p(v_i|v_j, r_k)$ equals to $1/m$.

Estimating the Joint Probability

The joint probability $p(v_j, r_k)$ is in general not easy to evaluate, so we employ the assumption of independence, that is, $p(v_j, r_k) = p(v_j) \times p(r_k)$. By applying this assumption on Eq. 4, we can obtain follows.

$$p(v_i) = \sum_{j=1}^{m} \sum_{k=1}^{l} p(v_i|v_j, r_k) \times p(v_j) \times p(r_k). \tag{6}$$

Now the problem turns to be how to calculate $p(r_k)$. Without loss of generality, $p(r_k)$ represents the summarization of the probabilities that a random surfer moves from a node v_j to one of its neighbor v_i using the relation r_k, that is,

$$p(r_k) = \sum_{i=1}^{m} \sum_{k=1}^{l} p(r_k|v_i, v_j) \times p(v_i, v_j) = \sum_{i=1}^{m} \sum_{k=1}^{l} p(r_k|v_i, v_j) \times p(v_i) \times p(v_j). \tag{7}$$

We also employ the assumption of independence in Eq. 7 during calculating $p(v_i, v_j)$.

$p(r_k|v_i, v_j)$ in Eq. 7 is called inter-relation transitional probability. Due to the heterogeneous relations in united graph G, it is not appropriate to calculate $p(r_k|v_i, v_j)$ in a similar way as intra-relation transitional probability $p(v_i|v_j, r_k)$ in Eq. 5, since the physical meaning of $\sum_{k=1}^{l} a_{i,j,k}$ is not clear and $a_{i,j,k}$ may has various distributions on different relations. We will explain in detail how to estimate $p(r_k|v_i, v_j)$ in the remaining of this section.

Given an input document D, and a corpus $\mathcal{D}$, let $N(D)$ denote its nearest neighbor documents under the bag-of-word model based on cosine similarity. Each document is represented by a vector d, where each dimension in d is the tf-idf value of the corresponding word in the bag-of-word model. Let $sim(d_i, d_j)$ denote the similarity of document pair $\langle D_i, D_j \rangle$, then $N(D) = \{D_k|sim(d, d_k) > \tau\}$, where τ is a pre-defined threshold.

The inter-relation transitional probability $p(r_k|v_i, v_j)$ tells us which relation a random surfer should favor during its random walks. Then we use the keyphrase information in $N(D)$ to help us estimate $p(r_k|v_i, v_j)$. The intuitive idea is to favor the relation that could generate correct importance score of each word in keyphrases. we build the co-occurrence graph and the semantic graph for each document in $N(D)$ as described in Sect. 4. Based on the keyphrase information of the documents in $N(D)$, let W denote the set of words contained by keyphrases, and $\overline{W}$ denote the complementary part of W, the set of words not in keyphrses. Then the random walk stationary probabilities of words in W should be larger than ones of words in $\overline{W}$, on both the co-occurrence graph and the semantic graph.

We utilize a loss function to establish the inter-relation transitional probability. Let h denote a loss function, and in this paper, we adopt Wilcoxon-Mann-Whitney (WMW) loss function with width b [20], which is

$$h(x) = \frac{1}{1 + e^{-x/b}}. \tag{8}$$

Let pr_w denote the random walk stationary probability of word w in a graph, then the quality of the co-occurrence graph and the semantic graph could be measured in terms of the summation of loss function values.

$$totalloss = \sum_{s \in W, t \in \overline{W}} h(pr_t - pr_s). \tag{9}$$

$h(\cdot)$ assigns a non-negative penalty according to the value $pr_t - pr_s$. If $pr_t - pr_s < 0$, then $h(pr_t - pr_s) = 0$, otherwise, $h(pr_t - pr_s) > 0$.

Let $totalloss_k$ denote the summation of loss function values on relation k, then $p(r_k|v_i, v_j)$ is

$$p(r_k|v_i, v_j) = 1 - \frac{totalloss_k}{\sum_{o=1}^{l} totalloss_o}, \tag{10}$$

where a random surfer favors the relations with smaller loss function values. Similar to intra-relation transitional probability, if $a_{i,j,k} = 0$ for all $1 \leq k \leq l$, then we set $p(r_k|v_i, v_j)$ equals to $1/l$. Recall that in this paper, l equals to 2.

5.2   The Biased Ranking Algorithm

With both the transitional probability $p(v_i|v_j, r_k)$ and the joint probability $p(v_j, r_k)$ calculated, we apply a similar way in the *MultiRank* [12] to evaluate the importance score of words given the input document D. With Eqs. 5, 6 and 7, we can estimate the stationary landing probability $p(v_i)$ of a random surfer on the united graph G. Let $\mathcal{Q}_{i,j,k}$ denote $p(v_i|v_j, r_k)$ and $\mathcal{R}_{i,j,k}$ denote $p(r_k|v_i, v_j)$, then the tensor form of Eqs. 6 and 7 is as follows.

$$\mathbf{x} = \mathcal{Q}\mathbf{x}\mathbf{y}, \quad \mathbf{y} = \mathcal{R}\mathbf{x}^2 \tag{11}$$

where $\mathbf{x} = (x_1, ..., x_m)$, $\mathbf{y} = (y_1, ..., y_l)$.

One insightful observation is that the word characteristics in a document play an import role in extracting keyphrases besides relations. According to several previous work, word frequency and the first occurrence position in the document are two essential features which have a large impact on the importance of a word. So we involving these two features in Eq. 11 to bias the ranking process. Let $f(w)$ and $pos(w)$ denote the word frequency and the first occurrence position respectively.

$$\mathbf{x}_b = \frac{f(v_i)}{pos(v_i)} / \sum_{i=1}^{m} \frac{f(v_i)}{pos(v_i)} \tag{12}$$

With Eq. 12, the biased form of Eq. 11 is

$$\mathbf{x} = (1 - \alpha)\mathcal{Q}\mathbf{x}\mathbf{y} + \alpha\mathbf{x}_b, \quad \mathbf{y} = \mathcal{R}\mathbf{x}^2 \tag{13}$$

By repeatedly applying Eq. 13 until it converges, we could obtain the stationary landing probabilities as the word important scores.

6   Generating Candidates and Selecting Keyphrases

In this section, we discuss how to generate the candidate phrases. Most of the previous works use syntactic rules based on the part-of-speech of words. Others use the adjacency relations of top ranking words. If several words are in a continuous sequence in the input document, these words form a keyphrase. We follow the approach in most previous research works which is to use the syntactic pattern to detect candidate phrases. If a sequence of words satisfy the pattern that zero or more adjectives are followed by one or more nouns, it is considered as a candidate phrase. Formally, we represent the syntactic pattern as "$(JJ)*(NN|NNS|NNP)+$". All candidate phrases are merged according to the following two rules, where $\mathcal{P}$ is the generated candidate phrase set.

(1) If the *NER*s of two candidate phrases are *PERSON*, and one is contained in the other, we merge the shorter one to the longer one. For example, "George Bush" and "Bush", we merge them as "George Bush" and add the frequency of "Bush" to "George Bush".

(2) If a candidate phrase is the acronym of another candidate phrase, then we merge them as one. For example, "FEMA" and "Federal Emergency Management Agency", we merge the former to the latter and add the frequency of the former to the latter.

Based on the word important score obtained in Sect. 5, the final ranking score of each candidate phrase $P_i \in \mathcal{P}$ is defined in Eq. 14 by summing up the important scores of the contained words.

$$PhraseScore(P_i) = \sum_{v_j \in P_i} p(v_j). \tag{14}$$

We further embed the frequency and the first occurrence position of phrases, denoted by $f(P_i)$ and $pos(P_i)$ respectively, into Eq. 14 to improve the quality of phrase scores, as shown in Eq. 15.

$$Score(P_i) = \frac{f(P_i)}{pos(P_i)} \times PhraseScore(P_i). \tag{15}$$

After we obtain the final ranking scores $Score(P_i)$ of all the candidate phrases $P_i \in \mathcal{P}$, we select the phrases with top-K largest scores as the keyphrases of the input document.

7   Experiments

In this section, we demonstrate our experimental results on news data by comparing with several other state-of-the-art methods. We denote our proposed methods as **HGRank** for convenience. We use the *DUC2001* dataset in the experimental evaluation. *DUC2001*[2] is a popular news dataset in many research tasks, such as keyphrase extraction and document summarization. There are 308 news articles in *DUC2001*, which are categorised into 30 topics. The average length of the news articles is about 700 words, and each article is manually assigned about 10 keyphrases. The manually labeled keyphrase dataset is provided by [17]. We evaluate the performance under three metrics, which are precision (P), recall (R) and F-measure (F), as Eq. 16 shows.

$$P = \frac{count_{correct}}{count_{output}}, R = \frac{count_{correct}}{count_{manual}}, F = \frac{2 \times P \times R}{P + R}. \tag{16}$$

7.1   Comparison with Other Methods

We compare *HGRank* with three unsupervised keyphrase extraction methods, which are *ExpandRank* [17], *SingleRank* [17], *TF-IDF*, and one supervised method *KEA* [19]. Note that all the unsupervised methods have the same candidate phrase set.

---

[2] http://www-nlpir.nist.gov/projects/duc/past_duc/duc2001/data.html.

TF-IDF: The method uses the *tf-idf* score as the ranking score of each keyterm. And the ranking score of each candidate phrase is the sum of the scores of words contained in it.

We also compare the two relations used in our work. One is the co-occurrence relation, the other is the *Word2Vec* semantic relation. We adopt *PageRank* on the co-occurrence graph and semantic graph respectively. The former is corresponding to the method *SingleRank*. The latter we refer to as **SimilarRank**.

In order to see the influence of phrase frequency with the first occurrence of phrase on keyphrase extraction, we use the same way to combine these two features as in our proposed method *HGRank* to rank candidate phrases and get the method **PTF-POS**. Meanwhile, we also compare with the method **HGNWRank** which follows the same framework with *HGRank* except it does not adopt phrase features.

Figure 1(a)–(c) shows the precision, recall and F-measure curves of all the above methods when the keyphrase number K ranges from 5 to 20 respectively. We use the variables which achieve the best performances. For *HGRank*, we set $W = 20$, $\theta = 0.4$. For *ExpandRank*, we set $W = 19$. For *SingleRank*, we set $W = 18$. For *SimilarRank*, we set $\theta = 0.4$. For *HGNWRank*, we set $W = 14$, $\theta = 0.3$.

From Fig. 1(a)–(c), we can see our method *HGRank* performs the best, which indicates integrating different word relations with phrase features improves the performance of keyphrase extraction, followed by *ExpandRank*, which is the state-of-art method on *DUC2001* dataset. Note that we use the same steps as the original work [17] of *ExpandRank*, the only differences are we use the form of each term after lemmatization and we merge the candidate phrases according to the two rules in Sect. 6. *ExpandRank* performs better than *SingleRank*, which indicates that documents with similar topics are helpful for keyphrase extraction. *HGNWRank* performs a little better than *TF-IDF* and *SingleRank*, which indicates that the combination of co-occurrence relations and semantic relations is helpful for keyphrase extraction. *TF-IDF* performs a little better than *SingleRank*, which indicates *IDF* is also helpful since *PageRank* score is proportional to degree (i.e., *TF*) in the undirected graph. Note that *PTF-POS* performs worse than *SingleRank*, which indicates from words perspective to measure the importance of phrases is meaningful. *SimilarRank* performs worse than *SingleRank*, which indicates the information from the document itself is more important and semantic relations are better to learn from the documents of the same domain. *KEA* performs worse than *PTF-POS*, which indicates supervised method is hard to transfer to different domains since the training documents used in *KEA*[3] are not in the same domain with the evaluation data.

7.2   Effect of Parameters

In order to investigate the influence of windowsize W on keyphrase extraction, we conduct experiments with different values of W. Figure 2 shows the effect of

---

[3] We use the original version of *KEA* provided in the public website created by its author, http://www.nzdl.org/Kea/download.html.

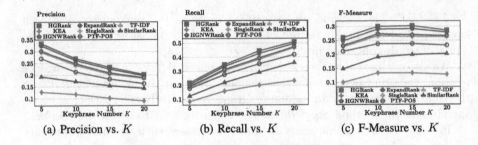

(a) Precision vs. K (b) Recall vs. K (c) F-Measure vs. K

Fig. 1. Comparison with other algorithms.

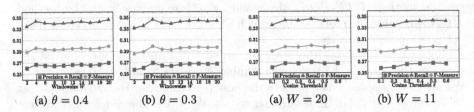

(a) $\theta = 0.4$ (b) $\theta = 0.3$ (a) $W = 20$ (b) $W = 11$

Fig. 2. Effect of windowsize W **Fig. 3.** Effect of cosine threshold θ

W when W ranges from 2 to 20. In Fig. 2(a) the cosine similarity threshold θ is set to 0.4. In Fig. 2(b) the cosine similarity threshold θ is set to 0.3. From the figures, we can see W has an obviously influence on the performances when it is small, however, as it becomes larger, performances change little. We also conduct experiments to learn the impact of cosine similarity threshold θ on keyphrase extraction. Figure 3 shows the effect of θ when θ ranges from 0.1 to 0.6. In Fig. 3(a) the windowsize W is set to 20 and in Fig. 3(b) W is set to 11. From the figures, We can see θ has the same influence as W, performances increase obviously when θ is small and keep almost constant when θ becomes larger.

8   Conclusions

In this paper, we construct a united graph based on the co-occurrence relation and semantic relation of the words in the input documents. We propose a novel graph-based approach to extract keyphrases by considering these heterogeneous latent word relations. The underlying random walk model behind our graph-based approach is made possible and reasonable by exploiting nearest neighbor documents. The combination of global features and phrase features further improve the performance. We demonstrate the effectiveness of our proposed approach by extensive experiments.

Acknowledgements. This work is supported by Research Grant Council of Hong Kong SAR No. 14221716, Natural Science Foundation of Jiangsu Province (BK20171447) and Nanjing University of Posts and Telecommunications (NY215045).

References

1. Blei, D.M., Ng, A.Y., Jordan, M.I.: Latent Dirichlet allocation. J. Mach. Learn. Res. **3**, 993–1022 (2003)
2. Boudin, F.: A comparison of centrality measures for graph-based keyphrase extraction. IJCNLP **2013**, 834–838 (2013)
3. Hammouda, K.M., Matute, D.N., Kamel, M.S.: CorePhrase: keyphrase extraction for document clustering. In: Perner, P., Imiya, A. (eds.) MLDM 2005. LNCS (LNAI), vol. 3587, pp. 265–274. Springer, Heidelberg (2005). doi:10.1007/11510888_26
4. Hasan, K.S., Ng, V.: Conundrums in unsupervised keyphrase extraction: making sense of the state-of-the-art. In: COLING, pp. 365–373 (2010)
5. Huang, C., Tian, Y., Zhou, Z., Ling, C.X., Huang, T.: Keyphrase extraction using semantic networks structure analysis. In: ICDM 2006, pp. 275–284 (2006)
6. Hulth, A.: Improved automatic keyword extraction given more linguistic knowledge. In: EMNLP 2003, pp. 216–223 (2003)
7. Jiang, X., Hu, Y., Li, H.: A ranking approach to keyphrase extraction. In: SIGIR 2009, pp. 756–757 (2009)
8. Liu, Z., Huang, W., Zheng, Y., Sun, M.: Automatic keyphrase extraction via topic decomposition. In: EMNLP, pp. 366–376 (2010)
9. Liu, Z., Li, P., Zheng, Y., Sun, M.: Clustering to find exemplar terms for keyphrase extraction. In: EMNLP 2009, pp. 257–266 (2009)
10. Manning, C.D., Surdeanu, M., Bauer, J., Finkel, J., Bethard, S.J., McClosky, D.: The Stanford CoreNLP natural language processing toolkit. In: ACL, pp. 55–60 (2014)
11. Mihalcea, R., Tarau, P.: Textrank: bringing order into text. In: EMNLP 2004, pp. 404–411 (2004)
12. Ng, M.K., Li, X., Ye, Y.: Multirank: co-ranking for objects and relations in multi-relational data. In: SIGKDD 2011, pp. 1217–1225 (2011)
13. Page, L., Brin, S., Motwani, R., Winograd, T.: The PageRank citation ranking: bringing order to the web. 1999
14. Tsatsaronis, G., Varlamis, I., Nørvåg, K.: SemanticRank: ranking keywords and sentences using semantic graphs. In: COLING 2010, pp. 1074–1082 (2010)
15. Turney, P.D.: Learning algorithms for keyphrase extraction. CoRR, cs.LG/0212020 (2002)
16. Turney, P.D.: Learning to extract keyphrases from text. CoRR, cs.LG/0212013 (2002)
17. Wan, X., Xiao, J.: Exploiting neighborhood knowledge for single document summarization and keyphrase extraction. ACM Trans. Inf. Syst. **28**(2) (2010)
18. Wan, X., Yang, J., Xiao, J.: Towards an iterative reinforcement approach for simultaneous document summarization and keyword extraction. In: ACL 2007 (2007)
19. Witten, I.H., Paynter, G.W., Frank, E., Gutwin, C., Nevill-Manning, C.G.: KEA: practical automatic keyphrase extraction. In: ACM DL 1999, pp. 254–255 (1999)
20. Yan, L., Dodier, R.H., Mozer, M., Wolniewicz, R.H.: Optimizing classifier performance via an approximation to the Wilcoxon-Mann-whitney statistic. In: ICML 2003, pp. 848–855 (2003)
21. Youn, E., Jeong, M.K.: Class dependent feature scaling method using naive bayes classifier for text datamining. Pattern Recogn. Lett. **30**(5), 477–485 (2009)

Mining High-Dimensional CyTOF Data: Concurrent Gating, Outlier Removal, and Dimension Reduction

Sharon X. Lee[✉]

School of Mathematics and Physics, University of Queensland, Brisbane, Australia
s.lee11@uq.edu.au

Abstract. Cytometry is a powerful tool in clinical diagnosis of health disorders, in particular, immunodeficiency diseases and acute leukemia. Recent technological advancements have enabled up to 100 measurements to be taken simultaneously on each cell, thus generating high-throughput and high-dimensional datasets. Current analysis, relying on manual segmentation of cell populations (gating) on sequential low-dimensional projections of the data, is subjective, time consuming and error-prone. It is also known that these multidimensional cytometric data typically exhibit non-normal features, including asymmetry, multimodality, and heavy tails. This present a great challenge to traditional clustering methods which are typically based on symmetric distributions.

In recent years, non-normal distributions have received increasing interest in the statistics literature. In particular, finite mixtures of skew distributions have emerged as a promising alternative to the traditional normal mixture modelling. However, these models are not well suited to high-dimensional settings.

This paper describes a flexible statistical approach designed for performing, at the same time, unsupervised clustering, dimension reduction, and outlier removal for cytometric data. The approach is based on finite mixtures of multivariate skew normal factor analyzers (SkewFA) with threshold pruning. The model can be fitted by maximum likelihood (ML) via an expectation-maximization (EM) algorithm. An application to a large CyTOF data is presented to demonstrate the usefulness of the SkewFA model and to illustrate its effectiveness relative to other algorithms.

1 Introduction

Flow cytometry is routinely used in clinical and research immunology. It allows the fluorescence expression of different markers on each cell to be measured. In a flow cytometer, samples stained with fluorophore-conjugated antibodies (or markers) are passed through a laser beam and the light that emerges from each cell are captured and quantitated. A more recent technology is mass cytometry, also known as cytometry by time-of-flight mass spectrometry (CyTOF), which uses metal isotopes in place of fluorescent antibodies. This latter advancement enables much more variables to be measured at the same time [7].

© Springer International Publishing AG 2017
Z. Huang et al. (Eds.): ADC 2017, LNCS 10538, pp. 178–189, 2017.
DOI: 10.1007/978-3-319-68155-9_14

One of the major tasks in the analysis of cytometric data is the identification of cell populations from these multidimensional data, currently performed manually by visually separating regions (gates) of interests on a series of sequential bivariate projections of the data. This process is known as gating. Due to the subjective and time-consuming nature of this approach, and the difficulty in detecting higher-dimensional inter-marker relationships, many efforts have been made to develop computational methods to automate the gating process; see [1] for a recent account.

Cytometric data can be challenging to model. It is well known that they are often heterogeneous (that is, contain more than one cell populations) and that these cell clusters are typically asymmetrically distributed as well as having heavier-tails than normal. Among the recent computational tools for flow cytometric data, many were based on the mixture model-based approach due to its powerfulness and flexibility. Mixture model is a well-established statistical framework and have enjoyed numerous applications in many fields. The traditional approach is to adopt the normal distribution as component densities of the mixture model, that is, the so-called normal mixture model. However, this is inadequate for cytometric data as the normal distribution is symmetric and hence is not suitable for modelling the asymmetrically distributed cell clusters. To address this, some authors have recently considered adopting skew mixture models, showing promising results [10,14,16,17,22,23,26]. For high-dimensional cytometric data such as those acquired from mass cytometry, these methods cannot be directly applied due to computational complexity. In view of this, we consider a factor analytic approach as an alternative to these fully specified mixture models.

In this paper, we consider the analysis of large and high-dimensional CyTOF data sets for which recent computational tools that are based on skew mixture models (such as [17,22]) find it challenging to analyse due to the large number of markers and cells. Our method is based on a mixture of skew normal of factor analyzers, thus allowing simultaneous gating and dimension reduction. Moreover, to allow for the identification and removal of outlying observations, we adopt a threshold-based pruning approach where a small number of observations that are deemed as very unlikely to occur are automatically pruned or discarded during estimation. Thus, this three-in-one approach can perform automated gating, dimension reduction, and outlier removal at the same time.

The rest of this paper is organized as follows. In Sect. 2, we present the main ideas of our approach, discussing each of the four parts of our model. In Sect. 3, we describe the model fitting process which is via a closed-form ECM algorithm. Finally, the usefulness of the proposed approach is demonstrated in Sect. 4 on a large high-dimensional CyTOF data containing 24 major immune cell populations.

2   The SkewFA Methodology

Our methodology adopt a three-in-one approach comprising of four parts: (1) skew component distributions, (2) mixture modelling, (3) factor analysis, and (4) pruning of observations. They allow for the modelling of multiple clusters that may exhibit non-normal features, perform implicit dimension reduction, and outlier removal, respectively. This is achieved by fitting a modified version of mixtures of skew normal factor analyzers, referred to as the SkewFA model. We now walk through each of these building blocks used to construct our SkewFA model.

2.1 Non-normal Clusters: Multivariate Skew Normal Distributions

A multivariate skew normal (MSN) distribution is a flexible generalization of the multivariate normal distribution, suitable for capturing skewness in data. The MSN distribution (1) is the basic part of SkewFA, which plays the role of modelling a single cell population. To ease discussion, we start with the definition of the MSN distribution to be adopted by SkewFA. Let Y be a p-dimensional random vector that follows a p-dimensional skew normal distribution [6] with a $p \times 1$ location vector μ, a $p \times p$ scale matrix Σ, and a $p \times 1$ skewness vector δ. Then its probability density function (pdf) can be expressed as the product of a multivariate normal density and a (univariate) normal distribution function, given by

$$f_p(y; \mu, \Sigma, \delta, \nu) = 2\phi_p(y; \mu, \Omega)\Phi(y^*; 0, \lambda), \tag{1}$$

where $\Omega = \Sigma + \delta\delta^T$, $y^* = \delta^T\Omega^{-1}(y - \mu)$, and $\lambda = 1 - \delta^T\Omega^{-1}\delta$. Here, we let $\phi_p(.; \mu, \Sigma)$ be the p-dimensional normal distribution with mean vector μ and covariance matrix Σ, and $\phi(.; \mu, \Sigma)$ is the corresponding (cumulative) distribution function. The notation $Y \sim \text{MSN}_p(\mu, \Sigma, \delta)$ will be used. Note that when $\delta = 0$, (1) reduces to the symmetric normal density $\phi_p(y; \mu, \Sigma)$. The vector δ specifies how skewness can be characterized, that is, its direction and magnitude.

It is worth noting that there exists many different versions of the multivariate skew normal density in the literature. This includes the well-known version proposed by [6] and the versions considered by [5,8,9], which involves univariate normal distribution function in their pdfs. Among other versions, there are also the skew family of distributions proposed by [3,4,24], which allows for a multivariate normal distribution function to be used in their densities. Here in (1), we are adopting the version as considered by [22], which is equivalent to the so-called classical form proposed by [6]; see [15] for the proof.

2.2 Multiple Clusters: Finite Mixtures of Skew Normal Distributions

The next part of SkewFA is the joint modelling of multiple cell populations or clusters. This can be conveniently formulated as a mixture of MSN distribution. More formally, a g-component finite mixture model is a convex linear combination of g component densities. In our case, the density of a mixture of g MSN distributions is given by

$$f\left(y; \Psi\right) = \sum_{i=1}^{g} \pi_i f_p\left(y; \mu_i, \Sigma_i, \delta_i\right), \tag{2}$$

where $f_p\left(y; \mu_i, \Sigma_i, \delta_i\right)$ denotes the ith MSN component of the mixture model as defined by (1), with location parameter μ_i, scale matrix Σ_i, and skewness parameter δ_i. The mixing proportions or weights satisfy $\pi_i \geq 0$ ($i = 1, \ldots, g$) and $\sum_{i=1}^{g} \pi_i = 1$. We shall refer to the model defined in (2) as the FMMSN (finite mixture of MSN) model. In the above, we let Ψ denote the vector containing all the unknown parameters of the FMMSN model; that is, $\Psi = \left(\pi_1, \ldots, \pi_{g-1}, \theta_1^T, \ldots, \theta_g^T\right)^T$ where now θ_i consists of the unknown parameters of the ith component density function which includes μ_i, δ_i, and the distinct elements of Σ_i. The FMMSN model has been shown to be effective in modeling flow cytometric data [22]. It provides a natural representation of the data where each cluster can have its own specific parametric model. In [22] and subsequent works, the FMMSN model have only been applied to data where p is relatively small. This may perhaps be due to the computational burden associated with fitting a high-dimensional FMMSN model. The next part of SkewFA attempts to alleviate this problem.

2.3 Dimension Reduction: Mixture of Skew Normal Factor Analyzers

The FMMSN model is a highly parametrized model and thus can quickly become computationally infeasible as the number of markers p and/or the number of clusters g increases. A popular and powerful alternative to a fully specified mixture model is a factor analysis (FA) model [13,19]. It assumes that the observations can be modelled by lower-dimensional latent representations. More formally, it models the distribution of the data using

$$y_j = \mu + Bu_j + e_j, \tag{3}$$

where u_j is a q-dimensional ($q < p$) random vector of latent variables known as factors, B is a $p \times q$ matrix of factor loadings, and e are independently distributed error variables. Traditionally, it is assumed that U follows a standard (multivariate) normal distribution, whereas e follows a centered normal distribution with a diagonal covariance matrix.

The third part of SkewFA adopts the FA framework for the FMMSN model, leading to a mixture of MSN factor analyzers. We proceed by letting each component of (2) to have a factor analytic representation similar to (3). More specifically, we have the component-specific factors and errors jointly following a MSN distribution. In addition, for ease of correspondence with the FA model, we use a parameterization of the MSN distribution such that the factors have mean being the zero vector and covariance matrix being the identity matrix. This enables the property of u in the FA model to be preserved. To achieve this, we let

$$u \sim \mathrm{MSN}_q\left(-c\Lambda^{\frac{1}{2}}, \Lambda, \Lambda^{\frac{1}{2}}\delta\right), \tag{4}$$

where $c = \frac{\pi}{2}$ and $\Lambda = \left(I_q + (1 - c^2)\delta\delta^T \right)^{-1}$. It follows that the marginal distribution of y is a FMMSN distribution given by

$$f(y_j; \Psi) = \sum_{i=1}^{g} \pi_i \, \mathrm{MSN}_p \left(\mu_i - cB_i\Lambda_i^{\frac{1}{2}}\delta_i, \ \Sigma_i, \ B_i\Lambda_i^{\frac{1}{2}} \right). \tag{5}$$

We shall refer to the above model as the SkewFA model.

2.4   Outlier Removal: Skew Normal Factor Analyzers with Pruning

The final part of SkewFA is the incorporation of an (automated) outlier removal technique. For this, we start with a commonly used technique known as the trimmed likelihood approach by [21]. This technique is quite simple to implement, but has been shown to be quite powerful and flexible; see [11,12] for its applications to mixtures of factor analyzers and skew mixture models, respectively. However, in addition to pruning a certain proportion of observations, we also discard observations in a component-wise manner based on a specified cutoff. The key idea is to (temporarily) remove a small number of observations that are deemed as least likely to occur, so that they (temporarily) do not contribute to (an iteration of) the model fitting procedure. By least likely to occur, we mean that the observation has the lowest (estimated) contribution to the likelihood function. To proceed, we attach a pruning label p_j to each observation y_j, which is a binary variable indicating whether the associated observation is to be pruned for the current iteration of the fitting procedure. The contribution of an observation is defined in terms of its density (5). These contributions are then ranked from smallest to largest, that is,

$$f(y_{(1)}; \Psi) \leq f(y_{(2)}; \Psi) \leq \cdots \leq f(y_{(n)}; \Psi), \tag{6}$$

where the subscript given in parenthesis denotes the ranked index. The pruning threshold is calculated based on the total number of observations and the pruning proportion ρ. Then observations with ranked index less than or equal to $\lfloor n\rho \rfloor$ is pruned, that is,

$$p_j = \begin{cases} 1, \text{ if } (j) \geq \lfloor n\rho \rfloor \\ 0, \text{ otherwise.} \end{cases} \tag{7}$$

In addition, if the component-wise density value of on observation falls below a very small cutoff value, it is temporarily discarded (for that component) by setting the corresponding indicator variable to zero. Note that at each iteration of the fitting procedure, different observations can be pruned, hence the term *temporary* was used. At the end of the last iteration, if an observation is pruned, it is labeled as an outlier. However, for clustering purposes, a cluster label can still be provided for these pruned observations by applying the maximum *a posteriori* rule to their estimated posterior probabilities of component membership.

3 Parameter Estimation via EM Algorithm

To fit the SkewFA model described above, that is, to carry out parameter estimation, we follow the most commonly adopted approach of maximum likelihood estimation (MLE). It follows that the SkewFA admits a convenient hierarchical characterization that facilitates the development of an EM algorithm for parameter estimation, namely,

$$\boldsymbol{Y}_j \mid \boldsymbol{u}_{ij}, x_j, Z_{ij} = 1 \sim N_p\left(\boldsymbol{\mu}_i + \boldsymbol{B}_i \boldsymbol{\Lambda}_i^{\frac{1}{2}} \boldsymbol{u}_{ij}, \boldsymbol{D}_i\right),$$
$$\boldsymbol{U}_{ij} \mid x_j, Z_{ij} = 1 \sim N_q\left((x_j - c)\boldsymbol{\delta}_i, \boldsymbol{I}_q\right),$$
$$x_j \mid Z_{ij} = 1 \sim HN\left(0, 1; \mathbb{R}^+\right),$$
$$Z_j \sim \text{Multi}_g\left(1; \boldsymbol{\pi}\right), \tag{8}$$

assuming $\boldsymbol{y}_j$ is not pruned. In the above, $HN(0, 1; \mathbb{R}^+)$ denotes the standard half-normal distribution truncated to the positive region, Multi denotes the multinomial distribution, and $\boldsymbol{\pi} = (\pi_1, \ldots, \pi_g)$. The variables $Z_j = (Z_{i1}, \ldots, Z_{ig})$ $(j = 1, \ldots, n)$ have been introduced to denote the latent labels for the jth observation where Z_{ij} is one or zero depending on whether $\boldsymbol{y}_j$ belongs to the ith component of the mixture model. The complete-data in this case consists of the observations $\boldsymbol{y}_j$, the indicator variables Z_j, and the latent variables $\boldsymbol{U}_{ij}$ and x_j.

The EM algorithm is implemented by alternating repeatedly the P-, E-, and M-steps until convergence. The P-step stands for the pruning step, which precedes the usual E-step of the EM algorithm. During this step, the pruning labels are calculated as described in Sect. 2.4. Only observations with $p_j = 1$ are passed to the E- and M-step of the current iteration. Furthermore, if the density of the ith component of the skewFA model for an observation $\boldsymbol{y}_j$ falls below the specified cutoff, then z_{ij} is set to zero and hence does not contribute to the corresponding component in the E- and M-steps. The E-step calculates the expectation of the complete-data log likelihood given the observed data $\boldsymbol{y}$ using the current estimate of the parameters, known as the Q-function, which is given by

$$Q(\boldsymbol{\Psi}; \boldsymbol{\Psi}^{(k)}) = E_{\boldsymbol{\Psi}^{(k)}}\left\{\log L_{pc}(\boldsymbol{\Psi}) \mid \boldsymbol{y}\right\},$$

where $\log L_{pc}(\boldsymbol{\Psi})$ denotes the pruned complete-data log likelihood function, given by

$$\log L_{pc}(\boldsymbol{\Psi}) = \sum_{j=1}^{n} p_j \sum_{i=1}^{g} z_{ij} \left[\log \pi_i - \tfrac{1}{2} \log |\boldsymbol{D}_i| + tr\left(\boldsymbol{D}_i^{-1} \boldsymbol{R}_{ij}\right)\right.$$
$$\left. + (x_j - c)^2 \boldsymbol{\delta}_i \boldsymbol{\delta}_i^T - 2(x_j - c)\boldsymbol{\delta}_i^T \boldsymbol{u}_{ij}\right]. \tag{9}$$

Here, we let $\boldsymbol{R}_{ij} = (\boldsymbol{y}_j - \boldsymbol{\mu}_i - \boldsymbol{B}_i \boldsymbol{\Lambda}^{\frac{1}{2}} \boldsymbol{u}_{ij})(\boldsymbol{y}_j - \boldsymbol{\mu}_i - \boldsymbol{B}_i \boldsymbol{\Lambda}^{\frac{1}{2}} \boldsymbol{u}_{ij})^T$ for notational convenience. It can be shown that on the $(k+1)$th iteration, the E-step requires the calculation of the following conditional expectations.

$$z_{ij}^{(k)} = E\left[Z_{ij} \mid \boldsymbol{y}_j\right] \tag{10}$$

$$e_{1ij}^{(k)} = E[W_j \mid \boldsymbol{y}_j, Z_{ij} = 1] \tag{11}$$

$$e_{2ij}^{(k)} = E[W_j^2 \mid \boldsymbol{y}_j, Z_{ij} = 1] \tag{12}$$

$$e_{3ij}^{(k)} = E[W_j \boldsymbol{U}_{ij} \mid \boldsymbol{y}_j, Z_{ij} = 1] \tag{13}$$

$$e_{4ij}^{(k)} = E[\boldsymbol{U}_{ij} \mid \boldsymbol{y}_j, Z_{ij} = 1] \tag{14}$$

$$e_{5ij}^{(k)} = E[\boldsymbol{U}_{ij} \boldsymbol{U}_{ij}^T \mid \boldsymbol{y}_j, Z_{ij} = 1] \tag{15}$$

The M-step of the EM algorithm maximizes the Q-function with respect to the parameters $\boldsymbol{\Psi}$. It follows that on the $(k+1)$th iteration of the EM algorithm, an updated estimate of the parameters of the SkewFA model is given by

$$\pi_i^{(k+1)} = \frac{n_i^{(k)}}{n}, \tag{16}$$

$$\mu_i^{(k+1)} = \frac{\sum_{j=1}^n z_{ij}^{(k)} (\boldsymbol{y}_j - \boldsymbol{B}_i^{(k)} \boldsymbol{\Lambda}_i^{(k)\frac{1}{2}} e_{4ij}^{(k)})}{n_i^{(k)}}, \tag{17}$$

$$\boldsymbol{B}_i^{(k+1)} = \left[\sum_{j=1}^n z_{ij}^{(k)} (\boldsymbol{y}_j - \mu_i^{(k)}) e_{4ij}^{(k)}\right] \left[\sum_{j=1}^n z_{ij}^{(k)} e_{5ij}^{(k)}\right] \boldsymbol{\Lambda}_i^{(k)-\frac{1}{2}}, \tag{18}$$

$$\boldsymbol{D}_i^{(k+1)} = \frac{1}{n_i^{(k)}} \text{diag}\left(\sum_{j=1}^n z_{ij}^{(k)} \left[\left(\boldsymbol{y}_j - \mu_i^{(k+1)} - \boldsymbol{B}_i^{(k+1)} \boldsymbol{\Lambda}_i^{(k)-\frac{1}{2}} e_{4ij}^{(k)}\right)\right.\right.$$
$$\left(\boldsymbol{y}_j - \mu_i^{(k+1)} - \boldsymbol{B}_i^{(k+1)} \boldsymbol{\Lambda}_i^{(k)-\frac{1}{2}} e_{4ij}^{(k)}\right)^T$$
$$\left.\left. + \boldsymbol{B}_i^{(k+1)} \boldsymbol{\Lambda}_i^{(k)-\frac{1}{2}} (e_{5ij}^{(k)} - e_{4ij}^{(k)} e_{4ij}^{(k)T}) \boldsymbol{\Lambda}_i^{(k)-\frac{1}{2}} \boldsymbol{B}_i^{(k)T}\right]\right), \tag{19}$$

$$\delta_i^{(k+1)} = \frac{\sum_{j=1}^n z_{ij}^{(k)} (e_{3ij}^{(k)} - c\, e_{4ij}^{(k)})}{\sum_{j=1}^n z_{ij}^{(k)} (e_{2ij}^{(k)} - 2c\, e_{1ij}^{(k)} + c^2)}, \tag{20}$$

$$\boldsymbol{\Lambda}_i^{(k+1)} = \boldsymbol{I}_q + (1 - c^2) \delta_i^{(k)} \delta_i^{(k)T}. \tag{21}$$

The factor loading matrices $\boldsymbol{B}_i$ carry important information about the dimension reduction part of the SkewFA model. They give an indication of the importance of each markers in the composition of the latent space. Accordingly, the factor scores can be interpreted as the lower-dimensional representation of the data. This can be easily retrieved by calculating

$$\sum_{i=1}^g \pi_i^{(k)} \boldsymbol{\Lambda}_i^{(k)\frac{1}{2}} e_{4ij}^{(k)}. \tag{22}$$

4   Analysis of High-Dimensional CyTOF Data

To demonstrate the usefulness of the SkewFA approach, we consider the tasks of automated gating, dimension reduction, and outlier detection, of a 13-dimensional CyTOF data analyzed in [18]. The data contain measurements on 167,044 cells derived from a sample donated by a healthy human individual. There are 24 major immune cell populations identified by manual analysis on this data, and their relative population size varies considerably (Fig. 1). Further description of the data, including the details of each label in Fig. 1, can be found in [18] and is not repeated here due to space limitation. This data is difficult to analyse [27] due to the large span in the abundance of the populations, which ranges from the smallest Platelet population of 5 cells to the largest population of 13,964 mature CD4+ T-cells. Our main task is to identify and model all these cell populations and provide a predicted class label for each cell. Our secondary tasks includes dimension reduction and identifying outlier cells.

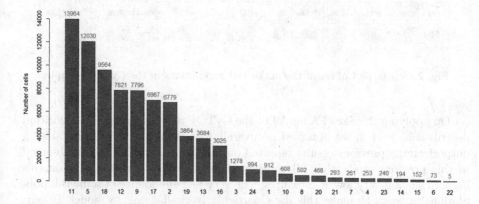

Fig. 1. Abundance of cell populations varies largely across the 24 clusters, ranging from as few as 5 cells to close to 14,000 cells. Details of the population labels (shown as #1 to #24 above) can be found in [18].

An initial inspection of the distribution of the cell populations reveals that some are evidently not symmetric. For example, the CD11b− monocyte cell population (labeled #1 in Fig. 1) is clearly not symmetrically distributed on the CD45 and CD45RA markers (see the red population within the top left plot in Fig. 2). To assess the performance of the SkewFA model, we take the manual gating results as the 'true' solution and measure how close the solution given by SkewFA is to this 'true' results. For this, we report the F-measure score as is typically used in evaluating the performance of cytometric data analysis algorithms. The F-measure is defined as the harmonic mean of precision and recall. It ranges between 0 and 1, with 1 indicating a perfect match with the true class labels and 0 is the worse match. In calculating the F-measure, we choose among the possible permutations of the cluster labels the one that gives the highest value.

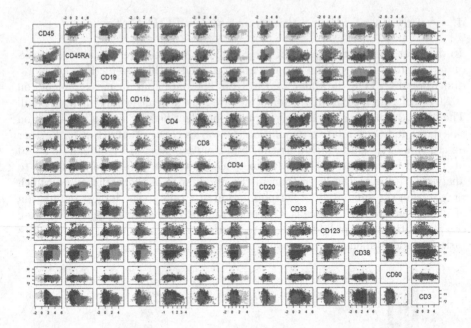

Fig. 2. Scatterplot of six of the major cell populations in the CyTOF sample.

On applying the SkewFA model to the CyTOF data using the EM algorithm described in Sect. 3, we obtained an overall F-measure of 0.830 (see Table 1), outperforming previous results reported on this data (see [27]). It was observed in this previous study that all considered methods performed poorly on this data, achieving very low F-measures. The very low abundance of some cell populations appears to make this data particularly challenging to model. Briefly, [27] reported a comparison of 13 unsupervised algorithms on this CyTOF data. This includes some well-known and/or commonly used algorithms specialised for cytometric data, such as flowMeans [2], immunoClust [25], and SWIFT [20]. The reported best results was 0.52, obtained by flowMeans. A comparison of the performance of the SkewFA approach relative to these methods is shown in Table 1, where the F-measure for other methods were reproduced from [27]. It can be observed that SkewFA provides a notable improvement in gating performance across most cell populations as evident by a higher F-measure, especially for low abundance populations. In particular, Table 1 reveals that SkewFA is the only method that had correctly identified all cells in the least abundant population (#22), achieving a perfect F-measure. The SWIFT algorithm is the only other algorithm that was able to identify at least some cells in this population, but with a F-measure of 0.01 it is far from the performance of SkewFA. The other algorithms failed to discriminate any cells from this population.

Furthermore, none of the competing methods considered have reported facilities to identify and eliminate outliers. For this data, SkewFA identified approximately 10% of outliers. To pursuit this further, we inspected all of these pruned

Table 1. The *F*-measure *per cell population* of various methods applied to the CyTOF data. The final row shows the *overall* *F*-measure results of the methods.

ID	SkewFA	FlowMeans	FlowSOM	k-means	FLOCK	ImmunoClust	SWIFT
1	0.98	0.61	0.39	0.64	0.00	0.00	0.43
2	0.82	0.95	0.86	0.64	0.81	0.95	0.30
3	0.71	0.19	0.30	0.13	0.00	0.37	0.29
4	0.96	0.03	0.42	0.44	0.43	0.00	0.43
5	0.82	0.49	0.46	0.57	0.80	0.81	0.14
6	0.74	0.39	0.28	0.30	0.39	0.04	0.14
7	0.51	0.16	0.02	0.11	0.00	0.00	0.14
8	0.49	0.19	0.05	0.19	0.14	0.00	0.17
9	0.99	0.62	0.65	0.86	0.96	0.94	0.28
10	0.77	0.51	0.57	0.05	0.00	0.00	0.32
11	0.86	0.86	0.97	0.63	0.94	0.80	0.20
12	0.99	0.97	0.96	0.74	0.08	0.00	0.24
13	0.96	0.88	0.55	0.69	0.00	0.00	0.15
14	0.98	0.42	0.36	0.22	0.00	0.10	0.45
15	0.62	0.04	0.04	0.02	0.04	0.00	0.17
16	0.43	0.32	0.23	0.36	0.15	0.10	0.16
17	0.99	0.94	0.92	0.78	0.68	0.00	0.28
18	0.97	0.96	0.96	0.90	0.72	0.71	0.17
19	0.92	0.82	0.93	0.90	0.93	0.70	0.32
20	1.00	0.35	0.61	0.00	0.78	0.00	0.77
21	1.00	0.64	0.62	0.63	0.59	0.55	0.61
22	1.00	0.00	0.00	0.00	0.00	0.00	0.01
23	0.90	0.45	0.00	0.00	0.00	0.00	0.72
24	0.97	0.66	0.72	0.62	0.64	0.00	0.49
All	0.83	0.52	0.50	0.44	0.38	0.31	0.18

observations (not shown). Surprisingly, it reveals that they all corresponds to cells that are deemed as dead by expert analysts. This suggests that SkewFA has the ability to discriminate between live and dead cells, which may have contributed to its superior performance in this data.

Finally, concerning the computation time, we note that the SkewFA approach is very competitive to other algorithms. In this experiment, the total computation time for SkewFA was only 30 s, placing it the fourth fastest among the 11 algorithms considered. The three faster algorithms were *k*-means, FlowSOM, and FLOCK, requiring 2, 15, and 29 s, respectively. The remaining algorithms have computation time ranging from 249 to 29,469 s. It is worth noting that SkewFA is a single-threaded implementation in this experiment, whereas for the other competing algorithms the parallel version is used if available.

5   Conclusion

We introduced a new computational tool, SkewFA, for mining large and high-dimensional CyTOF data. It allows for simultaneously clustering, dimension reduction, and removal of outliers. Based on an enhanced version of finite mixtures of skew normal factor analyzers, the approach can effectively accommodate clusters that are asymmetrically distributed. Moreover, by adopting a factor analytic characterization of the component densities, it enables SkewFA to perform implicit dimension reduction within very reasonable time. An illustration on a large CyTOF data shows that the SkewFA model compares favourably to other state-of-the-art specialized algorithms, achieving a higher F-measure than those reported in other analyses of these data. It is also the only approach among the competing algorithms that allows for automatic identification and removal of outlying observations.

References

1. Aghaeepour, N., Finak, G., The FlowCAP Consortium, The DREAM Consortium, Hoos, H., Mosmann, T., Gottardo, R., Brinkman, R.R., Scheuermann, R.H.: Critical assessment of automated flow cytometry analysis techniques. Nat. Methods **10**, pp. 228–238 (2013)
2. Aghaeepour, N., Nikoloc, R., Hoos, H.H., Brinkman, R.R.: Rapid cell population identification in flow cytometry data. Cytom. A **79**, 6–13 (2011)
3. Arellano-Valle, R.B., Azzalini, A.: On the unification of families of skew-normal distributions. Scand. J. Stat. **33**, 561–574 (2006)
4. Arellano-Valle, R.B., Genton, M.G.: On fundamental skew distributions. J. Multivar. Anal. **96**, 93–116 (2005)
5. Azzalini, A., Capitanio, A.: Distributions generated by perturbation of symmetry with emphasis on a multivariate skew t-distribution. J. Royal Stat. Soc. B **65**, 367–389 (2003)
6. Azzalini, A., Dalla Valle, A.: The multivariate skew-normal distribution. Biometrika **83**, 715–726 (1996)
7. Bendall, S.C., Simonds, E.F., Qiu, P., Amir, E.D., Krutzik, P.O., Finck, R.: Single-cell mass cytometry of differential immune and drug responses across a human hematopoietic continuum. Science **332**, 687–696 (2011)
8. Branco, M.D., Dey, D.K.: A general class of multivariate skew-elliptical distributions. J. Multivar. Anal. **79**, 99–113 (2001)
9. Cabral, C.R.B., Lachos, V.H., Prates, M.O.: Multivariate mixture modeling using skew-normal independent distributions. Comput. Stat. Data Anal. **56**, 126–142 (2012)
10. Frühwirth-Schnatter, S., Pyne, S.: Bayesian inference for finite mixtures of univariate and multivariate skew-normal and skew-t distributions. Biostatistics **11**, 317–336 (2010)
11. García-Escudero, L.A., Gordaliza, A., Ingrassia, S., Mayo-Iscar, A.: The joint role of trimming and constraints in robust estimation for mixtures of gaussian factor analyzers. Comput. Stat. Data Anal. **99**, 131–147 (2016)
12. García-Escudero, L.A., Greselin, F., Mayo-Iscar, A., McLachlan, G.J.: Robust estimation of mixtures of skew-normal distributions. In: Proceedings of the 48th Scientific Meeting of the Italian Statistical Society (SIS2016) (2016)

13. Ghahramani, Z., Beal, M.: Variational inference for bayesian mixture of factor analysers. In: Solla, S., Leen, T., Muller, K.R. (eds.) Advances in Neural Information Processing System, pp. 449–455. MIT Press, Cambridge (2000)

14. Lee, S.X., McLachlan, G.J.: Model-based clustering and classification with non-normal mixture distributions. Stat. Methods Appl. **22**, 427–454 (2013)

15. Lee, S.X., McLachlan, G.J.: On mixtures of skew-normal and skew t-distributions. Adv. Data Anal. Classif. **7**, 241–266 (2013)

16. Lee, S.X., McLachlan, G.J.: Finite mixtures of canonical fundamental skew t-distributions: The unification of the restricted and unrestricted skew t-mixture models. Stat. Comput. **26**, 573–589 (2016)

17. Lee, S.X., McLachlan, G.J., Pyne, S.: Modelling of inter-sample variation in flow cytometric data with the Joint Clustering and Matching (JCM) procedure. Cytom. A **89**, 30–43 (2016)

18. Levine, J.H., Simonds, E.F., Bendall, S.C., Davis, K.L., Amir, E.D., Tadmor, M.D., Nolan, G.P.: Data driven phenotypic dissection of aml reveals progenitor-like cells that correlate with prognosis. Cell **162**, 184–197 (2015)

19. McLachlan, G.J., Peel, D.: Mixtures of factor analyzers. In: Proceedings of the Seventeenth International Conference on Machine Learning. pp. 599–606. Morgan Kaufmann, San Francisco (2000)

20. Mosmann, T.R., Naim, I., Rebhahn, J., Datta, S., Cavenaugh, J.S., Weaver, J.M.: SWIFT - scalable clustering for automated identification of rare cell populations in large, high-dimensional flow cytometry datasets. Cytom. A **85A**, 422–433 (2014)

21. Neykov, N., Filzmoser, P., Dimova, R., Neytchev, P.: Robust fitting of mixtures using the trimmed likelihood estimator. Comput. Stat. Data Anal. **52**, 299–308 (2007)

22. Pyne, S., et al.: Automated high-dimensional flow cytometric data analysis. In: Berger, B. (ed.) RECOMB 2010. LNCS, vol. 6044, pp. 577–577. Springer, Heidelberg (2010). doi:10.1007/978-3-642-12683-3_41

23. Pyne, S., Lee, S.X., Wang, K., Irish, J., Tamayo, P., Nazaire, M.D., Duong, T., Ng, S.K., Hafler, D., Levy, R., Nolan, G.P., Mesirov, J., McLachlan, G.: Joint modeling and registration of cell populations in cohorts of high-dimensional flow cytometric data. PLoS ONE **9**, e100334 (2014)

24. Sahu, S.K., Dey, D.K., Branco, M.D.: A new class of multivariate skew distributions with applications to bayesian regression models. Can. J. Stat. **31**, 129–150 (2003)

25. Sorensen, T., Baumgart, S., Durek, P., Grutzkau, A., Haaupl, T.: immunoClust - an automated analysis pipeline for the identification of immunophenotypic signatures in high-dimensional cytometric datasets. Cytom. A **87A**, 603–615 (2015)

26. Wang, K., Ng, S.K., McLachlan, G.J.: Multivariate skew t mixture models: applications to fluorescence-activated cell sorting data. In: Shi, H., Zhang, Y., Bottema, M.J., Lovell, B.C., Maeder, A.J. (eds.) Proceedings of Conference of Digital Image Computing: Techniques and Applications, pp. 526–531. IEEE, Los Alamitos, California (2009)

27. Weber, L.M., Robinson, M.D.: Comparison of clustering methods for high-dimensional single-cell flow and mass cytometry data. Cytom. A **89A**, 1084–1096 (2016)

AI for Big Data

Locality-Constrained Transfer Coding
for Heterogeneous Domain Adaptation

Jingjing Li[1,2]([✉]), Ke Lu[1], Lei Zhu[2], and Zhihui Li[3]

[1] School of Computer Science and Engineering,
University of Electronic Science and Technology of China, Chengdu 610054, China
lijin117@yeah.net, kel@uestc.edu.cn
[2] School of Information Technology and Electrical Engineering,
University of Queensland, Queensland, QLD 4067, Australia
leizhu0608@gmail.com
[3] Beijing Etrol Technologies Co., Ltd., Beijing, China
zhihuilics@gmail.com

Abstract. Currently, most of widely used databases are label-wise. In other words, people organize their data with corresponding labels, e.g., class information, keywords and description, for the convenience of indexing and retrieving. However, labels of the data from a novel application usually are not available, and labeling by hand is very expensive. To address this, we propose a novel approach based on transfer learning. Specifically, we aim at tackling heterogeneous domain adaptation (HDA). HDA is a crucial topic in transfer learning. Two inevitable issues, feature discrepancy and distribution divergence, get in the way of HDA. However, due to the significant challenges of HDA, previous work commonly focus on handling one of them and neglect the other. Here we propose to deploy locality-constrained transfer coding (LCTC) to simultaneously alleviate the feature discrepancy and mitigate the distribution divergence. Our method is powered by two tactics: feature alignment and distribution alignment. The former learns new transferable feature representations by sharing-dictionary coding and the latter aligns the distribution gaps on the new feature space. By formulating the problem into a unified objective and optimizing it via an iterative fashion, the two tactics are reinforced by each other and the two domains are drawn closer under the new representations. Extensive experiments on image classification and text categorization verify the superiority of our method against several state-of-the-art approaches.

Keywords: Domain adaptation · Transfer learning · Knowledge discovery

1 Introduction

From the perspective of general users, a database is a set of well organized data. For a piece of data in a database, it is usually organized with several keywords associated with it for the convenience of indexing and retrieving. These keywords

© Springer International Publishing AG 2017
Z. Huang et al. (Eds.): ADC 2017, LNCS 10538, pp. 193–204, 2017.
DOI: 10.1007/978-3-319-68155-9_15

can be seen as labels of the corresponding data. In real-world applications, however, labels are not always accessible. What should one do in this situation? Most people would say we can train a classifier to automatically label samples, and some may say why bother, we can label them by hand. Unfortunately, both of them are not practical in specific circumstances, because either we have insufficient training samples to train an accurate classifier for a new application, or labeling by hand is too expensive to be afforded. The endless stream of novel applications and pervasive scarcity of well-labeled data have stimulated a remarkable wave of research on transfer learning [1]. Transfer learning handles the problem where available labeled samples in the target domain are too scarce to train an accurate model. It borrows knowledge from related domains, i.e., the source domain, to facilitate training. However, a majority of machine learning approaches might fail to transfer knowledge because they commonly assume that training and test data are drawn from the same probability distribution. Unfortunately, it is always hard to find a semantically matched source domain which happens to share the same data distribution with the target domain.

Domain adaptation [2,3,5,6] is proposed with the mission of knowledge transfer beyond the distribution gaps among different domains. Most of existing work [2,7,8] focus on the homogeneous domain adaptation problem where the two domains are sampled from different data distributions but the same feature representation. In practice, however, the two domains are often drawn from not only different probability distributions but also different feature representations. For instance, the target domain is sampled from convolutional neural network (CNN) features of images, whilst the source domain is drawn from bag-of-words (BoW) features of texts. Accordingly, HDA [9–11] is proposed and has received increasing attention.

To transfer knowledge among heterogeneous domains, HDA has to overcome two inevitable obstacles: feature discrepancy and distribution divergence. However, due to the intrinsic challenges of HDA, existing work generally select only one of them to tackle. They optimize either the feature discrepancy or the distribution divergence individually. For instance, previous work [8–10,12] focus on minimizing the distribution divergence. A fraction of existing approaches [11,13,14] propose to learn new feature representations to mitigate the feature discrepancy. Nevertheless, it is easy to know that the distribution gaps would be smaller if the two domains are represented by more similar feature representations. Meanwhile, minimizing the distribution gaps also leads to more optimal feature representations (feature representations which can alleviate the distribution divergence are preferred). In a nutshell, the two objectives can reinforce each other and jointly optimizing them would be more beneficial.

Motivated by above discussions, we propose a novel approach, named as locality-constrained transfer coding (LCTC). It simultaneously alleviates the feature discrepancy and mitigates the distribution divergence in a unified optimization problem. Specifically, our objective is formulated by the following motivations: (1) To alleviate the feature discrepancy, we learn new transferable feature representations for the two domains with a shared codebook. By sharing the codebook, the learned new features are interconnected and, thus, the knowledge can be transfered to the target domain. (2) To mitigate the distribution divergence, we further minimize the marginal distribution gaps between the two domains on the learned new feature space. (3) It is worth noting that samples with the same semantic label tend to stay close (as shown in Fig. 1). Therefore, we also exploit to preserve the manifold structure and local consistency in our formulation. An iterative updating algorithm is presented to optimize the objective. Experimental results on image classification and text categorization verify the superiority of our method.

2   Related Work

Many approaches [7,19,20] have been proposed to handle domain adaptation problems. A majority of them focus on homogeneous domain adaptation. For instance, Pan et al. [8] propose to learn transferable components across domains in a reproducing kernel Hilbert space (RKHS) by using maximum mean discrepancy (MMD) [15]. Ding et al. [5] deploy low-rank coding in a deep structure to learn a latent space shared by the two domains. However, those work concentrate on either minimizing the data distribution gaps or mining the shared factors among the two domains. They did not give much attention to the feature discrepancy.

Recently, some HDA methods [21,22] are proposed to handle the feature discrepancy problem. Wang and Mahadevan [9] align the data manifold of the two domains for adaptation. Li et al. [11] propose using augmented feature representations to effectively utilize the data from both domains with a SVM similar formulation. Thai et al. [10] re-weight the samples and select landmarks for classification.

As stated before, the current issue is that there are two inevitable problems of which need to be taken care, but previous work exploit to optimize either the feature discrepancy or the distribution divergence separately. They catch one of them and lose another. This paper proposes a novel approach which aims to simultaneously alleviate the feature discrepancy and mitigate the distribution divergence in a unified optimization problem.

3   Locality-Constrained Transfer Coding

3.1   Notations

In this paper, we use bold lowercase letters to represent vectors, bold uppercase letters to represent matrices. A sample is denoted as a vector, e.g., $\mathbf{x}$, and the i-th sample in a set is represented by the symbol $\mathbf{x}_i$. For a matrix $\mathbf{M}$, its Frobenius norm is defined as $\|\mathbf{M}\|_F = \sqrt{\sum_i \delta_i(\mathbf{M})^2}$, where $\delta_i(\mathbf{M})$ is the i-th singular value of the matrix $\mathbf{M}$. The trace of matrix $\mathbf{M}$ is denoted by $\mathbf{tr}(\mathbf{M})$. For clarity, the frequently used notations and corresponding descriptions are shown in Table 1.

Table 1. Frequently used notations and their descriptions.

Notation	Description	Notation	Description
$\mathbf{X}_s \in \mathbb{R}^{d_s * n_s}$, $\mathbf{y_s}$	Source samples/labels	$\mathbf{X} \in \mathbb{R}^{m*n}$	$\mathbf{X}_s$ and $\mathbf{X}_t$
$\mathbf{X}_t \in \mathbb{R}^{d_t * n_t}$, $\mathbf{y_t}$	Target samples/labels	$\mathbf{B} \in \mathbb{R}^{m*k}$	Codebook
$\mathbf{M} \in \mathbb{R}^{n*n}$	MMD matrix	$\mathbf{S} \in \mathbb{R}^{k*n}$	Coding matrix
$\mathbf{\Lambda}$	Lagrange multipliers	α, β	Penalty parameters

3.2 Definitions

Definition 1. *A domain $\mathbb{D}$ consists of two parts: feature space $\mathcal{X}$ and its probability distribution $P(\mathbf{X})$, where $\mathbf{X} \in \mathcal{X}$.*

We use subscripts s and t to indicate the source domain and the target domain, respectively. This paper focuses on the following problem:

Problem 1. *Given a well-labeled source domain $\mathbb{D}_s$ and a mostly unlabeled target domain $\mathbb{D}_t$, where $\mathbb{D}_s \neq \mathbb{D}_t$, $\mathcal{X}_s \neq \mathcal{X}_t$ and $P(\mathbf{X}_s) \neq P(\mathbf{X}_t)$. Simultaneously align the feature discrepancy and distribution divergence between $\mathbb{D}_s$ and $\mathbb{D}_t$.*

3.3 Problem Formulation

As we stated in the introduction, our model supposes to learn new transferable feature representations for $\mathbf{X}_s$ and $\mathbf{X}_t$ with a shared codebook $\mathbf{B}$. It also minimizes the distribution gaps between the two domains on the new feature space. Thus, our objective can be formulated as follows:

$$\min_{\mathbf{B},\mathbf{S}} \underbrace{\mathcal{C}_1(\mathbf{X_s}, \mathbf{X_t}, \mathbf{B}, \mathbf{S})}_{feature\ alignment} + \underbrace{\alpha \mathcal{C}_2(\mathbf{S_s}, \mathbf{S_t})}_{distribution\ alignment} + \underbrace{\beta \Omega(\mathbf{S})}_{constraint} \tag{1}$$

where $\mathbf{S}$ ($\mathbf{S}_s$ and $\mathbf{S}_t$ for $\mathbf{X}_s$ and $\mathbf{X}_t$ respectively) is the new feature representation, $\mathcal{C}_1$ is the feature alignment part, $\mathcal{C}_2$ is the distribution alignment part and Ω is the constraint. Notice that $\mathcal{C}_2$ is deployed on $\mathbf{S}$ rather than $\mathbf{X}$. It is a progressive alignment based on the results of $\mathcal{C}_1$. $\alpha > 0$ and $\beta > 0$ are two hyper-parameters. In the remainder of this section, we will present each part in detail and show how to optimize Eq. (1).

Feature Alignment. Suppose that we can learn a new feature representation $\mathbf{S}$ by which the feature discrepancy between the two domains can be alleviated. For the source domain data $\mathbf{X}_s$, we can learn $\mathbf{S}_s$ through

$$\min_{\mathbf{b}_s, \mathbf{s}_s} \sum_{i=1}^{n_s} \left\| \mathbf{x}_{s,i} - \sum_{j=1}^{k} \mathbf{b}_{s,j} \mathbf{s}_{s,i}^j \right\|_2^2 + \beta \sum_{i=1}^{n_s} \left\| \mathbf{s}_{s,i} \right\|_1,$$
$$s.t. \ \left\| \mathbf{b}_{s,j} \right\|^2 \leq c, \quad \forall j, \tag{2}$$

where c is a constant, we keep it as 1 in this paper. Equation (2) can be rewritten as the following form in matrix,

$$\min_{\mathbf{B_s},\mathbf{S_s}} \|\mathbf{X_s} - \mathbf{B_s}\mathbf{S_s}\|_F^2 + \beta \sum_{i=1}^{n_s} \|\mathbf{s}_{s,i}\|_1, \ s.t. \ \|\mathbf{b}_{s,j}\|^2 \le c, \ \forall j, \tag{3}$$

where $\mathbf{B_s}$ is the codebook, also known as dictionary matrix, learned on $\mathbf{X}_s$, and $\mathbf{S}_s$ is the coding matrix. $\mathbf{s}_{s,i}$ is a sparse representation for the corresponding data point in $\mathbf{X}_s$. In a similar fashion, we can learn a new feature representation for the target domain data by optimizing:

$$\min_{\mathbf{B_t},\mathbf{S_t}} \|\mathbf{X_t} - \mathbf{B_t}\mathbf{S_t}\|_F^2 + \beta \sum_{i=1}^{n_t} \|\mathbf{s}_{t,i}\|_1, \ s.t. \ \|\mathbf{b}_{t,j}\|^2 \le c, \ \forall j. \tag{4}$$

In order to transfer knowledge from the source domain to the target domain, we advocate $\mathbf{X_s}$ and $\mathbf{X_t}$ sharing the same codebook $\mathbf{B}$. Thus, their corresponding new feature representations $\mathbf{S_s}$ and $\mathbf{S_t}$ would be interconnected and can be directly compared. It means the new feature representations are transferable. To this end, we have the following problem:

$$\min_{\mathbf{B},\mathbf{S_s},\mathbf{S_t}} \|\mathbf{X_s} - \mathbf{B}\mathbf{S_s}\|_F^2 + \|\mathbf{X_t} - \mathbf{B}\mathbf{S_t}\|_F^2 + \beta \sum_{i=1}^{n} \|\mathbf{s}_i\|_1,$$
$$s.t. \ \|\mathbf{b}_j\|^2 \le c, \ \forall j. \tag{5}$$

Note that $\mathbf{S_s}$ and $\mathbf{S_t}$ may have different dimensionalities in practice. In this paper, to reduce the computational costs and filter out high-dimensional noises, samples are aligned to the same dimensionality by PCA. One can also align them in a RKHS, or by learning two projections, one for each domain. Equation (5) can be rewritten as the following equivalent equation after some algebraic manipulations,

$$\min_{\mathbf{B},\mathbf{S}} \|\mathbf{X} - \mathbf{B}\mathbf{S}\|_F^2 + \beta \sum_{i=1}^{n} \|\mathbf{s}_i\|_1,$$
$$s.t. \ \|\mathbf{b}_j\|^2 \le c, \ \forall j, \tag{6}$$

where $\mathbf{X}$ and $\mathbf{S}$ are defined as,

$$\mathbf{X} = [\mathbf{X}_s \ \mathbf{X}_t], \ \mathbf{S} = [\mathbf{S}_s \ \mathbf{S}_t].$$

Distribution Alignment. Please notice that in HDA tasks, one needs to take care of not only feature discrepancy but also distribution divergence. Recently, MMD [15] has been introduced to estimate the distance between distributions because of its non-parametric merit. The MMD between two datasets $\mathbf{X}_s$ and $\mathbf{X}_t$ can be computed as:

$$\left\| \frac{1}{n_s} \sum_{i=1}^{n_s} \phi(\mathbf{x}_{s,i}) - \frac{1}{n_t} \sum_{j=1}^{n_t} \phi(\mathbf{x}_{t,j}) \right\|^2, \tag{7}$$

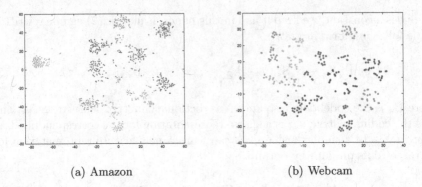

(a) Amazon   (b) Webcam

Fig. 1. Visualization of the Amazon and Webcam dataset from Office dataset [4]. The figure is generated by t-SNE [23] with DeCAF$_6$ features [24]. Each color denotes one class. (Color figure online)

where $\phi(\cdot)$ is a feature mapping. Since we have learned the new feature representations $\mathbf{S_s}$ and $\mathbf{S_t}$ for $\mathbf{X_s}$ and $\mathbf{X_t}$ respectively, we further mitigate the distribution divergence on the new feature space. The minimization of distribution divergence between $\mathbf{S}_s$ and $\mathbf{S}_t$ can be formulated as follows:

$$\min_{\mathbf{S_s},\mathbf{S_t}} \left\| \frac{1}{n_s}\sum_{i=1}^{n_s}\mathbf{S}_{s,i} - \frac{1}{n_t}\sum_{j=1}^{n_t}\mathbf{S}_{t,j} \right\|_2^2 = \min_{\mathbf{S}} \operatorname{tr}(\mathbf{SMS}^\top), \tag{8}$$

where $\mathbf{M}$ is the MMD matrix computed as:

$$\mathbf{M}_{ij} = \begin{cases} \frac{1}{n_s n_s}, & if\, \mathbf{x}_i, \mathbf{x}_j \in \mathbf{X}_s \\ \frac{1}{n_t n_t}, & if\, \mathbf{x}_i, \mathbf{x}_j \in \mathbf{X}_t. \\ \frac{-1}{n_s n_t}, & otherwise \end{cases} \tag{9}$$

Local Consistency. Suppose that the label information of the target domain is known, we visualize the target samples as shown in Fig. 1. An interesting observation is that the data samples with the same label stay in a compact cluster. More formally, a sample tends to have the same label with its k-nearest neighbors. Traditionally, the local consistency is formulated as a graph Laplacian regularization [25,26]. However, to formulate a compact and efficient objective, we advocate a locality constraint instead of the graph regularization in this paper. The locality constraint can be seamlessly incorporated into our sharing-dictionary coding framework. And, fortunately, it gives rise to a serendipity that the locality constraint must lead to sparsity [32]. Thus, we can remove the ℓ_1 norm from Eq. (6). The locality constraint is defined as follows:

$$\|\mathbf{d}_i \odot \mathbf{s}_i\|^2,\ with\ \ \mathbf{d}_i = \exp(\tfrac{\operatorname{dist}(\mathbf{x}_i,\mathbf{B})}{\sigma}) \tag{10}$$

where $\odot$ denotes the element-wise multiplication, $\mathbf{d}_i$ is the locality adaptor which measures the distance between an instance $\mathbf{x}_i$ and the codebook. $\operatorname{dist}(\mathbf{x}_i, \mathbf{b}_j)$ is the Euclidean distance between $\mathbf{x}_i$ and $\mathbf{b}_j$. σ is used for adjusting the weight decay speed [32]. We set $\sigma = 1$ in this paper.

At last, by taking all the considerations into account, we have the overall objective shown as follows:

$$\min_{\mathbf{B},\mathbf{S}} \|\mathbf{X} - \mathbf{BS}\|_F^2 + \alpha \mathrm{tr}(\mathbf{SMS}^\top) + \beta \sum_{i=1}^n \|\mathbf{d}_i \odot \mathbf{s}_i\|^2,$$
$$s.t. \ \|\mathbf{b}_j\|^2 \le c, \ \ \forall j. \tag{11}$$

3.4 Problem Optimization

It is easy to know that Eq. (11) is not convex for $\mathbf{B}$ and $\mathbf{S}$ simultaneously. However, it is convex for each of them when the other is fixed. We, therefore, deploy an alternative strategy to optimize it as shown in the following steps.

Step 1. Optimize coding matrix S. Optimizing Eq. (11) with respect to $\mathbf{S}$ when $\mathbf{B}$ is fixed can be reformulated as optimizing the following problem:

$$\min_{\mathbf{S}} \|\mathbf{X} - \mathbf{BS}\|_F^2 + \alpha \mathrm{tr}(\mathbf{SMS}^\top) + \beta \sum_{i=1}^n \|\mathbf{d}_i \odot \mathbf{s}_i\|^2. \tag{12}$$

To make the problem easier to be optimized, we introduce an auxiliary variable $\mathbf{D}_i = \mathrm{diag}(\mathbf{d}_i)$. As a result, Eq. (12) can be rewritten as,

$$\min_{\mathbf{S}} \|\mathbf{X} - \mathbf{BS}\|_F^2 + \alpha \mathrm{tr}(\mathbf{SMS}^\top) + \beta \sum_{i=1}^n \|\mathbf{D}_i \cdot \mathbf{s}_i\|^2. \tag{13}$$

To solve $\mathbf{S}$, by taking the derivative of Eq. (13) with respect to $\mathbf{S}$, and setting the derivative to zero, we have,

$$\mathbf{s}_i = \left(\mathbf{B}^\top \mathbf{B} + \alpha \mathbf{M}_{ii} \mathbf{I} + \beta \mathbf{D}_i^\top \mathbf{D}_i\right)^{-1} \mathbf{B}^\top \mathbf{x}_i. \tag{14}$$

Step 2. Optimize codebook B. Optimizing Eq. (11) with respect to $\mathbf{B}$ when $\mathbf{S}$ is fixed can be reformulated as optimizing the following problem:

$$\min_{\mathbf{B}} \|\mathbf{X} - \mathbf{BS}\|_F^2 + \beta \sum_{i=1}^n \|\mathbf{d}_i \odot \mathbf{s}_i\|^2, \ \ s.t. \ \|\mathbf{b}_j\|^2 \le c, \ \ \forall j. \tag{15}$$

The problem has been well investigated in previous work [32]. Limited by spaces, we do not present the details here. For clarity, we sketch out the main steps of our approach in Algorithm 1.

3.5 Complexity Analysis

Here we present the theoretical complexity of Algorithm 1 by big O notation. For the initialization part, the KMeans operation costs $O(n)$ and the construction of MMD matrix costs $O(n^2)$. For the iteratively updating part, solving the coding matrix $\mathbf{S}$ is some matrix multiplications, it generally costs $O(m^3)$, optimizing the codebook $\mathbf{B}$ costs $O(m + k^2)$. In sum, the overall time costs of Algorithm 1 is $O(n + n^2 + T(m^3 + m + k^2))$, where T is the number of iterations.

Algorithm 1. *HDA via locality-constrained transfer coding*

Input: Sample sets $\mathbf{X}_s$ and $\mathbf{X}_t$, parameters α, β, σ and c
Output: Label information of $\mathbf{X}_t$

Initialize
1. Initialize $\mathbf{d}_i := \mathbf{0}$, $\mathbf{s}_i := \mathbf{0}$, σ=1 and c=1
2. Use KMeans clustering [32] to initialize $\mathbf{B}$
3. Compute MMD matrix $\mathbf{M}$ by Eq. (9)
Repeat
  4. Optimize the coding matrix $\mathbf{S}$
  5. Optimize the codebook $\mathbf{B}$
  6. Update the locality adaptor $\mathbf{d}$
until *Convergence or max iteration*
7. Classify $\mathbf{S}_t$ with $\mathbf{S}_s$ used as reference

---

4   Experiments

4.1   Data Preparation

Office+Caltech-256 dataset [4] consists of 4 sub-dataset from **Amazon** (A), **Webcam** (W), **DSLR** (D), and **Caltech-256** (C). Samples in Amazon are downloaded from amazon.com. Webcam consists of low-resolution images captured by a web camera. On the contrary, images in DSLR are high-resolution ones captured by a digital SLR camera. In our experiments, we follow the same settings with previous work [2]. Ten common classes shared by four dataset are selected. There are 8 to 151 samples per category per domain, and 2,533 images in total. Furthermore, 800-dimensional SURF features and 4,096-dimensional DeCAF$_6$ features [24] are extracted as our low-level input.

   Multilingual Reuters Collection [33] is a cross-lingual text dataset. It consists of 11,000 articles from 6 categories in 5 languages, i.e., English, French, German, Italian, and Spanish. Here we follow the same settings in previous work [10,11]. Specifically, all the articles are represented by BoW with TF-IDF. Then, the BoW features are processed by PCA with dimensionality of 1,131, 1,230, 1,417, 1,041, and 807 for different language categories English, French, German, Italian, and Spanish, respectively.

4.2   Experimental Protocols

To fully evaluate our model, we perform three experiments. For instance, image classification across features, image classification across features and datasets and heterogeneous text categorization. Office+Caltech-256 are used in the first two experiments and Multilingual Reuters Collection are used in the last.

   For fair comparison, we follow the same settings with previous work. Specifically, for **image classification** tasks, the source domain consists of 20 samples per category for training, and 3 labeled target samples per category are randomly selected as reference for classification. For **text categorization** tasks, we have Spanish as the target domain and the others as the source domain.

Table 2. Accuracy (%) of HDA across features.

Method	A→A	C→C	W→W	Avg.
SVM$_t$	44.2 ± 1.1	30.1 ± 0.9	58.3 ± 1.2	44.2 ± 1.1
DAMA	39.5 ± 0.7	19.5 ± 0.8	47.5 ± 1.6	35.5 ± 1.1
MMDT	40.7 ± 1.0	31.5 ± 1.1	60.3 ± 0.8	44.2 ± 1.0
SHFA	43.4 ± 0.9	29.8 ± 1.3	62.4 ± 0.9	45.2 ± 1.0
SHFR	44.5 ± 1.1	33.4 ± 1.0	54.3 ± 0.9	44.1 ± 1.0
LCTC	**45.5 ± 1.4**	**34.5 ± 1.2**	**64.3 ± 1.1**	**48.1 ± 1.2**

Table 3. Accuracy (%) of HDA across features and datasets. The source domain and the target domain are represented by DeCAF$_6$ features and SURF features respectively.

Method	SVM$_t$	MMDT	SHFA	SHFR	LCTC
A→C	30.1 ± 0.9	28.5 ± 1.4	29.6 ± 1.5	27.4 ± 1.7	**34.7 ± 1.5**
A→W	58.3 ± 1.2	58.1 ± 1.3	58.8 ± 1.3	52.5 ± 1.4	**60.4 ± 1.4**
C→A	44.2 ± 1.1	44.7 ± 1.0	45.9 ± 1.3	43.6 ± 1.3	**50.1 ± 1.2**
C→W	58.3 ± 1.2	57.8 ± 1.1	59.1 ± 1.2	52.7 ± 1.5	**61.9 ± 2.0**
W→A	44.2 ± 1.1	45.3 ± 1.5	46.6 ± 1.3	43.1 ± 1.6	**52.3 ± 1.4**
W→C	30.1 ± 0.9	29.9 ± 1.3	29.7 ± 1.6	27.1 ± 1.3	**32.9 ± 1.6**
Avg.	44.2 ± 1.1	44.1 ± 1.3	45.0 ± 1.4	41.1 ± 1.5	**48.7 ± 1.5**

We randomly select 100 articles per category for the source domain and 500 articles for the target domain. Then, we select 10 labeled target samples as reference. For simplicity and without loss of generality, we set the hyper-parameters $\alpha = 1$ and $\beta = 1$ in our experiments.

We compare our method with several state-of-the-art HDA approaches, e.g., domain adaptation using manifold alignment (DAMA) [9], maximum margin domain transform (MMDT) [12], semi-supervised heterogeneous feature augmentation (SHFA) [11], sparse heterogeneous feature representation (SHFR) [14] and cross-domain landmark selection (CDLS) [10]. SVM trained on the labeled reference samples (SVM$_t$) is used as baseline. SVM is also used as the final classifier for the tested approaches. We report the accuracy rate [2,6] on the target domain, i.e., the ratio between the number of correctly predicted samples and the number of total samples in the target domain. Since the evaluated instances are randomly selected, each of the reported results of our algorithm is the average of 10 runs.

4.3   Experimental Results and Discussions

The image classification results of **HDA across features** on Office+Caltech-256 are shown in Table 2. The two domains are sampled from different feature representations but from the same dataset. DSLR is not tested for the limited

Table 4. Accuracy (%) of HDA on text categorization.

Method	English	French	German	Italian
SVM$_t$	67.1 ± 0.8			
DAMA	67.8 ± 0.7	68.3 ± 0.8	67.7 ± 1.0	66.5 ± 1.1
MMDT	68.9 ± 0.6	69.1 ± 0.7	68.3 ± 0.6	67.5 ± 0.5
SHFA	68.2 ± 0.9	68.7 ± 0.4	68.9 ± 0.5	68.5 ± 0.7
SHFR	67.7 ± 0.4	68.5 ± 0.7	68.1 ± 0.8	67.2 ± 1.0
CDLS	71.1 ± 0.7	71.2 ± 0.9	70.9 ± 0.7	**71.5 ± 0.6**
LCTC	**73.7 ± 0.5**	**74.0 ± 0.6**	**72.5 ± 0.4**	71.3 ± 0.5

number of samples. It can be seen from Table 2 that HDA approaches generally perform better than baseline SVM. However, DAMA not always can outperform the baseline. A possible explanation is that DAMA only considers the manifold matching and topology structure preservation between two domains. The further knowledge transfer after domain alignment is ignored in DAMA. Our approach considers not only the topology structure (by locality constraint) and the distribution alignment (by minimizing MMD), but also the knowledge transfer by a shared codebook. As a result, our approach performs better than state-of-the-art methods.

Table 3 shows the results of **HDA across features and datasets** on Office+Caltech-256. The source domain and the target domain are drawn not only from different feature representations but also different datasets. We can see that our model stays ahead of the evaluations. It further verifies the effectiveness of our model. It is worth noting that DAMA and MMDT are approaches that emphasize on distribution matching, whilst SHFA and SHFR mainly consider learning new feature representations and new classifiers. Both strategies are important and effective in some ways. However, the two obstacles of HDA are inevitable when the domain difference is substantially large. Thus, jointly optimizing both of them, as our approach does, can further improve the performance.

From Tables 2 and 3, it is clear that our approach performs well on image classification tasks. Now, we further test it on text categorization tasks. Specifically, we evaluate it on Multilingual Reuters Collection dataset. Following the previous work [10,11], we use 'Spanish' as the target domain, 'English', 'French', 'German' and 'Italian' as the source domain respectively. The experimental results are reported in Table 4.

It can be seen from Table 4 that our algorithm also performs well on **text categorization** tasks. Limited by space, we only report the results of 10 labeled samples as reference. It is worth noting that although HDA methods outperform the baseline, the performance superiority between HDA methods and SVM would get smaller with the increasing number of labeled target samples. It means transfer learning is especially suitable for tasks where the target domain has just a few or even no labeled data. Besides, CDLS proposes to minimize the distribution gaps and re-weight samples for better adaptation. However, it does not preserve the local structure of data samples. That is the reason that our method outperforms CDLS.

5 Conclusion

This paper proposes a novel approach for HDA, which takes both alleviating the feature discrepancy and mitigating the distribution divergence into consideration. By sharing a dictionary, the source domain and the target domain are coded in shared new feature representations. The probability distributions of the two domains are further aligned on the new feature space. A locality constraint is deployed to preserve the local structure and to reduce the computational costs. Extensive experiments on image classification and text categorization tasks verify the superiority of our approach.

Acknowledgements. This work was supported in part by the National Natural Science Foundation of China under Grant 61371183, ARC under Grant FT130101530 and DP170103954, the Applied Basic Research Program of Sichuan Province under Grant 2015JY0124, and the China Scholarship Council.

References

1. Pan, S.J., Yang, Q.: A survey on transfer learning. IEEE TKDE **22**(10), 1345–1359 (2010)
2. Gong, B., Shi, Y., Sha, F., Grauman, K.: Geodesic flow kernel for unsupervised domain adaptation. In: CVPR, pp. 2066–2073. IEEE (2012)
3. Si, S., Tao, D., Geng, B.: Bregman divergence-based regularization for transfer subspace learning. IEEE TKDE **22**(7), 929–942 (2010)
4. Saenko, K., Kulis, B., Fritz, M., Darrell, T.: Adapting visual category models to new domains. In: Daniilidis, K., Maragos, P., Paragios, N. (eds.) ECCV 2010. LNCS, vol. 6314, pp. 213–226. Springer, Heidelberg (2010). doi:10.1007/978-3-642-15561-1_16
5. Ding, Z., Shao, M., Fu, Y.: Deep low-rank coding for transfer learning. In: AAAI, pp. 3453–3459. AAAI Press (2015)
6. Li, J., Zhao, J., Lu, K.: Joint feature selection and structure preservation for domain adaptation. In: IJCAI, pp. 1697–1703 (2016)
7. Fernando, B., Habrard, A., Sebban, M., Tuytelaars, T.: Unsupervised visual domain adaptation using subspace alignment. In: ICCV, pp. 2960–2967 (2013)
8. Pan, S.J., Tsang, I.W., Kwok, J.T., Yang, Q.: Domain adaptation via transfer component analysis. IEEE TNN **22**(2), 199–210 (2011)
9. Wang, C., Mahadevan, S.: Heterogeneous domain adaptation using manifold alignment. In: IJCAI, vol. 22, no. 1, p. 1541 (2011)
10. Hubert Tsai, Y.-H., Yeh, Y.-R., Frank Wang, Y.-C.: Learning cross-domain landmarks for heterogeneous domain adaptation. In: CVPR, pp. 5081–5090 (2016)
11. Li, W., Duan, L., Xu, D., Tsang, I.W.: Learning with augmented features for supervised and semi-supervised heterogeneous domain adaptation. IEEE TPAMI **36**(6), 1134–1148 (2014)

12. Hoffman, J., Rodner, E., Donahue, J., Kulis, B., Saenko, K.: Asymmetric and category invariant feature transformations for domain adaptation. IJCV 109(1–2), 28–41 (2014)
13. Raina, R., Battle, A., Lee, H., Packer, B., Ng, A.Y.: Self-taught learning: transfer learning from unlabeled data. In: ICML, pp. 759–766 (2007)
14. Zhou, J.T., Tsang, I.W., Pan, S.J., Tan, M.: Heterogeneous domain adaptation for multiple classes. In: AISTATS, pp. 1095–1103 (2014)
15. Gretton, A., Borgwardt, K.M., Rasch, M., Rasch, Schölkopf, B., Smola, A.J.: A kernel method for the two-sample-problem. In: NIPS, pp. 513–520 (2006)
16. Long, M., Wang, J., Ding, G., Sun, J., Yu, P.S.: Transfer joint matching for unsupervised domain adaptation. In: CVPR, pp. 1410–1417. IEEE (2014)
17. Cortes, C., Vapnik, V.: Support-vector networks. Mach. Learn. 20(3), 273–297 (1995)
18. Long, M., Wang, J., Ding, G., Shen, D., Yang, Q.: Transfer learning with graph co-regularization. IEEE TKDE 26(7), 1805–1818 (2014)
19. Bruzzone, L., Marconcini, M.: Domain adaptation problems: a DASVM classification technique and a circular validation strategy. IEEE TPAMI 32(5), 770–787 (2010)
20. Gopalan, R., Li, R., Chellappa, R.: Domain adaptation for object recognition: an unsupervised approach. In: ICCV, pp. 999–1006. IEEE (2011)
21. Zhu, Y., Chen, Y., Lu, Z., Pan, S.J., Xue, G.-R., Yu, Y., Yang, Q.: Heterogeneous transfer learning for image classification. In AAAI (2011)
22. Shi, X., Liu, Q., Fan, W., Philip, S.Y., Zhu, R.: Transfer learning on heterogenous feature spaces via spectral transformation. In: ICDM, pp. 1049–1054 (2010)
23. van der Maaten, L., Hinton, G.: Visualizing data using t-SNE. JMLR 9, 2579–2605 (2008)
24. Donahue, J., Jia, Y., Vinyals, O., Hoffman, J., Zhang, N., Tzeng, E., Darrell, T.: Decaf: a deep convolutional activation feature for generic visual recognition. arXiv preprint arXiv:1310.1531 (2013)
25. Cai, D., He, X., Han, J.: Spectral regression: a unified approach for sparse subspace learning. In: ICDM, pp. 73–82. IEEE (2007)
26. Yan, S., Xu, D., Zhang, B., Zhang, H.-J., Yang, Q., Lin, S.: Graph embedding and extensions: a general framework for dimensionality reduction. IEEE TPAMI 29(1), 40–51 (2007)
27. Lee, H., Battle, A., Raina, R., Ng, A.Y.: Efficient sparse coding algorithms. In: NIPS, vol. 19, p. 801 (2007)
28. Censor, Y., Zenios, S.A.: Parallel Optimization: Theory, Algorithms, and Applications. Oxford University Press on Demand, New York (1997)
29. Long, M., Ding, G., Wang, J., Sun, J., Guo, Y., Yu, P.S.: Transfer sparse coding for robust image representation. In: CVPR, pp. 407–414 (2013)
30. Zheng, M., Bu, J., Chen, C., Wang, C., Zhang, L., Qiu, G., Cai, D.: Graph regularized sparse coding for image representation. IEEE TIP 20(5), 1327–1336 (2011)
31. Tsai, Y.-H.H., Hou, C.-A., Chen, W.-Y., Yeh, Y.-R., Wang, Y.-C.F.: Domain-constraint transfer coding for imbalanced unsupervised domain adaptation. In: AAAI, pp. 3597–3603. AAAI Press (2016)
32. Wang, J., Yang, J., Yu, K., Lv, F., Huang, T., Gong, Y.: Locality-constrained linear coding for image classification. In: CVPR, pp. 3360–3367. IEEE (2010)
33. Amini, M., Usunier, N., Goutte, C.: Learning from multiple partially observed views-an application to multilingual text categorization. In: NIPS, pp. 28–36 (2009)

Training Deep Ranking Model
with Weak Relevance Labels

Cheng Luo, Yukun Zheng, Jiaxin Mao, Yiqun Liu[✉], Min Zhang,
and Shaoping Ma

Tsinghua National Laboratory for Information Science and Technology,
Department of Computer Science and Technology,
Tsinghua University, Beijing 100084, China
{chengluo,yiqunliu}@tsinghua.edu.cn
http://www.thuir.cn

Abstract. Deep neural networks have already achieved great success
in a number of fields, for example, computer vision, natural language
processing, speech recognition, and etc. However, such advances have
not been observed in information retrieval (IR) tasks yet, such as ad-hoc
retrieval. A potential explanation is that in a particular IR task, training
a document ranker usually needs large amounts of relevance labels which
describe the relationship between queries and documents. However, this
kind of relevance judgments are usually very expensive to obtain. In this
paper, we propose to train deep ranking models with weak relevance
labels generated by click model based on real users' click behavior. We
investigate the effectiveness of different weak relevance labels trained
based on several major click models, such as DBN, RCM, PSCM, TCM,
and UBM. The experimental results indicate that the ranking models
trained with weak relevance labels are able to utilize large scale of behavior data and they can get similar performance compared to the ranking
model trained based on relevance labels from external assessors, which
are supposed to be more accurate. This preliminary finding encourages
us to develop deep ranking models with weak supervised data.

Keywords: Ranking model · Click model · Deep learning

1 Introduction

Deep neural networks have already delivered great improvements in many
machine learning tasks, such as speech recognition, computer vision, natural
language processing, and etc. This line of research is often referred to as *deep
learning*, as these neural networks usually comprise multiple interconnected layers. A number of "deep models" have been proposed to address the challenges
in IR tasks, in particular ad-hoc search. For example, Huang et al. proposed
DSSM [1], which is a feed forward neural network to predict the click probability given a query string and a document title.

© Springer International Publishing AG 2017
Z. Huang et al. (Eds.): ADC 2017, LNCS 10538, pp. 205–216, 2017.
DOI: 10.1007/978-3-319-68155-9_16

Learning such kind of deep models requires large amount of labeled data. In IR tasks, relevance judgments, i.e. query-document pairs, often provide supervision for the training process of ranking models. However, large scale of labeled data can be very expensive and time consuming to obtain. To circumvent the lack of labeled samples, researchers proposed to use unsupervised learning methods or weak relevance labels to train ranking models.

Unsupervised neural models aim to describe the implicit internal structure of the textual contents. Several methods for distributed text representations (for example, word2vec [2], GloVe [3], Paragraph2vec [4], and etc.) have been shown to be effective in various tasks such as text classification, recommendation, as well as Web search. The pre-trained distribution of text can be fed into document ranking algorithms to capture the relationship between query string and target documents.

Another line of research attempts to utilize weak labels for model training. The weak labels can be generated based on heuristic methods or users' behavior. Yin et al. proposed to use a 10-slot windows where the first document is treated as a positive sample while the remaining ones are treated as negative ones [6]. Dehghani et al. developed a neural network using the output of an unsupervised ranking model, BM25, as the weak supervision signal [5]. During a user's search session, the click-through data can be collected by the search engine, which is often treated as pseudo feedback from users. Huang et al. used the clicked result as a positive document and randomly selected unclicked results as negative documents. Compared to relevance judgments assessed by people, it is able to obtain much more weak labels by utilizing large scale of behavior data, i.e. much more queries and documents. This proves to be vital to the success of a lot of deep models [5,7].

In this paper, we try to train deep ranking models based on the relevance labels estimated by click model. Click model is widely used nowadays in commercial search engine to model user clicks on a search engine result page (SERP). Different click models are actually based on different user behavior assumptions. One of the key functions of click model is to predict the click probability of a result given the users' behavior on the corresponding query [8]. This probability is shown to be strongly related to the relevance score of the result. We first train several click models with query logs of a commercial search engine to generate weak relevance labels. Then we learn ranking models based on both weak relevance labels and actual relevance judgments assessed people. We adopt several major evaluation metrics to compare the ranking performance.

In summary, the main contributions of our study are as following:

- We investigate the effectiveness of weak relevance labels estimated by several major click models in training deep ranking models.
- We compare the performance of the ranking model trained based on weak relevance labels to that trained on relevance judgments made by assessors.

The remaining of this paper is organized as follows: we review related work in Sect. 2 and describe the weak relevance label generation procedure in Sect. 3. The training of deep ranking models is illustrated in Sect. 4 and the performances of different models are presented in Sect. 4.3. Finally, we conclude our research in Sect. 5.

2 Related Work

2.1 Ranking with Deep Models

Deep neural networks have achieved dramatic improvements in multiple fields of computer sciences. IR community has begun to apply neural methods to advance state of the art retrieval technology. The central IR task can be typically formalized as a matching problem [9]. Guo et al. suggested that most of recent neural models in IR application can be generally partitioned into two categories [9] according to the model architecture.

The first category is the representation-focused model, which tries to construct a representation for the text in both queries and candidate documents with deep neural networks. Then the similarity between a query and a candidate document can be measured between the two representations with a similarity function. This line of research includes DSSM [1], C-DSSM [10] and ARC-I [11]. These approaches are also referred to as "late combination methods" since the representations of queries and documents are learnt separately. Guo et al. argued that the shortcoming of representation-focused model is that the *semantic* matching is not necessarily appropriate for *relevance* matching in IR tasks.

The second category is the interaction-focused model including Deep-Match [12], ARC-II [11], DRMM [9], and MatchPyramid [13]. In these models, the interactions between queries and candidate documents are fed into neural networks. Thus, the neural networks get the opportunity to capture various matching patterns between pieces of text. In a typical ad-hoc search scenario, the query is usually very brief while the candidate documents can be much longer. The information of matched terms and matched positions is very valuable to learn a good ranker. Therefore, recently more effort has been spent on the interactions-focused models [14].

In this study, we conduct a preliminary study based on a deep ranking model, Duet, which is proposed by Mitra et al. [15]. In this approach, the local and distributional representations (early combination model and late combination model) are learnt simultaneously to take advantage of both relevance matching and semantic matching. More details about our experiment will be presented in Sect. 4.

2.2 Click Model

Modern search engines exploit user's interaction logs to improve search quality. However, although the click-through-rate (CTR) of a result can be regarded as an implicit relevance feedback from real users, it is systematically affected by some biases. For example, Joachims et al. [17] showed that the CTR could be affected by the *position bias* and top results could attract more user clicks than results in lower positions. Wang et al. [19] found that the presentation style of search results could influence their CTRs.

To distill accurate relevance labels from the noisy and biased query logs, a series of *click models* were proposed in previous studies. Most click models are probabilistic models that follow the *examination hypothesis* [24]: a search result will be clicked ($C_i = 1$) only if it is examined ($E_i = 1$) and it is relevant to the query ($R_i = 1$):

$$C_i = 1 \rightarrow E_i = 1 \land R_i = 1 \tag{1}$$

Under this hypothesis, the click probability is given by:

$$P(C_i = 1) = P(E_i = 1)P(R_i = 1) \tag{2}$$

Most click models assume that $P(R_i)$ only depends on the query and result (URL): $P(R_i = 1) = r_{qu}$ and incorporate the behavior biases in the estimation of $P(E_i)$. By inferring r_{qu} from the query log, we can estimate the relevance score between query q and result u.

Different click models make different assumptions of how users browse and interact with SERPs, and therefore, have different estimation of $P(E_i)$. For example, the cascade model proposed by Craswell et al. [24] assumes the user will examine the results sequentially until he or she finds and clicks a relevant result:

$$P(E_1 = 1) = 1 \tag{3}$$

$$P(E_{i+1} = 1|E_i = 1, C_i) = 1 - C_i \tag{4}$$

While the cascade model assumes that the user will always be satisfied with a single click, the Dynamic Bayesian Network model (DBN) model proposed by Chapelle and Zhang [21] uses a separate variable (S_i) to model whether the user will be satisfied after a click.

$$P(S_i = 1|C_i = 1) = s_{qu} \tag{5}$$

$$P(E_{i+1} = 1|E_i = 1, S_i = 0) = \lambda \tag{6}$$

$$P(E_{i+1} = 1|E_i = 1, S_i = 1) = 0 \tag{7}$$

The assumption that the user scans the results on the SERP one-by-one might be too strong. Therefore, Dupret and Piwowarski [22] proposed the User Browsing Model (UBM), which allows the user to skip some of the results. The examination probability of UBM depends on the position of last click (r_i) and the its distance to current result (d_i):

$$P(E_i = 1|C_{1...i-1}) = \gamma_{r_i,d_i} \tag{8}$$

Recently, Wang et al. [19] further found that the user does not always browse the SERP in the top-to-bottom order and there is revisiting behavior in user's interaction with SERPs. So they incorporated these non-sequential behaviors into the Partially Sequential Click Model (PSCM), in which the examination probability is determined by the position of current result (i), the position of previous clicks (m), and the position of next click (n):

$$P(E_i = 1 | C_{1...N}) = \gamma_{i,m,n} \tag{9}$$

The CTR of the result can also be influenced by the current search context. Therefore, Zhang et al. [23] built a Task-Centric Click Model (TCM), which incorporates the *query bias* (i.e. whether the query actually matches user's information need) and *duplicate bias* (i.e. whether the result is examined before), to model the click probability at a session level.

In this study, we will use a series of click models, including DBN, UBM, PSCM, and TCM, along with the RCM in which the examination probability $P(E_i = 1)$ is always set to 1, to generate weak relevance labels deep ranking models training. The detailed process of weak relevance label generation is described in Sect. 3.

3   Weak Relevance Label Generation

Click-through behavior during Web search provides implicit feedback of users' click preferences [16]. Joachims et al. looked into the reliability of implicit feedback and found that the click-through information is "informative yet biased" [17]. User clicks are biased toward many aspects: *position bias*: users tend to prefer the documents higher in the ranking list [17]; *novelty bias*: previously unseen documents are more likely to be clicked [18]; *attention bias* states that the impact of visually salient documents [19].

The central of a click model is to predict the clicked probability (Click-through rate) of a search result. Although the click possibility is not defined as the document relevance, it is closely related to document relevance. It is intuitive that the more relevant a document is, the more likely that a user will click it. Therefore, we can infer the document relevance based on the click probability predicted by click models.

In this study, we adopt several popular click models including DBN, RCM, PSCM, TCM, and UBM. We use an open-source implementation of these models [20].

We trained these click models with a real-world dataset collected by a commercial search engine in China. We removed all the queries that appeared less than 10 times (i.e. less than 10 sessions) since it seem unlikely to train a reliable click model with insufficient behavior data. For each query, at most 500 search sessions are selected for click model training to keep a balance between model precision and the amount of calculation. The statistics of our behavior dataset is shown in Table 1.

The distributions of click probability for each click model is illustrated in Fig. 1. The x-axis is the click probability ranged from 0 to 1 while the y-axis is the number of corresponding documents in logarithmic scale. We can see that the distributions of DBN, UBM and RCM are quite similar to each other, i.e. the documents with relatively low click probability are much more that with high

Table 1. The statistics of user behavior dataset

Num of queries	64,169
Num of documents	747,792
Date	From Apr. 1st 2015 to Apr. 18 2015
Language	Chinese

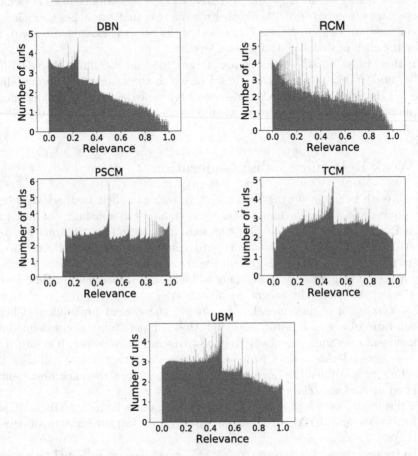

Fig. 1. Click probability distributions of different click models

click probability. The distributions of PSCM and TCM are close to uniform distribution.

Though we have the predicted click probability for each URL, it is not obvious how to map the click probability to document relevance. We adopt two different strategies to organize document pairs. More detailed will be discussed in the Sect. 4.

In this section, we discuss how to generate weak relevance labels with click models and users' click-through data. In previous study by Huang et al. [1], they proposed to use a clicked document as the positive (relevant) sample and randomly select an unclicked document as the negative sample. The click-through action can also be treated as a kind of "weak supervision". Although their approach may generate more document pairs for training as each search session can be utilized to generate training pairs, we argue that their methods are more likely to be affected by the noise and bias from individual user's actions. For example, if we have a document pair $\langle D^+, D^- \rangle$ sampled from a search session, it is possible that D^- is as relevant as D^+, even more relevant than D^-. The reason that the user did not click D^- might be that the D^- is ranked at lower positions and the user got satisfied by D^+, i.e. "position bias" in Web search. Our weak relevance label generation method is able to utilize the behavior of a group of users to reduce the impact of position bias. Our method may need more behavior data to train click model compared to directly sampling from search sessions. We argue that it is possible to construct a dataset with millions of queries with public available query logs (Sogou or Yandex) [20] in lab environment.

4 Deep Ranking Models with Weak Relevance Label

In this section, we describe how we train the deep ranking model with weak relevance labels estimated by click model. The performances of ranking models based on the output of several click models are compared. We also investigate whether the ranking model based on weak labels can get similar performance compared to that based on strong labels which are assessed by human.

4.1 Ranking Model

In this study, we choose to train our ranking model based on one of the most recent approaches, Duet, which was proposed by Mitra et al. [15]. According to Guo et al. taxonomy [9], most neural ranking models can be classified into two categories: *representation-focused* methods which try to get a good representation for query and document and *interaction-focused* methods which put emphasis on capturing the textual matching pattern between query and document.

Duet model actually combines these two lines of research. It composed of two separate neural networks, a local one and a distributed one. The two networks are jointly learnt as part of a single network.

The local model estimates the document relevance based on the exact matches of query terms in the document. It uses a local term representation, i.e. the one-hoc vectors which are widely used in traditional retrieval models. The local model focuses on capturing the exact matches on term level and terms are considered to be distinct entities. As suggested by Guo et al., the exact matching between query and document is valuable to measure the document relevance [9]. The distributed model first learns low-dimensional vector representations for both query and document. Then it estimates the positional similarity between query and document. Instead of the higher-dimensional one-hot representations, distributed model projects n-graph vectors of query and document into an lower-dimensional embedding space. This would be helpful to solve the vocabulary mismatch problem. The Duet model linearly combines the local model and the distributed model, which are jointly trained on labeled query-document pairs:

$$f^{duet}(\mathbf{Q}, \mathbf{D}) = f^l(\mathbf{Q}, \mathbf{D}) + f^d(\mathbf{Q}, \mathbf{D}) \tag{10}$$

where $\mathbf{Q}$ is the query and $\mathbf{D}$ is the document pair. f^l and f^d denotes the local model and the distributed model respectively.

In our experiment, we use the implementation of Duet model which was released by the authors[1]. The original Duet model was trained on an English corpus. In our experiment, we did some necessary data pre-processing to make the model appropriate for Chinese environment: First, all queries and documents are segmented into words. The original Duet model used 2000 most frequent n-grams for n-graph. We put 5000 most frequent Chinese n-grams into the vocabulary. We adopt the other parameters in the original Duet mode, including the dropout rate and the learning rate. The model were trained based on a single GPU.

4.2   Dataset

The dataset for model training includes the following parts:

1. Weak Relevance Label: as mentioned in Sect. 3. We have estimated click probability for query-document pairs.
2. Strong Relevance Label: we have 200 queries which are released for NTCIR WWW task [25]. For each query, there are some documents whose relevances are judged by professional assessors in a five level scale (from irrelevant to high relevant). The number of max/avg/min judged documents is 424/170/120 respectively.

We want to investigate the performance of ranking models trained with strong relevance label. Therefore, we randomly split the Strong Relevance Label dataset into two parts: Training Set contains 150 queries while the Test Set contains 50 queries.

In the remaining of this paper, we evaluate all the ranking models based on the Test Set. We use AP, ERR, nDCG@10, P@10, Q-measure and RBP, which is calculated with an open-source tool NTCIREVAL[2]. We also introduce a widely used baseline method BM25.

---

[1] https://github.com/bmitra-msft/NDRM/blob/master/notebooks/Duet.ipynb.
[2] http://research.nii.ac.jp/ntcir/tools/ntcireval-en.html.

4.3 Comparison Between Models Based on Different Weak Labels

We first look into the performance of the rankers based on different click models. Recall that the Duet model was trained based on document pairs, e.g. $\langle D^+, D^- \rangle$. We design two methods to organize training samples.

The first method is called Absolute method (ABS): we can map the click probability to relevance score by using a map function $rel\,(p)$. In our approach, we simply split the probability into 4 segments and each segments represents a relevance level respectively, e.g. the relevance score is 1 if the click probability is between 0.0 to 0.25. Then we adopt the method in Mitra et al.'s study [15] to organize document pairs. For a document pair $\langle D^+, D^- \rangle$, the relevance scores of two documents can be 3 vs. 1/0, or 2 vs. 0.

The second method is called Relative methods (REL): Assume we have two documents, d_a and d_b, their relevance scores are s_a and s_b respectively. If $s_a - s_b$ is greater than a predefined threshold t. Then $\langle d_a, d_b \rangle$ can be viewed as a valid training sample. In our experiments, we use $t = 0.42$ to make sure that the number of training pairs are comparable to that with Absolute method.

Table 2. Comparison between ranking models based on different weak relevance labels (The ranker with best performance in FIX is marked in bold while that in REL is marked with underline.)

	Model	AP	ERR	nDCG@10	P@10	Q	RBP	#Pair
ABS	DBN	**0.6283**	0.5180	0.5374	0.6540	**0.6385**	**0.3925**	11,251
	RCM	0.5569	0.4924	0.4922	0.6080	0.5761	0.3446	18,554
	PSCM	0.6251	0.5151	0.5364	0.6640	0.6344	0.3829	59,296
	TCM	0.6264	0.5124	0.5412	0.6680	0.6359	0.3889	42,268
	UBM	0.6240	**0.5302**	**0.5542**	**0.6840**	0.6333	0.3855	34,537
	DBN	0.6271	0.5197	0.5387	0.6600	<u>0.6379</u>	<u>0.3880</u>	8,339
	RCM	0.5662	0.5083	0.5107	0.6200	0.5863	0.3528	20,719
REL	PSCM	<u>0.6276</u>	<u>0.5251</u>	<u>0.5469</u>	0.6720	0.6366	0.3866	70,682
	TCM	0.6265	0.5012	0.5454	<u>0.6760</u>	0.6348	0.3872	43,088
	UBM	0.6221	0.5097	0.5427	0.6660	0.6333	0.3871	32,919
Baseline	BM25	0.5591	0.4657	0.4405	0.5560	0.5772	0.3386	747,792

The performance ranking models based on different weak relevance labels is presented in Table 2. We can see that the methods based on the training samples which are generated by ABS method is slightly better than that generated by REL method. It is potentially due to ABS method is able to produce samples with higher quality. For example, the REL sample may generate a document pair whose click probability is $\langle 0.42, 0.0 \rangle$. This sample will not be accepted by ABS method. We find that the click models which are most helpful are different for ABS and REL. For ABS, DBN and UBM are more effective while for REL,

PSCM, TCM and DBN are more beneficial for model training. The more complex models (DNB, PSCM, TCM and UBM) are more likely to generate training samples of high quality than naive model like RCM, since the more complex models can better estimate the click probability of documents.

4.4   Comparison Between Strong/Weak Relevance Labels

We further investigate the performances of rankers based on strong and weak relevance labels.

For strong relevance labels, we adopt a similar approach like ABS method to differentiate positive documents and negative ones. The smaller the threshold (t) is, the more training samples we will get. The evaluation results are shown in Table 3.

Table 3. Comparison between models which are based on strong/weak relevance labels (The ranker with best performance trained on weak labels is marked in bold while that trained on strong labels is marked with underline.)

	Model	AP	ERR	nDCG@10	P@10	Q	RBP	#Pair
Strong label	Duet(t = 1)	0.6416	0.5418	0.5578	0.6800	0.6466	0.3897	469,790
	Duet(t = 2)	0.6293	0.5214	0.5403	0.6560	0.6395	0.3821	95,985
	Duet(t = 3)	0.6203	0.5118	0.5084	0.6400	0.6263	0.3781	2,557
Weak label+ABS	DBN	0.6283	0.5180	0.5374	0.6540	0.6385	0.3925	11,251
	RCM	0.5569	0.4924	0.4922	0.6080	0.5761	0.3446	18,554
	PSCM	0.6251	0.5151	0.5364	0.6640	0.6344	0.3829	59,296
	TCM	0.6264	0.5124	0.5412	0.6680	0.6359	0.3889	42,268
	UBM	0.6240	0.5302	0.5542	0.6840	0.6333	0.3855	34,537

We can see that the models which are based on strong labels are slightly better than that are based on weak labels. This conclusion is consistent across different evaluation metrics. The reason may be due to that in our experiments, rankers with strong labels have the opportunity to utilize much more training samples. If we look into the rankers with different threshold in strong label group, we find that the Duet(t = 1) performs much better that the remaining two models. The number of training document pair in Duet(t = 1) is also much larger than that in the other two models. This observation suggest that it is necessary to feed large amounts of training samples, even they contains more noise, to train a good ranking sample. In this pilot study, the scale of data we used is relatively small due to the limit of calculation resource. We would like to leave exploration with much more data in our future work.

All the neural ranking models (except that for RCM) performs significantly better than BM25 (p<0.01). This encourages us to continue applying neural network in IR tasks.

5 Conclusions and Future Work

In this study, we present a novel neural ranking model training method based on weak relevance labels. We propose to generate weak relevance labels for documents by training click models with users' click behavior. Experiments based on a real-world user behavior dataset demonstrate that the ranking models trained with weak labels can get similar performance compared to that with relevance judgments. We also find that the more data (even more noisy) fed into the neural model, the better performance the model can achieve.

Our work has a few limitations. First, deep learning for IR is developing rapidly and a number of neural methods have been proposed recently. We should validate the effectiveness training methods with various neural models. Second, compared to previous attempts [5,9,10] based on millions of queries, the dataset in our experiment is too small. We would like to explore if we will get better performance with a larger dataset in our future work.

Acknowledgement. This work is supported by Natural Science Foundation of China (Grant No. 61622208, 61732008, 61532011) and National Key Basic Research Program of China (2015CB358700).

References

1. Huang, P.-S., et al.: Learning deep structured semantic models for web search using clickthrough data. In: Proceedings of the 22nd ACM International Conference on Conference on Information and Knowledge Management. ACM (2013)
2. Mikolov, T., et al.: Distributed representations of words and phrases and their compositionality. In: Advances in Neural Information Processing Systems (2013)
3. Pennington, J., Socher, R., Manning, C.D.: Glove: global vectors for word representation. In: EMNLP, vol. 14 (2014)
4. Le, Q.V., Mikolov, T.: Distributed representations of sentences and documents. In: ICML, vol. 14 (2014)
5. Dehghani, M., Zamani, H., Severyn, A., Kamps, J., Croft, W.B.: Neural ranking models with weak supervision. arXiv preprint arXiv:1704.08803 (2017)
6. Yin, D., Hu, Y., Tang, J., Daly, T., Zhou, M., Ouyang, H., Chen, J., et al.: Ranking relevance in Yahoo search. In: Proceedings of the 22nd ACM SIGKDD International Conference on Knowledge Discovery and Data Mining, pp. 323–332. ACM (2016)
7. Salakhutdinov, R., Hinton, G.: Semantic hashing. Int. J. Approx. Reason. **50**(7), 969–978 (2009)
8. Chuklin, A., Markov, I., de Rijke, M.: Click models for web search. Synth. Lect. Inf. Concepts Retr. Serv. **7**(3), 1–115 (2015)
9. Guo, J., Fan, Y., Ai, Q., Croft, W.B.: A deep relevance matching model for ad-hoc retrieval. In: Proceedings of the 25th ACM International on Conference on Information and Knowledge Management, pp. 55–64. ACM (2016)
10. Shen, Y., He, X., Gao, J., Deng, L., Mesnil, G.: Learning semantic representations using convolutional neural networks for web search. In: Proceedings of the 23rd International Conference on World Wide Web, pp. 373–374. ACM (2014)

11. Hu, B., Lu, Z., Li, H., Chen, Q.: Convolutional neural network architectures for matching natural language sentences. In: Advances in Neural Information Processing Systems, pp. 2042–2050 (2014)
12. Lu, Z., Li, H.: A deep architecture for matching short texts. In: Advances in Neural Information Processing Systems, pp. 1367–1375 (2013)
13. Pang, L., Lan, Y., Guo, J., Xu, J., Wan, S., Cheng, X.: Text matching as image recognition. arXiv preprint arXiv:1602.06359 (2016)
14. Hui, K., Yates, A., Berberich, K., de Melo, G.: A position-aware deep model for relevance matching in information retrieval. arXiv preprint arXiv:1704.03940 (2017)
15. Mitra, B., Diaz, F., Craswell, N.: Learning to match using local and distributed representations of text for web search. arXiv preprint arXiv:1610.08136 (2016)
16. Agichtein, E., Brill, E., Dumais, S., Ragno, R.: Learning user interaction models for predicting web search result preferences. In: Proceedings of the 29th Annual International ACM SIGIR Conference on Research and Development in Information Retrieval, pp. 3–10. ACM (2006)
17. Joachims, T., Granka, L., Pan, B., Hembrooke, H., Gay, G.: Accurately interpreting clickthrough data as implicit feedback. In: Proceedings of the 28th Annual International ACM SIGIR Conference on Research and Development in Information Retrieval, pp. 154–161. ACM (2005)
18. Zhang, Y., Chen, W., Wang, D., Yang, Q.: User-click modeling for understanding and predicting search-behavior. In Proceedings of the 17th ACM SIGKDD International Conference on Knowledge Discovery and Data Mining, pp. 1388–1396. ACM (2011)
19. Wang, C., Liu, Y., Zhang, M., Ma, S., Zheng, M., Qian, J., Zhang, K.: Incorporating vertical results into search click models. In: Proceedings of the 36th International ACM SIGIR Conference on Research and Development in Information Retrieval, pp. 503–512. ACM, July 2013
20. Wang, C., Liu, Y., Wang, M., Zhou, K., Nie, J., Ma, S.: Incorporating non-sequential behavior into click models. In: Proceedings of the 38th International ACM SIGIR Conference on Research and Development in Information Retrieval, pp. 283–292. ACM (2015)
21. Chapelle, O., Zhang, Y.: A dynamic Bayesian network click model for web search ranking. In: Proceedings of the 18th International Conference on World Wide Web, pp. 1–10. ACM (2009)
22. Dupret, G.E., Piwowarski, B.: A user browsing model to predict search engine click data from past observations. In: Proceedings of the 31st Annual International ACM SIGIR Conference on Research and Development in Information Retrieval, pp. 331–338. ACM (2008)
23. Xu, W., Manavoglu, E., Cantu-Paz, E.: Temporal click model for sponsored search. In: Proceedings of the 33rd International ACM SIGIR Conference on Research and Development in Information Retrieval, pp. 106–113. ACM (2010)
24. Craswell, N., Zoeter, O., Taylor, M., Ramsey, B.: An experimental comparison of click position-bias models. In: WSDM 2008, pp. 87–94. ACM (2008)
25. Luo, C., Zheng, Y., Liu, Y., Wang, X., Xu, J., Zhang, M., Ma, S.: SogouT-16: a new web corpus to embrace IR research. In: The 40th ACM SIGIR International Conference on Research and Development in Information Retrieval, SIGIR 2017 (2017)

A Deep Approach for Multi-modal User Attribute Modeling

Xiu Huang, Zihao Yang, Yang Yang$^{(\boxtimes)}$, Fumin Shen, Ning Xie,
and Heng Tao Shen

School of Computer Science and Technology and Center for Future Media,
University of Electronic Science and Technology of China, Chengdu, China
hxiu321@163.com, zivon396@163.com, dlyyang@gmail.com, fumin.shen@gmail.com,
seanxiening@gmail.com, shenhengtao@hotmail.com

Abstract. With the explosive growth of user-generated contents (e.g.,
texts, images and videos) on social networks, it is of great significance
to analyze and extract people's interests from the massive social media
data, thus providing more accurate personalized recommendations and
services. In this paper, we propose a novel multimodal deep learning
algorithm for user profiling, dubbed multi-modal User Attribute Model
(mmUAM), which explores the intrinsic semantic correlations across different modalities. Our proposed model is based on Poisson Gamma Belief
Network (PGBN), which is a deep learning topic model for count data in
documents. By improving PGBN, we succeed in addressing the problem
of learning a shared representation between texts and images in order
to obtain textual and visual attributes for users. To evaluate the effectiveness of our proposed method, we collect a real dataset from Sina
Weibo. Experimental results demonstrate that the proposed algorithm
achieves encouraging performance compared with several state-of-the-art
methods.

Keywords: User profiling · Deep learning · Multi-model · Social media

1 Introduction

With the rapid development of social networks, massive information (e.g., texts,
images and videos) generated by users is emerging on various social media platforms. The activities people participating in and the contents people producing
play a significant role of analyzing people's interests and preferences, which are of
great importance to provide personalized recommendation and on-line retrieval
for them. In particular, microblogging is now one of the most popular social
media services, where people are keen on posting daily activities, sharing opinions and focusing on hot and interesting topics. For example, Sina Weibo[1], a
commonly used social media platform in China, has attracted a great amount
of users to participate in. Released by Sina Weibo Data Center, the number of

---

[1] http://www.weibo.com.

Z. Huang et al. (Eds.): ADC 2017, LNCS 10538, pp. 217–230, 2017.
DOI: 10.1007/978-3-319-68155-9_17

monthly active users approaches to 222 million up to October 2015. Besides, the social media applications involve multi-modal data, where the visual information is vital to strengthen the description of short texts.

In order to explore user attributes, prior works construct topic modeling from users' previous behaviors and preferences. For example, Latent Dirichlet Allocation (LDA) [7] is a widely used generative probabilistic model for text corpora. By modifying LDA, there are other traditional topic models to tackle the problem of short texts from social media data, such as author topic model [19] and twitter-user model [24]. Besides, previous works also focus on dynamic topic models to analyze the change of topics in data streams, such as Dynamic User Attribute Model (DUAM) [12] which models the dynamics using time windows, and dynamic User Clustering Topic model (UCT) [27] to capture the dynamics of users' interests by integrating the interests at previous time periods with newly collected data in text streams. In addition, there are topic models proposed to explore the correlations among different modalities. For instance, mm-LDA [2] and corr-LDA [6] are presented to learn the correspondence between textual and visual information. Cross-Media-LDA (CMLDA) [5] is also proposed to discover the intrinsic correlations among multiple media types for social event summarization. Some similar methods proposed by Bian et al. are demonstrated in [3,4].

Recently, there is a great interest in deep learning, which succeeds in many applications. The Deep Belief Network (DBN) [11] and the Deep Boltzmann Machine (DBM) [20] are deep networks both designed to model binary observations, whose hidden units are also typically restricted to be binary. However, different from conventional deep networks, the Poisson Gamma Belief Network (PGBN) [29] is proposed to construct a deep networks architecture with nonnegative real hidden units to automatically tune both the width of each layer and the depth of the network. Despite PGBN learns the representation of count observations, it is a unimodal network and not applicative to short texts of social data. To deal with multi-modal social data, we propose a novel multi-modal User Attribute Model (mmUAM). Different from traditional methods of constructing user interest model that only take account of one layer of topic modelling, our model is designed to capture the correlations among multiple modalities. To facilitate this study, we collect a real dataset from Sina Weibo, on which extensive experiments show the superiority over state-of-the-art methods.

The main contributions of our work are summarized as follows.

1. We propose a novel multi-modal deep learning approach, named multi-modal User Attribute Model (mmUAM), through which we manage to automatically infer user attributes.
2. The proposed mmUAM captures the semantic correlations between texts and images, which enables us to learn effective textual and visual representation for more comprehensive user profiling.
3. We construct a Sina Weibo microblog dataset with multi-modality information. The promising results on this dataset demonstrate the efficacy of our proposed approach.

2   Related Work

Text-based User Profiling. With the tremendous growth of social networks, how to provide more accurate services for users is tough challenging. Previous works have been studied to explore users interests through extracting users' characteristics and preferences from user-generated texts on social media platforms. Generative topic models, such as LDA [7], provide an explicit representation of a document. However, such topic models fail to tackle the sparsity problem of short texts. Many variations of LDA have been proposed. For example, He et al. [10] propose a modified topic model, named Bi-labeled LDA, which utilizes users' relationship information to learn interest tags. Rosen-Zvi et al. [19] extend LDA to propose the author-topic model, which models the content of documents, including the author information. While, Xu et al. [24] introduce a modified author-topic model, twitter-user mode, which outperforms LDA and author-topic model. Besides, some other studies also make attempts to exploit external knowledge to enrich the s of short texts. Abel et al. [1] analyze Twitter activities in semantic way by integrating Twitter posts with related news articles. Instead of introducing external knowledge, Cheng et al., [8] model the generation of word co-occurrence patterns for topic modeling in order to address the sparsity problem of short texts. However, the above topic models are mostly applied to text corpus.

Image-based User Profiling. Deep Convolutional Neural Networks (CNNs) [13] have recently achieved a great success in large scale image feature learning. Consequently, many researches focus on building user profiling by extracting visual information. For instance, Geng et al. [9] propose a deep learning strategy to learn visual features for user profiling on Pinterest[2] in fashion domain. A Socially Embedded Visual Representation learning (SEVIR) approach [15] has been proposed to capture the semantics and user intentions based on learning image representation, which tackles the sparsity and unreliability problems. Li et al. [14] construct a Gaussian relational topic model by utilizing user-shared images to infer users' interests. Moreover, a pinboard recommendation system for Twitter users is presented in [25], which combines two different social media platforms in order to recommend users for more relevant and interesting topics. Also, the way of exploiting user-tagged Web images for video indexing can be learned in [26]. Despite the visual information exploiting user interests is definitely significant, more works should take account of multiple modalities.

Multi-modal User Profiling. As more and more social media data is integrated with texts, images and videos, most of the works have shifted their focus to dealing with multi-modal data. In [6], the correspondence Latent Dirichlet Allocation (corr-LDA) is a three hierarchical probabilistic mixture model to describe the correlations between images and annotations. Similarly, multi-modal Latent Dirichlet Allocation (mm-LDA) [2] is proposed to learn the joint distribution of images and their associated texts, which is used for social relation mining.

---

[2] http://www.pinterest.com.

In addition, Bian et al. [5] present a novel probabilistic modal, named Cross-Media-LDA (CMLDA), which aims to explore intrinsic correlations between texts and images for multimedia microblog summarization.

Besides, some deep networks are constructed to learn features among multiple modalities. Ngiam et al. [17] propose a cross-modality deep learning methods based on Restricted Boltzmann Machines (RBMs). Subsequently, Srivastava et al. [23] propose a multi-modal Deep Boltzmann Machine (DBM) model for images and texts. They construct multi-modal DBM by building an image-specific two-layer DBM that uses Gaussian RBM and a text-specific two-layer DBM that utilizes Replicated Softmax model. Similarly, Pang et al. [18] use multi-modal DBN to learn joint representation of the visual, auditory and textual features for user-generated web videos. In addition, the Deep Belief Network (DBN) presented in [22] to create a joint representation for texts and images, is different from DBM in that DBN is a directed model.

Nevertheless, the hidden units of DBM and DBN are typically restricted to be binary. These multi-modal deep learning approaches are not successfully applied to a real dataset for social service. We work on learning features of multiple modalities input data from large-scale real dataset and construct multi-modal deep networks to tackle the sparsity of short texts to explore more relevant interests that meet users' demand. To achieve better inference of our proposed deep topic modal, we employ upward-downward Gibbs sampling.

3   The Proposed Model

3.1   Overview

We employ the conventional bag-of-word method to deal with the texts and images to automatically infer user attributes. To construct our model, we utilize Sina Weibo data with user-generated short texts and corresponding images. Firstly, we extract both texts and images raw features as bag of words and bag of visual words, respectively. Formally, each document is under two different topic distributions. Note that Θ, which is a shared latent variable between visual and textual modalities, is concatenated by textual hidden unit θ_{w-j} and visual hidden unit θ_{v-j}. Topics Φ_w are specific to textual modality and topics Φ_v are unique to visual modality. Then, we build our proposed mmUAM in deep networks with five layers. The performance of multi-modal fusion in five different layers is presented in Sect. 4.3. As we use probabilistic models, upward-downward Gibbs sampling [29] is adopted to infer various parameters.

As we all know, microblogs always consist of short texts and relevant images, in which each text is restricted to 140 characters. Thus, each document is a piece of microblog and is composed of textual content, visual content, or the mixing of textual and visual information. In particular, the observation of an image is represented as a multivariate vector of visual words, which is denoted as v_j in jth document. Similarly, the observation of a text is defined as a vector w_j in jth document. The correlations between the K_0 features of $(v_1, v_2, ..., v_J)$ can be represented by the columns of Φ_v. In the same way, the correlations between

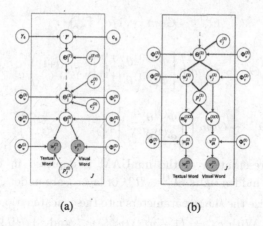

(a)   (b)

Fig. 1. The graphical illustration of the proposed mmUAM. (a) the mmUAM hierarchical model; (b) a presentation of layer t = 1 in the mmUAM.

the K_0 features of $(w_1, w_2, ..., w_J)$ are captured by the columns of Φ_w. Note that $\Theta_j \in R_+^{K_t}$ is the K_t hidden units of sample the jth document. We use the Poisson likelihood to connect the observed textual content $w_j^{(1)} \in Z^{K_0}$ (visual content $v_j^{(1)} \in Z^{K_0}$) to the product $\Phi_w^{(1)}\theta_{w-j}^{(1)}$ ($\Phi_v^{(1)}\theta_{v-j}^{(1)}$) at layer one as follows

$$w_j^{(1)} \sim Pois\left(\Phi_w^{(1)}\theta_{w-j}^{(1)}\right), \quad v_j^{(1)} \sim Pois\left(\Phi_v^{(1)}\theta_{v-j}^{(1)}\right).$$

3.2 Multi-modal User Attribute Model

Our proposed model is based on PGBN [29], a deep networks architecture that is designed only for text analysis. Then, Zhou et al. [30] propose augmentable gamma belief networks to learn multilayer deep representations for high-dimensional sparse count vectors and nonnegative real vectors. Nevertheless, the augmentable gamma belief networks are not adapted to social data. As a result, we propose a multi-modal user attribute model for multiple modalities data on social media. For microblogging document, we make an assumption that the generated topics are composed of two domains, including textual topics generated from microblog texts, and visual topics generated from posted images. In order to capture correlations of these two modalities, we learn a shared representation between textual and visual information. We use Θ_j to represent the shared gamma distribution between textual and visual information in the jth microblogging document. With T hidden layers, we give the example of our proposed mmUAM fusing in the first hidden layer as follows

$$\boldsymbol{\Theta}_j^{(T)} \sim Gam\left(\boldsymbol{r}, 1/c_j^{(T+1)}\right),$$

$$\cdots$$

$$\boldsymbol{\Theta}_j^{(t)} \sim Gam\left(\begin{bmatrix} \boldsymbol{\Phi}_w^{(t+1)}\boldsymbol{\theta}_{w-j}^{(t+1)} \\ \boldsymbol{\Phi}_v^{(t+1)}\boldsymbol{\theta}_{v-j}^{(t+1)} \end{bmatrix}, 1/c_j^{(t+1)}\right), \tag{1}$$

$$\cdots$$

$$\boldsymbol{\Theta}_j^{(1)} \sim Gam\left(\begin{bmatrix} \boldsymbol{\Phi}_w^{(2)}\boldsymbol{\theta}_{w-j}^{(2)} \\ \boldsymbol{\Phi}_v^{(2)}\boldsymbol{\theta}_{v-j}^{(2)} \end{bmatrix}, p_j^{(2)}/(1-p_j^{(2)})\right).$$

The graphic representation of the mmUAM is depicted in Fig. 1. For $t = 1, 2, ..., T-1$, the hidden units $\boldsymbol{\Theta}_j^{(t)} \in R_+^{K_t}$ of layer t are under gamma distribution, which factorize the shape parameters into the concatenation of $\boldsymbol{\Phi}_w^{(t+1)}\boldsymbol{\theta}_{w-j}^{(t+1)}$ and $\boldsymbol{\Phi}_v^{(t+1)}\boldsymbol{\theta}_{v-j}^{(t+1)}$. With $c_j^{(2)} = (1-p_j^{(2)})/p_j^{(2)}$, $p_j^{(2)}$ and $\{1/c^{(t)}\}_{3,T+1}$ are probability parameters and gamma scale parameters respectively. For the top layer, the gamma shape parameters of hidden units are vector $\boldsymbol{r} = (r_1, ..., r_K^{(T)})'$. The columns of $\phi_w^{(t+1)}$ and $\phi_v^{(t+1)}$ decide the correlations between the K_t latent features of $(\boldsymbol{\Theta}_1^{(t)}, ..., \boldsymbol{\Theta}_J^{(t)})$.

In order to simplify parameter inference, we impose the constraints on $\boldsymbol{\Phi}_w^{(t)}$ and $\boldsymbol{\Phi}_v^{(t)}$ that every column of $\boldsymbol{\Phi}_w^{(t)}$ and $\boldsymbol{\Phi}_v^{(t)}$ has a unit L_1 norm. Thus, for $t \in \{1, ..., T-1\}$, the hierarchical model is completed as follows

$$\phi_{w-k}^{(t)} \sim Dir(\eta^t, ..., \eta^t), \quad \phi_{v-k}^{(t)} \sim Dir(\xi^t, ..., \xi^t),$$

$$c_0 \sim Gam(e_0, 1/f_0), \quad \gamma_0 \sim Gam(a_0, 1/b_0), \quad r_k \sim Gam(\gamma_0/K_T, 1/c_0).$$

For $t \in \{3, ..., T+1\}$, we have

$$p_j^{(2)} \sim Beta(a_0, b_0), \quad c_j^{(t)} \sim Gam(e_0, 1/f_0). \tag{2}$$

we divide T hidden layers into T related subproblems, thus every subproblem has the similar way of solution.

Lemma 1 *(augment-and-conquer the mmUAM).* *With $p_j^{(1)} = 1 - e^{-1}$ and*

$$p_j^{(t+1)} = -ln(1-p_j^{(t)})/\left[c_j^{(t+1)} - ln(1-p_j^{(t)})\right]. \tag{3}$$

For $t \in \{1, ..., T\}$, we can define that the observed (if $t = 1$) or some latent (if $t \geq 2$) textual contents $w_j^t \in Z^{K_t-1}$ are under the Poisson distribution with the product $\boldsymbol{\Phi}_w^t\boldsymbol{\theta}_{w-j}^t$, and the observed (if $t = 1$) or some latent (if $t \geq 2$) visual word counts $v_j^t \in Z^{K_t-1}$ are under the Poisson distribution with the product $\boldsymbol{\Phi}_w^t\boldsymbol{\theta}_{v-j}^t$.

$$w_j^{(t)} \sim Pois\left[-\boldsymbol{\Phi}_w^{(t)}\boldsymbol{\theta}_{w-j}^{(t)}ln\left(1-p_j^{(t)}\right)\right], \tag{4}$$

$$v_j^{(t)} \sim Pois\left[-\boldsymbol{\Phi}_v^{(t)}\boldsymbol{\theta}_{v-j}^{(t)}ln\left(1-p_j^{(t)}\right)\right]. \tag{5}$$

Proof. The definition (4), (5) are absolutely true for layer one. Assume that (4), (5) are true for layer $t \geq 2$, then each textual count $w_{ij}^{(t)}$ and visual count $v_{ij}^{(t)}$ are separately augmented into the summation of K_t latent textual and visual counts. Thus, the summation of K_t two different latent counts is smaller than or equal to $w_{ij}^{(t)}$ and $v_{ij}^{(t)}$.

$$w_{ij}^{(t)} = \sum_{k=1}^{K_t} w_{ijk}^{(t)}, \quad w_{ijk}^{(t)} \sim Pois\left[-\phi_{w-ik}^{(t)}\theta_{w-kj}^{(t)}ln(1-p_j^{(t)})\right],$$

$$v_{ij}^{(t)} = \sum_{k=1}^{K_t} v_{ijk}^{(t)}, \quad v_{ijk}^{(t)} \sim Pois\left[-\phi_{v-ik}^{(t)}\theta_{v-kj}^{(t)}ln(1-p_j^{(t)})\right].$$

where $i \in \{1, ..., K_{t-1}\}$. Then, we have

$$m_{kj}^{(t)(t+1)} = w_{\cdot jk}^{(t)} = \sum_{i=1}^{K_{t-1}} w_{ijk}^{(t)}, m_j^{(t)(t+1)} = \left(w_{\cdot j1}^{(t)}, ..., w_{\cdot jK_t}^{(t)}\right)',$$

$$n_{kj}^{(t)(t+1)} = v_{\cdot jk}^{(t)} = \sum_{i=1}^{K_{t-1}} v_{ijk}^{(t)}, n_j^{(t)(t+1)} = \left(v_{\cdot j1}^{(t)}, ..., v_{\cdot jK_t}^{(t)}\right)'.$$

$m_{kj}^{(t)(t+1)}$ denotes the counts in layer t that factor $k \in \{1, ...K_t\}$ appears in document j, and $v_{kj}^{(t)(t+1)}$ represents the counts in layer t that factor $k \in \{1, ...K_t\}$ appears in document j. On account of $\sum_{i=1}^{K_{t-1}} \phi_{w-ik}^{(t)} = 1$ and $\sum_{i=1}^{K_{t-1}} \phi_{v-ik}^{(t)} = 1$, we utilize the method in [31] to marginalize out Φ_w^t and Φ_v^t. As a result,

$$m_j^{(t)(t+1)} \sim Pois\left[-\theta_{w-j}^{(t)}ln(1-p_j^{(t)})\right], \quad n_j^{(t)(t+1)} \sim Pois\left[-\theta_{v-j}^{(t)}ln(1-p_j^{(t)})\right].$$

Then, by employing the above Poisson likelihood, we further marginalize out $\theta_{w-j}^{(t)}$ and $\theta_{v-j}^{(t)}$ that follows the gamma distribution.

$$m_j^{(t)(t+1)} \sim NB\left[\Phi_w^{(t+1)}\theta_{w-j}^{(t+1)}, p_j^{(t+1)})\right], \tag{6}$$

$$n_j^{(t)(t+1)} \sim NB\left[\Phi_v^{(t+1)}\theta_{v-j}^{(t+1)}, p_j^{(t+1)})\right]. \tag{7}$$

As demonstrated in [28], (6) and (9) can also be generated from their compound Poisson distribution as

$$m_{kj}^{(t)(t+1)} = \sum_{x=1}^{w_{kj}^{(t+1)}} u_x, u_x \sim Log(p_j^{(t+1)}), w_{kj}^{(t+1)} \sim Pois\left[\phi_{w-k:}^{(t+1)}\theta_{w-j}^{(t+1)}ln(1-p_j^{(t+1)})\right],$$

$$n_{kj}^{(t)(t+1)} = \sum_{y=1}^{v_{kj}^{(t+1)}} u_y, u_y \sim Log(p_j^{(t+1)}), v_{kj}^{(t+1)} \sim Pois\left[\phi_{v-k:}^{(t+1)}\theta_{v-j}^{(t+1)}ln(1-p_j^{(t+1)})\right].$$

Hence, if (4), (5) are true for layer t, they are also true for layer $t+1$.

Inspired by the lemmas and theorems in [28,31], we propagate the latent textual counts $w_{ij}^{(t)}$ and visual counts $v_{ij}^{(t)}$ of layer t upward to layer $t+1$ as

$$\left\{ \left(w_{ij1}^{(t)}, ..., w_{ijK_t}^{(t)} \right) | w_{ij}^{(t)}, \phi_{w-i:}^{(t)}, \theta_{w-j}^{(t)} \right\}$$
$$\sim Mult \left(w_{ij}^{(t)}, \frac{\phi_{w-i1}^{(t)} \theta_{w-1j}^{(t)}}{\sum_{k+1}^{K_t} \phi_{w-ik}^{(t)} \theta_{w-kj}^{(t)}}, ..., \frac{\phi_{w-iK_t}^{(t)} \theta_{w-K_t j}^{(t)}}{\sum_{k+1}^{K_t} \phi_{w-ik}^{(t)} \theta_{w-kj}^{(t)}} \right), \tag{8}$$

$$\left(w_{kj}^{(t+1)} | m_{kj}^{(t)(t+1)}, \phi_{w-k:}^{(t+1)}, \theta_{w-j}^{(t+1)} \right) \sim CRT \left(m_{kj}^{(t)(t+1)}, \phi_{w-k:}^{(t+1)} \theta_{w-j}^{(t+1)} \right), \tag{9}$$

$$\left\{ \left(v_{ij1}^{(t)}, ..., v_{ijK_t}^{(t)} \right) | v_{ij}^{(t)}, \phi_{v-i:}^{(t)}, \theta_{v-j}^{(t)} \right\}$$
$$\sim Mult \left(v_{ij}^{(t)}, \frac{\phi_{v-i1}^{(t)} \theta_{v-1j}^{(t)}}{\sum_{k+1}^{K_t} \phi_{v-ik}^{(t)} \theta_{v-kj}^{(t)}}, ..., \frac{\phi_{v-iK_t}^{(t)} \theta_{v-K_t j}^{(t)}}{\sum_{k+1}^{K_t} \phi_{v-ik}^{(t)} \theta_{v-kj}^{(t)}} \right), \tag{10}$$

$$\left(v_{kj}^{(t+1)} | n_{kj}^{(t)(t+1)}, \phi_{v-k:}^{(t+1)}, \theta_{v-j}^{(t+1)} \right) \sim CRT \left(n_{kj}^{(t)(t+1)}, \phi_{v-k:}^{(t+1)} \theta_{v-j}^{(t+1)} \right). \tag{11}$$

3.3   Parameter Inference

In conventional topic models, variational inference and collapsed Gibbs sampling are often used for parameter inference. To estimate the latent variables under the multivariate observations, we utilize upward-downward Gibbs sampling [29] with the width of the first layer being restricted to K_{1max}. The sampling process of mmUAM is as below.

Sample $w_{ijk}^{(t)}$ and $v_{ijk}^{(t)}$. For all the layers, we can use (10) to sample $w_{ijk}^{(t)}$ and (12) to sample $v_{ijk}^{(t)}$. But for the first hidden layer, the observed counts $w_{ij}^{(1)}$ is considered as word tokens at the ith term in the jth document, where the size of textual vocabulary is denoted as $I = K_0$, and the observed counts $v_{ij}^{(1)}$ is treated as visual word tokens at the ith term (the size of visual vocabulary $I' = K_0'$) in the jth document. We define z_{w-js} and z_{v-js} as the topic index for i_{js} and $i_{js'}$ ($s \in \left\{ 1, ..., w_{\cdot j}^{(1)} \right\}$, $s' \in \left\{ 1, ..., v_{\cdot j}^{(1)} \right\}$).

$$P \left(z_{w-js} = k | - \right) \propto \frac{\eta^{(1)} + (w_{i_{js} \cdot k}^{(1)})_{-js}}{I \eta^{(1)} + (w_{\cdot \cdot k}^{(1)})_{-js}} \left((w_{\cdot jk}^{(1)})_{-js} + \phi_{w-k:}^{(2)} \theta_{w-j}^{(2)} \right), \tag{12}$$

$$P \left(z_{v-js'} = k | - \right) \propto \frac{\xi^{(1)} + (v_{i_{js'} \cdot k}^{(1)})_{-js'}}{I' \xi^{(1)} + (v_{\cdot \cdot k}^{(1)})_{-js'}} \left((v_{\cdot jk}^{(1)})_{-js'} + \phi_{v-k:}^{(2)} \theta_{v-j}^{(2)} \right). \tag{13}$$

where $k \in \{1, ..., K_{1max}\}$. We let $w_{ijk}^{(1)} = \sum_s \delta \left(i_{js} = i, z_{w-js} = k \right)$ and $v_{ijk}^{(1)} = \sum_{s'} \delta \left(i_{js'} = i, z_{v-js'} = k \right)$. $w_{ijk}^{(1)}$ and $v_{ijk}^{(1)}$ represent the number of times when term i is assigned to the topic k in document j. Besides, we use w_{-js} and v_{-js} to separately denote the count of w and v except when term i appears in document j. Especially when $T = 1$, we use Poisson Factor Analysis (PFA) with gamma-negative binomial process [28] to replace $\phi^{(2)}_{w-k}:\theta^{(2)}_{w-j}$ and $\phi^{(2)}_{v-k}:\theta^{(2)}_{v-j}$ with r_k. For simplification, if $T = 1$, we set K_{1max} factors and let $r_k \sim Gam(\gamma_0/K_{1max}, 1/c_0)$.

Sample $\phi^{(t)}_{w-k}$. We sample the textual topic $\phi^{(t)}_{w-k}$ as

$$\left(\phi^{(t)}_{w-k}|-\right) \sim Dir\left(\eta^{(t)} + w^{(t)}_{1\cdot k}, ..., \eta^{(t)} + w^{(t)}_{K_{t-1}\cdot k}\right).$$

Sample $\phi^{(t)}_{v-k}$. In the same way, we sample the visual topic $\phi^{(t)}_{v-k}$ as

$$\left(\phi^{(t)}_{v-k}|-\right) \sim Dir\left(\xi^{(t)} + v^{(t)}_{1\cdot k}, ..., \xi^{(t)} + v^{(t)}_{K_{t-1}\cdot k}\right).$$

Sample $w^{(t+1)}_{ij}$ and $v^{(t+1)}_{ij}$. We sample $\boldsymbol{w}^{(t+1)}_j$, $\boldsymbol{v}^{(t+1)}_j$ separately using (9), (11).

Sample r. c_0 and γ_0 are sampled using (3), whose detailed introduction is in [28].

$$(r_v|-) \sim Gam\left(\gamma_0/K_T + w^{(T+1)}_{i\cdot} + v^{(T+1)}_{i\cdot}, c_0 - \sum_j ln\left(1 - p^{(T+1)}_j\right)^{-1}\right).$$

Sample $\Theta^{(t)}_j$. Using the latent counts propagated upward and the gamma-Poisson conjugacy, we downward sample the hidden units Θ_j as

$$\left(\Theta^{(t)}_j|-\right) \sim Gam\left(\begin{bmatrix}\Phi^{(t+1)}_w\theta^{(t+1)}_{w-j}\\\Phi^{(t+1)}_v\theta^{(t+1)}_{v-j}\end{bmatrix} + \begin{bmatrix}m^{(t)(t+1)}_j\\n^{(t)(t+1)}_j\end{bmatrix}, c^{(t+1)}_j - ln\left(1 - p^{(t)}_j\right)^{-1}\right).$$

Sample $p^{(2)}_j$ and $c^{(t)}_j$. We calculate $p^{(t)}_j$ ($t \geq 3$) and $c^{(2)}_j$ using (6), and sample $p^{(2)}_j$ and $c^{(t)}_j$, where $t \geq 3$ as

$$\left(p^{(2)}_j|-\right) \sim Beta\left(a_0 + \begin{bmatrix}m^{(1)(2)}_{\cdot j}\\n^{(1)(2)}_{\cdot j}\end{bmatrix}, b_0 + \Theta^{(2)}_{\cdot j}\right),$$

$$\left(c^{(t)}_j|-\right) \sim Gam\left(e_0 + \Theta^{(t)}_{\cdot j}, \left[f_0 + \Theta^{(t-1)}_{\cdot j}\right]^{-1}\right).$$

4 Experiments

4.1 Dataset Construction

As we know, Sina Weibo is one of the most popular social media platforms in China, where we collect a dataset and conduct our experiments on the real

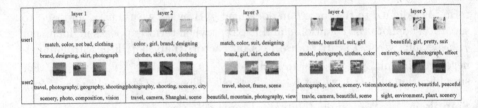

Fig. 2. The generated visual words and textual keywords for the mmUAM-1 of five different layers.

data. We crawl 1349 users, the data including users' basic information and their posted microblogs from January 2015 to December 2016, in which each microblog contains both texts and images. After filtering inactive users and separating each user's microblogs into two documents according to posting time, we have 193798 documents. In order to comprehensively evaluate the generated user attributes, we utilize both the crawled tags and the posted microblogs to manually label the users' interests.

For preprocessing the textual dataset, we firstly utilize jieba participle[3] to segment the Chinese words, and then we eliminate the non-Chinese characters, stop words, and the low-frequency words that appear less than five times. For visual feature description, Scale-Invariant Feature Transform (SIFT) [16] is used to extract discriminant local features of images, thus generating 128-dimensional SIFT descriptors. To construct a codebook of visual words, we utilize k-means with each descriptor being a cluster center quantized into a visual word. As a result, each image is represented with the count of visual words in the codebook.

4.2 Evaluation Metrics and Compared Methods

The standard classification algorithm evaluation methods like precision, recall and F1-measure would not be sufficient to understand the performance of multi-label problems. Thus, we adopt the following evaluation measures proposed in [21], where we set $X = \{x_1, x_2, ..., x_k\}$ as a set of output attributes and $Y = \{y_1, y_2, ..., y_k\}$ as a set of ground-truth attributes.

Average Precision. We use average precision to calculate the mean value of ranking H of ground-truth attributes in predicted attributes. $|N|$ is the number of documents.

$$ap(H) = \frac{1}{N}\sum_{i=1}^{N}\frac{1}{|Y_i|}\sum_{y\in Y_i}\frac{\left|\left\{y^{'}\in Y_i | rank_f\left(x_i, y^{'}\right) \leq rank_f\left(x_i, y\right)\right\}\right|}{rank_f\left(x, y\right)}. \quad (14)$$

One Error. The measure evaluates how many times the top-ranked predicted attributes were not in the set of possible attributes Y. We express one-error of a hypothesis f as $one - err(f)$.

---

[3] https://github.com/fxsjy/jieba.

$$oe\left(f\right) = \frac{1}{N}\sum_{i=1}^{N}\left\{[argmax_{y \in Y}f\left(x_i, y\right)] \notin Y_i\right\}. \tag{15}$$

Ranking Loss. This evaluation is to minimize the average fraction of crucial pairs which are misordered.

$$rl = \frac{1}{N}\sum_{i=1}^{N}\frac{1}{|Y_i|\,\overline{|Y_i|}}\left|\left\{\left(y, y'\right)\,|f\left(x_i, y\right) \leq f\left(x_i, y''\right), \left(y, y''\right) \in Y_i \times \overline{Y_i}\right\}\right|. \tag{16}$$

Coverage. The coverage measures in a sequence queue, to go down the list of predicted attributes in order to cover all the possible attributes assigned to a document.

$$co\left(f\right) = \frac{1}{N}\sum_{i=1}^{N}maxrank_f\left(x_i, y\right) - 1. \tag{17}$$

To demonstrate the effectiveness of our proposed mmUAM, we compare our algorithm with the following methods.

Poisson Gamma Belief Network (PGBN). The PGBN is applied on our crawled Sina Weibo dataset for short texts.

Multi-modal User Attribute Model (mmUAM). We adapt the multi-modal fusion in five layers, separately. The ways of five different fusion are denoted as mmUAM-1, mmUAM-2, mmUAM-3, mmUAM-4, mmUAM-5. Specifically, mmUAM-1 is expressed as the shared representation for texts and images learned in the first layer, and other mmUAMs are denoted in the same way.

4.3   Experimental Results

In this paper, we employ the layer-wise training method for the mmUAMs, with which we set a fixed budget on the width of layer one $K_{1max} = 400$ and the depth of the network $T = 5$. Besides, we set hyper-parameters as $e_0 = f_0 = 1$, $a_0 = b_0 = 0.01$, and $\eta^{(t)} = \xi^{(t)} = 0.05$ for all the layers.

As the mmUAM learns the shared representation of textual and visual information, we do the qualitative analysis to show the effectiveness of our proposed model. We randomly selected two users to analyze the topics generated by the mmUAM-1 of five different layers. As an example, we choose 3 generated visual words and the top 6 textual words of the specific topic showed in Fig. 2. We can see that the visual representation and textual representation strengthen the description of user attributes. Obviously, the mmUAM can effectively model the semantic correlations of social data in multiple modalities.

We compare our mmUAM with PGBN and different ways of layer-fusion to evaluate the quality of our generating user attributes. Figure 3 displays the performance in term of average precision, one error, ranking loss, coverage. On the crawled Sina Weibo dataset, our proposed mmUAM with five ways of multi-modal fusion all performs better than the PGBN. The results also confirm the fact that multi-modal topic modeling works better than unimodal

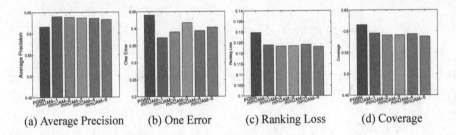

(a) Average Precision   (b) One Error   (c) Ranking Loss   (d) Coverage

Fig. 3. The performance is evaluated by average precision, one error, ranking loss, coverage, respectively.

approaches. As for mmUAMs, there is slight difference among results on the four evaluation metrics. Especially, mmUAM-1 achieves the best result of the average precision and the one error. For the evaluation of the ranking loss and the coverage, mmUAM-5 achieves the lowest results that represent the best quality of the classification. As a result, the first-layer-fusion and the last-layer-fusion capture better correlations between texts and images than other layer-fusion.

5   Conclusion

In this paper, we proposed a novel multi-modal user attribute (mmUAM) model which automatically generated user interests from multi-modal social media data, by capturing correlations between textual and visual information. In particular, we improved the PGBN network to extract topics of interests better in line with users' characteristics. We conducted experiments on our crawled microblog dataset, where the results demonstrated the superiority of our mmUAM as compared to the state-of-the-art methods.

Acknowledgments. This work was supported in part by the National Natural Science Foundation of China under Project 61572108 and Project 61502081, the National Thousand-Young-Talents Program of China, and the Fundamental Research Funds for the Central Universities under Project ZYGX2014Z007 and Project ZYGX2015J055.

References

1. Abel, F., Gao, Q., Houben, G.-J., Tao, K.: Semantic enrichment of Twitter posts for user profile construction on the social web. In: Antoniou, G., Grobelnik, M., Simperl, E., Parsia, B., Plexousakis, D., Leenheer, P., Pan, J. (eds.) ESWC 2011. LNCS, vol. 6644, pp. 375–389. Springer, Heidelberg (2011). doi:10.1007/978-3-642-21064-8_26
2. Barnard, K., Duygulu, P., Forsyth, D., de Freitas, N., Blei, D.M., Jordan, M.I.: Matching words and pictures. JMLR **3**, 1107–1135 (2003)
3. Bian, J., Yang, Y., Chua, T.S.: Multimedia summarization for trending topics in microblogs. In: ACM CIKM, pp. 1807–1812 (2013)

4. Bian, J., Yang, Y., Chua, T.S.: Predicting trending messages and diffusion participants in microblogging network. In: ACM SIGIR, pp. 537–546 (2014)
5. Bian, J., Yang, Y., Zhang, H., Chua, T.S.: Multimedia summarization for social events in microblog stream. IEEE Trans. Multimedia 17(2), 216–228 (2015)
6. Blei, D.M., Jordan, M.I.: Modeling annotated data. In: ACM SIGIR, pp. 127–134 (2003)
7. Blei, D.M., Ng, A.Y., Jordan, M.I.: Latent dirichlet allocation. JMLR 3, 993–1022 (2003)
8. Cheng, X., Yan, X., Lan, Y., Guo, J.: BTM: topic modeling over short texts. IEEE TKDE 26(12), 2928–2941 (2014)
9. Geng, X., Zhang, H., Song, Z., Yang, Y., Luan, H., Chua, T.S.: One of a kind: User profiling by social curation. In: ACM MM, pp. 567–576 (2014)
10. He, W., Liu, H., He, J., Tang, S., Du, X.: Extracting interest tags for non-famous users in social network. In: ACM CIKM, pp. 861–870 (2015)
11. Hinton, G.E., Osindero, S., Teh, Y.W.: A fast learning algorithm for deep belief nets. Neural Comput. 18(7), 1527–1554 (2006)
12. Huang, X., Yang, Y., Hu, Y., Shen, F., Shao, J.: Dynamic user attribute discovery on social media. In: Li, F., Shim, K., Zheng, K., Liu, G. (eds.) APWeb 2016. LNCS, vol. 9931, pp. 256–267. Springer, Cham (2016). doi:10.1007/978-3-319-45814-4_21
13. Krizhevsky, A., Sutskever, I., Hinton, G.E.: Imagenet classification with deep convolutional neural networks. In: NIPS, pp. 1097–1105 (2012)
14. Li, X., Cheung, M., She, J.: Connection discovery using shared images by gaussian relational topic model. arxiv:1612.03639 (2016)
15. Liu, S., Cui, P., Zhu, W., Yang, S.: Learning socially embedded visual representation from scratch. In: ACM MM, pp. 109–118 (2015)
16. Lowe, D.G.: Object recognition from local scale-invariant features. In: ICCV, vol. 2, pp. 1150–1157 (1999)
17. Ngiam, J., Khosla, A., Kim, M., Nam, J., Lee, H., Ng, A.Y.: Multimodal deep learning. In: ICML, pp. 689–696 (2011)
18. Pang, L., Ngo, C.W.: Mutlimodal learning with deep Boltzmann machine for emotion prediction in user generated videos. In: ICMR, pp. 619–622 (2015)
19. Rosen-Zvi, M., Griffiths, T., Steyvers, M., Smyth, P.: The author-topic model for authors and documents. In: UAI, pp. 487–494 (2004)
20. Salakhutdinov, R., Hinton, G.E.: Deep boltzmann machines. In: AISTATS, vol. 1, p. 3 (2009)
21. Schapire, R.E., Singer, Y.: Boostexter: a boosting-based system for text categorization. Mach. Learn. 39(2–3), 135–168 (2000)
22. Srivastava, N., Salakhutdinov, R.: Learning representations for multimodal data with deep belief nets. In: ICML Workshop (2012)
23. Srivastava, N., Salakhutdinov, R.R.: Multimodal learning with deep Boltzmann machines. In: NIPS, pp. 2222–2230 (2012)
24. Xu, Z., Ru, L., Xiang, L., Yang, Q.: Discovering user interest on twitter with a modified author-topic model. In: IEEE/WIC/ACM WI-IAT, vol. 1, pp. 422–429 (2011)
25. Yang, X., Li, Y., Luo, J.: Pinterest board recommendation for twitter users. In: ACM MM, pp. 963–966 (2015)
26. Yang, Y., Zha, Z.J., Gao, Y., Zhu, X., Chua, T.S.: Exploiting web images for semantic video indexing via robust sample-specific loss. IEEE Trans. Multimedia 16(6), 1677–1689 (2014)
27. Zhao, Y., Liang, S., Ren, Z., Ma, J., Yilmaz, E., de Rijke, M.: Explainable user clustering in short text streams. In: ACM SIGIR, pp. 155–164 (2016)

28. Zhou, M., Carin, L.: Negative binomial process count and mixture modeling. IEEE T PAMI **37**(2), 307–320 (2015)
29. Zhou, M., Cong, Y., Chen, B.: The poisson gamma belief network. In: NIPS, pp. 3043–3051 (2015)
30. Zhou, M., Cong, Y., Chen, B.: Augmentable gamma belief networks. JMLR **17**(163), 1–44 (2016)
31. Zhou, M., Hannah, L., Dunson, D.B., Carin, L.: Beta-negative binomial process and poisson factor analysis. In: AISTATS, vol. 22, pp. 1462–1471 (2012)

Potpourri

Reversible Fragile Watermarking for Fine-Grained Tamper Localization in Spatial Data

Hai Lan and Yuwei Peng$^{(\boxtimes)}$

Computer School, Wuhan University, Wuhan, Hubei, China
ywpeng@whu.edu.cn

Abstract. Spatial data has been widely used in many areas. Meanwhile, the increasingly simplified data access and manipulation methods in spatial data related applications make it important to verify data truthfulness. In this paper, we propose a reversible fragile watermarking scheme for the content authentication of spatial data. The proposed scheme embeds (detects) four kinds of watermarks, i.e., group, object, vertex-local and vertex-global watermarks, into (from) each object (vertex) by altering the digits of the coordinates after the *LSD* (Least Significant Digit). According to the detection results of those four kinds of watermarks, the scheme can not only locate tampers, but also recognize modification types on both object and vertex levels. In addition, the proposed scheme is reversible, that is the original data can be restored from the watermarked data. The feature of reversibility makes the scheme suitable for applications which require high-precision data.

Keywords: Spatial data · Content authentication · Reversible · Fragile watermarking

1 Introduction

Spatial data has been serving various applications, e.g. geographical information systems (GIS), location based services (LBS) and navigation systems, for many years. With the rapid development of Internet, the requirements for opening and sharing of spatial data are increasing significantly. Due to the nature of data sharing, spatial data in those applications is prone to be copied and tampered. Thus, data protection techniques are needed to ensure that spatial data has not been illegally copied or tampered.

Digital watermarking is the most popular data protection technique [2]. Initially, digital watermarking is intended for right protection of digital data. This type of watermarking is also known as so-called robust watermarking which is used to identify the ownership and illegal copy. Basically, digital watermarking is

---

Y. Peng—This work is supported by the National Natural Science Foundation of China under grant (No.61100019).

Z. Huang et al. (Eds.): ADC 2017, LNCS 10538, pp. 233–247, 2017.
DOI: 10.1007/978-3-319-68155-9_18

comprised of two elementary operations, embedding and detecting. Embedding operation stands for slightly modifying the digital data (host data) to embed a signal (watermark) in it. Detecting operation extracts a signal (detected watermark) from one data copy (suspicious data). Due to the special design of robust watermarking technique, the detected watermark should be very close to the original one. Then the detected watermark can be used for ownership identification of the suspicious data [5]. In past decades, digital watermarking has already been applied on right protection of multimedia data [9], relational data [15], spatial data [14], trajectory data [24], etc. Recently, digital watermarking has been also applied on content authentication [7], which is called fragile watermarking. In fragile watermarking, a damaged detected watermark means the host data has been tampered. To achieve that, the fragile watermark should be sensitive to modifications.

As a mainstream of content authentication technique, a mass of fragile watermarking schemes have been proposed for various applications, such as image [8], audio [10], video [11] and relational data [12]. Naturally, fragile watermarking has also been introduced in spatial data due to the increasing requirement of content authentication of spatial data [13]. Unlike watermark employed in robust watermarking, fragile watermark is typically generated from original host data rather than specified by the user. Embedding fragile watermark does not affect the regeneration of the fragile watermark itself. Since any data modification could result in difference between the watermarks regenerated and detected from the suspicious data respectively. Therefore, the suspicious data could be determined as modified if those two watermarks could not match each other [3,4,20]. According to whether the original host data can be restored, fragile watermarking can be divided into irreversible [3] and reversible [4,20] fragile watermarking.

Although the aforementioned fragile watermarking schemes can effectively detect modifications on the data, they can not determine the type of modifications, such as insertion or deletion. The fragile watermarking schemes for relational database can achieve such goal. Scheme proposed in [12] first divides tuples into groups according to the primary key. For each group, an attribute fragile watermark is generated by a hash function that takes all values of that attribute as the hash seed. The attribute fragile watermark is embedded into the LSBs of all values of the corresponding attribute. Finally, for each tuple, a tuple fragile watermark is generated in the similar way of attribute fragile watermark while the hash seed is all attribute values of that tuple. Similarly, the tuple fragile watermark is embedded into the second LSBs of all attribute values of the corresponding tuple. With the detection results of attribute and tuple fragile watermarks, the type of modification can be determined. In [24], a fragile watermarking scheme is proposed for spatial data, which is able to determine the type of modification. Basically, in that scheme, two kinds of watermarks are embedded into every object. The objects in a vector map are first divided into groups and assigned identifiers. A group fragile watermark is generated for each group based on all identifiers of the objects in that group, and then embedded in those objects. Also, an object fragile watermark is generated for each object based on all vertices of the object, and then embedded in that object. In the detection process, the groups are first restored. Then, the group and object fragile watermarks are detected from the objects. According to the detection result combination of those two watermarks, the modification type can be determined. However, in the scheme proposed in [24], only the tamper on object level can be located, e.g. object insertion and object deletion. The tamper on vertex level, such as vertex insertion, deletion and update, can not be located by that method.

In this paper, we propose a reversible fragile watermarking scheme for spatial data, which achieves fine-grained tamper localization. That is, the proposed watermark scheme can locate tampers on object level or vertex level and determine their modification types. Also, for the spatial data passed content authentication, its original data can be restored from it. In summary, the major contribution of this paper are: (1) developing a reversible fragile watermarking scheme that achieves fine-grained tamper localization and reversibility simultaneously; (2) the design of vertex level fragile watermark that ensures locating tampers on vertex level and determining their modification type; (3) an easy but effective choice of watermark embedding method that ensures the restoration of original spatial data while reducing the computational cost.

The remainder of this paper is organized as follows. In Sect. 2, some preliminaries are introduced. Then, the proposed reversible fragile watermarking scheme is described in Sect. 3, as well as the analysis of fragility and reversibility. Finally, experimental results and conclusions are given in Sects. 4 and 5 respectively.

2 Preliminaries

Before entering the detail of the proposed scheme, we first introduce some preliminaries of spatial data and related modification types as well as the generate-and-embed/detect watermarking framework.

2.1 Spatial Data and Modification Types

In general, a spatial dataset D is represented as a set of *geographical objects*. A geographical object O is a sequence of vertices (points): $\{P_1, P_2, ..., P_n\}(n > 0)$. If n equals 1, i.e. the object contains only one vertex, O is just a point. If $n > 1$ and $P_1 = P_n$, O is a polygon. Otherwise O is a line segment. In 2D space, a vertex P_i is a pair of coordinates (x_i, y_i), where x_i and y_i are normally double-precision floating-point numbers.

Due to the limitation of acquisition methods, each spatial data has its own precision. That is, the digits after a specific digit of x_i/y_i are meaningless (insignificant), and the specific digit is so-called *Least Significant Digit (LSD)*. In other words, the modifications on digits after *LSD* would not lower the usability of the spatial data. Hence *LSD* or digit after it has been widely used to carry watermark.

Commonly, modifications on spatial data can be categorized to two levels: object level and vertex level. Modifications on object level include (geographical) object addition, deletion and update. Vertex addition, deletion and update belong to vertex level. It is easy to understand that object addition and deletion means to add a newly created object to or remove an existing object from the data. As for object update, the situation is a little bit complicated. Generally, any change on any part of an object is considered as an update. Therefore, the vertex level modifications also incur object update. The object update can be further classified to vertex addition, deletion and update. Vertex addition inserts a vertex in an object while vertex deletion removes a vertex from an object, and vertex update means moving an existing vertex. In our proposed watermarking scheme, we focus on locating modifications both on object level and vertex level.

2.2   Generate-and-Embed/Detect Watermarking Framework

Like the work in [24], the proposed watermarking scheme for spatial data is based on the Generate-and-Embed/Detect Framework [7]. Therefore, we borrow some preliminaries of such framework and related concepts from [24].

Basically, the generate-and-embed/detect watermarking framework [7] is the foundation of almost all existing fragile watermarking schemes. In the embedding phase of the generate-and-embed/detect watermarking framework, a watermark is generated from the host data itself. The generated watermark is embedded into the host data. Then the watermarked data could be published. In the detection phase, two watermarks are extracted: *(a)* the first one is generated from the suspicious data in the same way used in the embedding phase, which is so-called the generated watermark; *(b)* the second one is detected from the suspicious data with the detection algorithm, which is so-called the detected watermark. The suspicious data is determined to be untampered if and only if the generated and detected watermarks are exactly the same. The key point of such framework is that the embedding of generated watermark should not disturb the re-generating of the watermark.

As a fragile watermarking framework, an essential property should be ensured that the embedded fragile watermark could never be forged. Almost all fragile watermarking schemes are dependent on an one-way hash function to achieve that. An one-way hash function, $H()$, gives a hash value V' for an input value V incorporated with a user-specified key K. It is computationally infeasible to obtain an input value V'' for a given hash value V' which satisfies both $H(V'') = V'$ and $V'' \neq V$. Since the embedding process is all controlled by the hash value, the attacker cannot tamper the watermarked data while keeping the same generated watermark. On the other hand, the fragility also relies on the hash function. Any modification, even if one bit change, on the input value certainly causes random change on the resulting hash value. For above properties, the application of hash function guarantees the overall security and fragility of a fragile watermarking framework scheme.

There are many potential hash functions including MD5 and SHA hash [21]. In this paper, we also employ the generate-and-embed/detect watermarking framework and hash functions.

3   The Proposed Scheme

Basically, the proposed watermarking scheme embeds various kinds of fragile watermarks into the host spatial data: *(a)* group watermark, *(b)* object watermark and *(c)* vertex watermark. The embedding process can be divided into following steps:

(1) Construct an identifier (ID) for each object based on $Area_T$ employed in [24], and then divide objects into a given number of groups according to their *ID*s.
(2) Generate one group watermark for each group from its member objects, then the group watermark is embedded into each of those objects.
(3) Generate one object watermark for each object from its all vertices, the object watermark is then embedded into each of those vertices.
(4) Generate one vertex-global watermark for each vertex of each object based on the number of vertices belong to the object, the vertex-global watermark is then embedded into the x-coordinate of each of those vertices.
(5) Generate one vertex-local watermark for each vertex of each object from its all coordinates, then embed the vertex-local watermark in its y-coordinate.

Like the embedding process, the first step of detection process is dividing objects into groups and generating object IDs generation. After the grouping operation, object and vertex watermarks are regenerated and detected from the suspicious spatial data respectively. The tampers and respective modification types can be determined according to the detection results of the watermarks.

The main components of the proposed watermarking scheme are detailed respectively in Sects. 3.1 and 3.2. Section 3.3 describes locating tampers and determining modification types according to the detection results. The fragility and reversibility of the proposed watermarking scheme is analyzed in Sects. 3.4 and 3.5.

3.1   Watermark Generating and Embedding

In the proposed scheme, four arguments should be supplied by the data owner before embedding and detection could be performed: g, K, λ and η. Here g is the number of groups obtained after dividing objects and K is a secret key. λ is an integral value indicates the position of LSD, that is λ stands for the precision of the data. η is the argument for controlling the sensitivity against vertex modifications, which is the length of vertex level watermarks.

Group Division. In order to divide objects into groups, each object is given an identifier which is constructed from its characteristics. In the proposed watermarking scheme, we adopt $Area_T$ used in [24], which is an integral value calculated from the turning function [22] of the object. Although $Area_T$ keeps same across embedding and detection theoretically, we adopt $ID_i = hb(Area_{T_i}, \gamma)$ as the identifier of O_i to avoid computational errors on different platforms, where $hb(a, b)$ is a function which returns the most significant b bits from the binary form of a. In the next step, each object O_i is assigned a group ID $G_j = H(ID_i, K) \bmod g$, where H is a hash function produces a hash code with two input arguments. In one word, the objects are divided into g groups in this step.

Group Watermark Embedding. The first watermark we are going to generate is group watermark. For each group G_j, a group watermark GW_j is generated from the identifiers of all its member objects, i.e. $GW_j = H((ID_1||...||ID_{n_j}), K)$, where n_j is the number of objects in G_j and '$||$' is a string concatenation operator. $||$ treats its both operands as strings and concatenates them to a single string which is consisted of decimal digits. The group watermark of G_j is then embedded into every object in G_j. For each vertex P_m of object O_i in G_j, the mth digit of GW_j is embedded into P_m's coordinates by replacing the $(\lambda + 1)$th digit of each coordinate with it. However, the length of GW_j might not be exactly the same with the m (the number of vertices of an object). Thus, in group watermark embedding, if the length of GW_j is larger than m, only the first m digits of GW_j are embedded. Or else GW_j is repeatedly embedded into the object.

Object Watermark Embedding. For each object O_i, an object watermark OW_i is generated as $OW_i = H((x_0||y_0||...||x_{l_i}||y_{l_i}), K)$. For each vertex P_j of object O_i, the mth digit of OW_i is embedded into P_j's mth vertex by replacing the $(\lambda + 2)$th digit of every coordinate of the vertex with it. If the length of OW_i does not equal the number of the vertices of P_j, like the group watermark, the object watermark would be partly or repeatedly embedded into the object.

Vertex Watermark Embedding. For each vertex P_s, a vertex-local watermark VWL_s is generated as $VWL_s = H((x_s||y_s), K)$. y_s is embedded with the vertex-local watermark by replacing its η digits after $(\lambda + 2)$th digit with $msd(VWL_s, \eta)$, where $msd(a, b)$ returns the first b digits of a. A vertex-global watermark VWG_s is also generated as $VWG_s = H(count(P_s), K)$, where $count(a)$ returns the number of vertices in object a. x_s is embedded with the vertex-global watermark by replacing its η digits after $(\lambda + 2)$th digit with $msd(VWG_s, \eta)$.

In the proposed watermarking scheme, watermark embedding only affects the digits after the LSD. Obviously, such watermarking scheme is not a robust watermarking. However, it is sufficient for tamper localization and modification type determination.

3.2 Watermark Detection

When a spatial data is suspected to be tampered (namely suspicious data), watermark detection process could be utilized to extract generated and detected watermarks respectively. By comparing those watermarks, the tampers could be located and the modification types could be determined.

During the detection process, groups are rebuilt on all objects of the suspicious data with the same identification-and-division method used in the watermark embedding. For each object, the generated (group, object and vertex) watermarks could be regenerated with the same method used in the embedding process, while the corresponding detected watermarks could be detected by assembling the digits after LSD. Thus, every generated watermark has respective detected watermark. The detection result for one kind of watermark (group, object or vertex) is true if its generated and detected watermarks match each other. Or else the detection result is false.

For each group G_i with m_i objects, two detection vectors can be constructed for recording the detection results of group and object watermarks respectively.

- $V_i^G = \{v_1^G, v_2^G, ..., v_{m_i}^G\}$, where $v_j^G (1 \leq j \leq m_i)$ is the boolean value representing the detection result of group watermark for object O_j. v_j^G is true if the generated and detected group watermarks of O_j are exactly the same. Otherwise it is false.
- $V_i^O = \{v_1^O, v_2^O, ..., v_{m_i}^O\}$, where $v_j^O (1 \leq j \leq m_i)$ is the boolean value representing the detection result of object watermark for object O_j. v_j^O is true if the generated and detected object watermarks of O_j are exactly the same. Otherwise it is false.

For an object O_i with n_i vertices, two detection vectors can be constructed for recording the detection results of vertex-global and vertex-local watermarks respectively.

- $V_i^{VG} = \{v_1^{VG}, v_2^{VG}, ..., v_{n_i}^{VG}\}$, where $v_j^{VG} (1 \leq j \leq n_i)$ is the boolean value representing the detection result of vertex-global watermark for vertex P_j. v_j^{VG} is true if the generated and detected vertex-global watermarks of P_j are exactly the same. Otherwise it is false.
- $V_i^{VL} = \{v_1^{VL}, v_2^{VL}, ..., v_{n_i}^{VL}\}$, where $v_j^{VL} (1 \leq j \leq n_i)$ is the boolean value representing the detection result of vertex-local watermark for vertex P_j. v_j^{VL} is true if the generated and detected vertex-local watermarks of P_j are exactly the same. Otherwise it is false.

3.3   Tamper Locating and Modification Type Determination

Based on aforementioned detection vectors, we can discover the tampers possibly occurred on the data and further determine their modification types.

Object Deletion. Suppose object O_i has been deleted from group G_j, since the generating of group watermark requires IDs of all objects belong to the same group, the generated group watermarks of objects in G_j are different from those generated from the unwatermarked data. Although the generated group watermark used in embedding phase can not be regenerated from the suspicious data (because the data has been tampered), one copy of it has already been embedded in each object of G_j, which are called detected group watermarks.

Therefore, the generated and detected group watermarks of any object in G_j are different from each other, i.e. $\forall v_k^G \in V_j^G, v_k^G = F$ (F means false and T means true). For any untampered object of G_j, the generated and detected object watermarks are the same since (generated or detected) object watermark only depends on vertices of object itself, that is $\forall v_k^O \in V_j^O, v_k^O = T$. Therefore, the group G_{del} has any deleted object is formulated as follows.

$$G_{del} = \{G_j | \forall O_i \in G_j, v_i^G = F \wedge v_i^O = T\}$$

Apparently, the object deletion could not be located if a whole group has been dropped. However, this situation rarely happens [12]. The experimental results illustrated in Sect. 4 also support such conclusion.

Object Addition. Suppose object O_i has been added into group G_j, as discussed above the detection result of group watermark for any object in G_j is false, i.e. $\forall v_k^G \in V_j^G, v_k^G = F$. The detection result of object watermark for any object other than O_i is true while the one of O_i is false, i.e. $\forall v_k^O \in V_j^O (k \neq i), v_k^O = T \wedge v_i^O = F$. Therefore, the group G_{add} with newly added object satisfies all following conditions.

(1) $\exists O_i \in G_{add}, v_i^O = F \wedge (\forall O_k \in G_{add} - \{O_i\}, v_k^O = T)$
(2) $\forall O_k \in G_{add}, v_k^G = F$

Furthermore, in any G_{add} satisfies above conditions, object O_i satisfies $v_i^O = F \wedge v_i^G = F$ could be located as an added object.

Object Update. Suppose object O_i is a modified object in group G_j, according to the definition of update in Sect. 2.1, the identifier of O_i is not changed while being updated. Therefore, the detection result of group watermark for any object in G_j is true since the set of IDs the object relies on keeps same, i.e. $\forall v_k^G \in V_j^G, v_k^G = T$. However, for object watermark, only the detection result of object watermark for O_i is false, i.e. $\forall v_k^O \in V_j^O (k \neq i), v_k^O = T \wedge v_i^O = F$. Therefore, the group G_{update} has updated object satisfies all following conditions.

(1) $\exists O_i \in G_{update}, v_i^O = F \wedge (\forall O_k \in G_{update} - \{O_i\}, v_k^O = T)$
(2) $\forall O_k \in G_{update}, v_k^G = T$

Furthermore, in any G_{update} satisfies above conditions, object O_i satisfies $v_i^O = F \wedge v_i^G = T$ could be located as an updated object.

Vertex Deletion. Once an object is determined to be updated, the tampers on vertex level can be located and recognized according to the vertex-level watermarks. Suppose vertex P_i is a deleted vertex in updated object O_j, the generated vertex-local watermark is not affected by the deletion of P_i since its generation only involves P_i's coordinates. Therefore, the detection result of vertex-local watermark for any vertex of O_j is true. The object O_{del} has vertex deleted is formulated as follows.

$$O_{del} = \{O_j | \forall \widetilde{P_i} \in O_j, v_i^{VL} = T\}$$

Vertex Addition and Update. For an object has one or more vertices whose detection result of vertex-local watermark is false, we can not determine whether the modification type is vertex addition or update. Since a new vertex has random detected vertex-local watermark and coordinates of an updated vertex are changed, vertex addition and update can both incur difference between generated and detected vertex-local watermarks. Thus, we need assistance from vertex-global watermark to distinguish vertex addition and update. Suppose vertex P_i is added into object O_j, the amount of vertices is changed, so the detection result of vertex-global watermark for any vertex (including P_i) of O_j is false since such kind of watermark relies on the amount. On the other hand, suppose vertex P_i is updated, the amount of vertices and all sequence numbers stay the same, thus the detection result of vertex-global watermark for any vertex (including P_i) of O_j is true. According to above analysis, vertex addition and update on O_j could be distinguished as follows.

- The modification on O_j is vertex addition if

$$\forall P_i \in O_j, v_i^{VG} = F \wedge \exists P_s \in O_j, v_s^{VL} = F$$

And the set of added vertices is

$$P_{add} = \{P_i | P_i \in O_j \wedge v_i^{VG} = F \wedge v_i^{VL} = F\}$$

- The modification on O_j is vertex update if

$$\forall P_i \in O_j, v_i^{VG} = T \wedge \exists P_s \in O_j, v_s^{VL} = F$$

And the set of added vertices is

$$P_{update} = \{P_i | P_i \in O_j \wedge v_i^{VG} = T \wedge v_i^{VL} = F\}$$

3.4   Analysis of Fragility

In this section, we give analysis on the probability of successfully locating tampers and recognizing the modification type. Like [24], we have some similar assumptions for the discussion: *(1)* objects uniformly distribute in the g groups, every group has m member objects and every object has l vertices; *(2)* every group has the same probability for being modified as well as every object; *(3)* after data modification, every decimal digit (i.e. 0 to 9) has the same probability to appear on every digit of the generated and detected watermarks, i.e. the corresponding generated and detected watermarks have the probability of $\frac{1}{10^l}$ to be exactly the same.

Object Deletion. As discussed in Sect. 3.2, locating of object detection only relies on the detection result of group watermark. The probability of locating one single object deletion is $1 - (\frac{1}{10^l})^{m-1}$. Suppose ω objects have been deleted from ρ $(1 \le \rho \le min(\omega, g))$ groups in the watermarked data, where group G_i has ω_i $(1 \le i \le \rho, 1 \le \omega_i \le m, \sum_{i=1}^{\rho} \omega_i = \omega)$ objects deleted. Then, the probability of locating all these object deletion is $p = \prod_{i=1}^{k}(1 - (\frac{1}{10^l})^{m-\omega_i})$.

Object Addition. Locating of object addition involves the detection results both of group and object watermarks. The probability of locating one single object addition is $(1 - \frac{1}{10^l})(1 - (\frac{1}{10^t})^m)$. Suppose ω objects have been added into ρ $(1 \leq \rho \leq min(\omega, g))$ groups in the watermarked data, where group G_i has ω_i $(1 \leq i \leq \rho, 1 \leq \omega_i \leq m, \sum_{i=1}^{\rho} \omega_i = \omega)$ objects added. Then, the probability of locating all these object addition is $p = \prod_{i=1}^{\rho}((1 - (\frac{1}{10^l})^{\omega_i})(1 - (\frac{1}{10^t})^m))$.

Object Update. In the situation of object update, the detection result of group watermark for any object is always true. However, only the detection result of object watermark for the updated object is false. Thus, the probability of locating one single object update is $1 - \frac{1}{10^t}$. Suppose there are ω objects have been updated, the probability of locating all updated objects is $p = (1 - \frac{1}{10^t})^{\omega}$.

Vertex Deletion. For one single vertex deletion, it can be located and characterized if: for all the remaining vertices, their detection results of vertex-local watermark are true (p_1), while the results of the vertex-global watermark are false (p_2). Since the results of vertex-local and vertex-global watermarks are naively true and false respectively, we have $p_1 = 1$ and $p_2 = 1$. The probability of locating one single object deletion is $p = p_1 * p_2 = 1$. For multiple vertex deletion, the probability of locating all these deletion is $p = 1$ no matter how many vertices have been deleted from the watermarked data.

Unlike the situation of object deletion, every object has at least one vertex left after deletion. It is rational since an object with all vertices deleted means the object itself has already been deleted, that situation could be located and recognized on object level.

Vertex Addition. For one single vertex addition, an added vertex can be located and characterized if: the detection results of its vertex-local and vertex-global watermarks are both false. That is, the probability of locating one single vertex addition is $(1 - \frac{1}{10^\eta})(1 - \frac{1}{10^\eta})$. For multiple vertex addition, suppose ω vertices are added into the watermarked spatial data, and the addition causes ρ $(1 \leq \rho \leq min(\omega, g))$ added vertices, with ω_i $(1 \leq i \leq \rho, \sum_{i=1}^{\rho} \omega_i = \omega)$ vertices added into object O_i. Then, the probability of locating all these added vertices is $p = \prod_{i=1}^{\rho}(1 - \frac{1}{10^\eta})^{2\omega_i} = (1 - \frac{1}{10^\eta})^{2\omega}$.

Apparently, the probability depends on the value of ρ. The larger ρ can help reduce the possibility of missing some vertex addition, that conclusion also follows the intuition.

Vertex Update. For one single vertex update, an updated vertex can be located and characterized if: the detection results of its vertex-local and vertex-global watermarks is false and true respectively. That is, the probability of locating one single vertex update is $(1 - \frac{1}{10^\eta})$. For multiple vertex update, suppose ω vertices are updated in the watermarked spatial data, and the update causes ρ $(1 \leq \rho \leq min(\omega, g))$ updated vertices, with ω_i $(1 \leq i \leq \rho, \sum_{i=1}^{\rho} \omega_i = \omega)$ vertices updated in object O_i. Then, the probability of locating all these updated vertices is $p = \prod_{i=1}^{\rho}(1 - \frac{1}{10^\eta})^{\omega_i} = (1 - \frac{1}{10^\eta})^{\omega}$.

Like the situation of vertex addition, the probability of locating vertex update relies on the value of ρ too.

3.5   Analysis of Reversibility

Reversibility is another feature of the proposed scheme. Unlike other existing reversible schemes embed watermark in LSB/LSD or bits/digits before it, the proposed scheme simply embeds several watermarks in the digits after the LSD. Such approach has following two advantages:

- Easy to restore the original host data. In the proposed scheme, the original host data can be restored from the watermarked spatial data by truncating all coordinates after the LSD, while the existing reversible watermarking schemes need to do relatively complicated calculation to revert the changes brought by watermark embedding.
- Better fidelity of the watermarked data. Since the watermark embedding process only changes the digits after LSD, the fidelity of the watermarked data can be guaranteed. To some extent, the watermarked data produced by our scheme is the original spatial data. This is very useful for the applications with requirement for high precision.

4   Experimental Results

In order to evaluate the performance of the proposed watermarking scheme, a series of experiments are performed on the county administrative boundary map of China provided by the National Administration of Surveying, Mapping and Geoinformation of China. This vector map has 3207 objects and 1128242 vertices, and its LSD is the 5th digit at the right of the decimal point. The programs used in the experiments is written by C and executed on a computer with a 1.8 GHz CPU and 4G RAM. From the analysis in Sect. 3.4, we find that the fragility of the proposed watermarking scheme is determined by g and η. In the experiments, we focus on demonstrating the influence of those arguments.

In the experiments, we fixed γ to 7 which allows the change on $Area_T$ of an updated object to be approximately 1% (which is a rational assumption [6,22]). It is worth to be noted that in practice $Area_T$ would never be influenced by the watermarking embedding since the watermarks are all embedded on the digits after the LSD.

4.1   Evaluation on Fragility

The first experiment aims to demonstrate the effectiveness of the proposed scheme on locating and recognizing the tampers. We applied $g = 150$ and $\eta = 5$ on the watermark embedding to obtain the watermarked map. Next, we applied

Table 1. Results of evaluation with various modification type and magnitude

Modification / Magnitude	Object Addition	Object Deletion	Object Update		
			Vertex Addition	Vertex Deletion	Vertex Update
25%	99.20	98.23	100		
			98.57	100	100
50%	98.31	99.17	100		
			98.78	100	100
75%	100	100	100		
			100	100	99.61

each type of modification with different magnitude (the ratio of objects or vertices randomly modified) on different copies of the watermarked map and then detected the modified maps to locate and recognize the modifications. With every combination of modification type and magnitude, the evaluation was performed for 50 times on different copies. The average success rate is illustrated in Table 1, where the value in each cell represents the percents of successfully located and recognized modifications. As shown in Table 1, the proposed scheme is useful for locating and recognizing tampers on both object and vertex levels.

4.2   Evaluation on g

Since m is roughly determined by g, the probability of locating and recognizing tampers on object level actually relies on g. The second experiment aims to evaluate the fragility with different value of g. In this experiment, η was fixed to 5, and g was set to 256, 512 and 1024 respectively (correspondingly, m was approximately 12, 6 and 3). In each pass of evaluation, modifications with magnitude 10%, 20%, 30%, 40% and 50% were applied on different copies of the watermarked map. According to the experimental results, all addition and update on object level were successfully located and recognized. However, for object deletion, the experimental results illustrated in Table 2 show that: *(1)* all object deletion could be successfully located and recognized only if a small portion of objects were deleted; *(2)* the ratio of deletion could be located and recognized decreases with increasing deletion magnitude and g. This is because high deletion ratio may cause some groups to be deleted entirely thus the deletion occurred in them can not be located and recognized. However, for a small g, the probability of removing whole group is very low.

On the vertex level, the experimental results also demonstrate that the locating and recognizing of vertex level modification is not affected by the value of g, since the watermarks on vertex level only rely on the object itself rather than the entire group.

Table 2. Results of evaluation with various object deletion magnitude and g

g \ Magnitude	10%	20%	30%	40%	50%
256	100	100	100	100	100
512	100	100	98.63	96.24	87.45
1024	100	96.23	87.82	72.86	68.19

4.3 Evaluation on η

From the analysis in Sect. 3.4, the probability of locating and recognizing tampers on vertex level highly relies on the value of η namely the length of vertex level watermarks. The last experiment aims to show the relationship between the probability and η. In this experiment, we used a fixed $g = 256$ and η varied from 1 to 5 in watermark embedding. Vertex modifications with magnitude 50% are performed on different copies of the watermarked map before the detection. The evaluation was executed 50 times for every combination of η and modification type. From the experimental results illustrated in Table 3, the larger value η gets, the higher success ratio we obtain. This is because that a small η gives higher possibility for a modified vertex has the same digits after the LSD with its unmodified version. Thus, the detection results of vertex-local and vertex-global watermarks are more likely to be true.

Table 3. Results of evaluation with various η

Modification \ η	1	2	3	4	5
Vertex Addition	97.64	98.36	99.91	100	100
Vertex Deletion	96.33	99.37	100	100	100
Vertex Update	88.82	90.33	93.71	98.67	100

5 Conclusion

In this paper, we discussed the problem of authenticating spatial data both on object level and vertex level. We proposed a reversible fragile watermarking scheme that can not only locate tampers, but also recognize modification types. Moreover, the proposed scheme is reversible so that the original (unwatermarked) spatial data can be restored from the watermarked one. We also theoretically analyzed fragility of the proposed scheme in a probabilistic way, and verified it experimentally on real data. Both the analysis and evaluation show that the proposed scheme is well suited for spatial data authentication. However, the proposed scheme currently can only locate single modification type, the improved scheme which can handle combination of various modification types is part of future work.

References

1. Lin, C.E., Kao, C.M., Lai, Y.C., Shan, W.L., Wu, C.Y.: Application of integrated GIS and multimedia modeling on NPS pollution evaluation. Environ. Monit. Assess. **158**(11),'319–331 (2009)
2. Petitcolas, F.A.P., Anderson, R.J., Kuhn, M.G.: Information hiding: a survey. Proc. IEEE **87**(7), 1062–1078 (1999)
3. Peng, F., Guo, R.S., Li, C.T., Long, M.: A semi-fragile watermarking algorithm for authenticating 2D CAD engineering graphics based on log-polar transformation. CAD **12**, 1207–1216 (2012)
4. Cao, L., Men, C., Gao, Y.: A recursive embedding algorithm towards lossless 2D vector map watermarking. DSP **23**, 912–918 (2013)
5. Guo, J.M., Liu, Y.F.: Continuous-tone watermark hiding in halftone images. In: Proceedings of the 16th International Asia-Pacific Web Conference on Web Technologies and Applications, pp. 411–414 (2010)
6. Guting, R.H.: An introduction to spatial database systems. VLDBJ **3**(4), 357–399 (1994)
7. Wu, C.-C., Chang, C.-C., Yang, S.-R.: An efficient fragile watermarking for web pages tamper-proof. In: Chang, K.C.-C., Wang, W., Chen, L., Ellis, C.A., Hsu, C.-H., Tsoi, A.C., Wang, H. (eds.) APWeb/WAIM -2007. LNCS, vol. 4537, pp. 654–663. Springer, Heidelberg (2007). doi:10.1007/978-3-540-72909-9_72
8. Kee, E., Johnson, M.K., Farid, H.: Digital image authentication from JPEG headers. IEEE Trans. Inf. Forensics Secur. **6**(3), 1066–1075 (2011)
9. Sarreshtedari, S., Akhaee, M.A.: A source-channel coding approach to digital image protection and self-recovery. IEEE Trans. Image Process. **24**(7), 2266–2277 (2015)
10. Li, W., Zhu, B., Wang, Z.: On the music content authentication. In: Proceedings of ACM International Conference on Multimedia, pp. 1101–1104 (2012)
11. Upadhyay, S., Singh, S.K.: Video authentication: issues and challenges. Int. J. Comput. Sci. **9**(1–3), 409–418 (2012)
12. Guo, H., Li, Y., Liu, A., Jajodia, S.: A fragile watermarking scheme for detecting malicious modifications of database relations. Inf. Sci. **176**, 1350–1378 (2006)
13. Niu, X.M., Shao, C.Y., Wang, X.T.: A survey of digital vector map watermarking. Int. J. Innov. Comput. Inf. Control **2**(6), 1301–1316 (2006)
14. Niu, X.M., Shao, C.Y., Wang, X.T.: GIS watermarking: hiding data in 2D vector maps. In: Pan, J.S., Huang, H.C., Jain, L.C., Fang, W.C. (eds.) Intelligent Multimedia Data Hiding. SCI, vol. 58, pp. 123–155. Springer, Heidelberg (2007). doi:10.1007/978-3-540-71169-8_6
15. Rao, U.P., Patel, D.R., Vikani, P.M.: Relational database watermarking for ownership protection. Procedia Technol. **6**, 988–995 (2012)
16. Zheng, L., You, F.: A fragile digital watermark used to verify the integrity of vector map. In: Proceedings of IEEE International Conference on E-Business and Information System Security, pp. 1–4 (2009)
17. Wang, N., Men, C.: Reversible fragile watermarking for locating tampered blocks in 2D vector maps. In: Multimedia Tools and Applications, pp. 1–31 (2013)
18. Zhang, H., Gao, M.: A Semi-fragile digital watermarking algorithm for 2D vector graphics tamper localization. In: Proceedings of IEEE International Conference on Multimedia Information Networking and Security, vol. 1, pp. 549–552 (2009)
19. Zheng, L., Li, Y., Feng, L., Liu, H.: Research and implementation of fragile watermark for vector graphics. In: Proceedings of IEEE International Conference on Computer Engineering and Technology, vol. 1, pp. V1-522–V1-525 (2010)

20. Wang, N., Men, C.: Reversible fragile watermarking for 2-D vector map authentication with localization. Comput. Aided Des. **44**(4), 320–330 (2012)
21. Schneier, B.: Applied Cryptography. Wiley, Chichester (1996)
22. Cohen, S.D., Guibas, L.J.: Partial matching of planar polylines under similarity transformations. In: Proceedings of the ACM-SIAM Symposium on Discrete Algorithms, pp. 777–786 (1997)
23. IEEE Standard for Binary Floating-point Arithmetic. ANSI/IEEE Standard 754–1985
24. Yue, M., Peng, Z., Peng, Y.: A fragile watermarking scheme for modification type characterization in 2D vector maps. In: Chen, L., Jia, Y., Sellis, T., Liu, G. (eds.) APWeb 2014. LNCS, vol. 8709, pp. 129–140. Springer, Cham (2014). doi:10.1007/978-3-319-11116-2_12

Jointly Learning Attentions with Semantic Cross-Modal Correlation for Visual Question Answering

Liangfu Cao, Lianli Gao$^{(\boxtimes)}$, Jingkuan Song, Xing Xu, and Heng Tao Shen

University of Electronic Science and Technology of China, Chengdu, China
lianli.gao@uestc.edu.cn

Abstract. Visual Question Answering (VQA) has emerged as a prominent multi-discipline research problem in artificial intelligence. A number of recent studies are focusing on proposing attention mechanisms such as visual attention ("where to look") or question attention ("what words to listen to"), and they have been proved to be efficient for VQA. However, they focus on modeling the prediction error, but ignore the semantic correlation between image attention and question attention. As a result, it will inevitably result in suboptimal attentions. In this paper, we argue that in addition to modeling visual and question attentions, it is equally important to model their semantic correlation to learn them jointly as well as to facilitate their joint representation learning for VQA. In this paper, we propose a novel end-to-end model to jointly learn attentions with semantic cross-modal correlation for efficiently solving the VQA problem. Specifically, we propose a multi-modal embedding to map the visual and question attentions into a joint space to guarantee their semantic consistency. Experimental results on the benchmark datasets demonstrate that our model outperforms several state-of-the-art techniques for VQA.

1 Introduction

Recently, VQA [1] task has become an increasing hot topic in the area of artificial intelligence (AI), which is involved in a new inter-discipline research field of computer vision (CV) and natural language processing (NLP) [2,3]. Unlike the previous tasks, visual captioning [7,15,29,33,35,36], machine translation [2,4] or text-based question answering [6,32], VQA system is designed for answering a natural language question from a related image. In order to answer the question correctly, we need to learn to understand semantic information of question and accurate visual object recognition as well. Meanwhile, it is rather challenging to integrate the question and image content in a unified framework. So, this is a more difficult and challenging task than the other CV tasks like image annotation [25] and retrieval [24,26,30]. In [28], they develop a so-called correlation component manifold space learning (CCMSL) to learn a common feature space by capturing the correlations between the heterogeneous databases. In [21], they propose a content similarity based fast reference frame selection algorithm for

© Springer International Publishing AG 2017
Z. Huang et al. (Eds.): ADC 2017, LNCS 10538, pp. 248–260, 2017.
DOI: 10.1007/978-3-319-68155-9_19

Q:what is the color of the bench ?   Q:what is the color of the dog ?
(a)   (b)

Fig. 1. The sample questions and images from COCO-QA. Usually, to answer the object color, we first locate where the object is, such as "bench" and "dog" in a figure. (Color figure online)

reducing the computational complexity of the multiple reference frames based inter-frame prediction.

Most state-of-the-art VQA models contain three component, the question encoding, image feature extraction and the answer generation [1,5,18]. A commonly method is to encode the question semantic information as a feature vector by using a recurrent neural network (RNN) [8], such as long short-term memory networks (LSTMs) [10]. And the deep convolutional neural networks (CNN) [37] are used to extract the global image features. Finally, through integrating the question feature and the image feature to generate the answer. The integrations between the question feature and image feature are various, such as element-wise dot or concatenation. However, this integration is not sufficient because the relationships between the question and image are complicated, and through simple operation may not make the best use of interaction between them. Most of the current models answer a single word by treating the VQA as a multi-class classification problem. Of course, we can also answer a complete sentence by a RNN decoder model [19].

When answering a question from the image, people tend to locate the related image region according to the question information before giving the final answer. For example, in Fig. 1, to answer the question "what is the color of the bench?" or "what is the color of the dog?", we first need to locate the object, "bench" or "dog", in the image, that means we focus more attention on the region of an image before draw a conclusion about the color of the "bench" or "dog". In addition, we hope that our location is as accurate as possible, in other words, the encoded question semantic should be more close to the image region features.

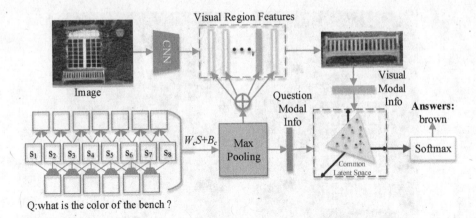

Fig. 2. The framework of our proposed method for VQA. There are mainly four parts in our method: (1) the image regions feature extraction layer; (2) the question encoding layer; (3) the attention with semantic consistency layer; and (4) the answer prediction layer.

Since attention mechanism has been proved effective in CV tasks [5, 16,17,33,34], in this paper, we adopt attention model with semantic consistency for VQA. The whole framework of our proposed method is shown in Fig. 2. Our method mainly including four components: (1) the image feature extraction layer, which is a CNN to extract image regions feature and one vector for a region of the image; (2) the question semantic encoding layer, which is a 1D CNN to extract the question feature after word embedding; (3) the attention with semantic consistency layer, which locates the specific image region related the question by a semantic consistency constraint; (4) the answer generation layer, which integrates the question and image features for answer prediction.

In summary, our main contributions are as following:

- We propose a new framework for VQA by using attention model with semantic consistency. With the semantic consistency constraint, the learned semantic of a question is more close to the image region features. To our best knowledge, we are the first to introduce semantic consistency for VQA.
- We evaluate our method on two benchmark datasets: COCO-QA and VQA datasets. The results of our experiments show superiority of our model over most of state-of-the-art methods.
- We conduct extensive experiment to show how much the semantic consistency influences our results and analyze our experimental results in detail.

2   Proposed Model

The whole framework of our model is shown in Fig. 2. There mainly four components are included in our model: the image feature extraction layer, the question semantic encoding layer, the attention with semantic consistency layer and the answer generation layer. Next, we will describe these four components in detail.

2.1 Image Feature Representation

There are many CV tasks employing CNN [23,27] to encode image content, including videos. In our work, we use the ResNet [9] to extract image features, where the ResNet has shown the best performance in image classification, object detection and image segmentation in 2015 ILSVRC competitions [12]. Differing from the most of previous VQA model using the output of last fully-connected layer as global image feature, we select features from res5c layer of ResNet, which contains more image spatial information. Before inputting the image into CNN, we resize the image to be 448×448. Therefore, then we obtain image feature with dimension of $2048 \times 14 \times 14$. The 14×14 is the number of image regions and 2048 is the dimension for each region vector. We define image region vector as $r_i, i \in \{0, 1, \ldots, 195\}$. Then we embed each region vector into a new vector v_i:

$$r_i = CNN_{Res}(I) \tag{1}$$
$$v_i = \phi(W_I r_i + b_I) \tag{2}$$

where ϕ is a nonlinear activation function, $W_I \in \mathbb{R}^{d \times 2048}$ and b_I are parameter, and d is designed for fitting the size of question vector.

2.2 Question Semantic Encoding

Most of current VQA models employ RNN to encode question semantic information, which has shown excellent performance in NLP and visual captioning. Abandoning the previous method, we employ 1D CNN to encode question semantic information. As demonstrated in [18], the relationships between image and question are complicated. Image feature extracted with high-level semantic information from CNN may not combine well with the question encoded from RNN. In other words, RNN cannot exploit this interactions effectively.

Given a question sentence $Q = [x_1, x_2, \ldots, x_T]$, where x_t is the tth word in question and represented as a one-hot vector which index the location in word vocabulary, and the T is max length of question. Before employing 1D CNN, we first embed each word x_t into a vector space by using an embedding matrix $s_t = W_e x_t$. Then we get the question representation by concatenating all the word vectors:

$$s_{1:T} = s_1 || s_2 || \ldots || s_T \tag{3}$$

where $||$ denotes the concatenation operation. Next, we employ 1D convolution operation on question which can be seen as a feature map. The tth unit output using convolution filter window size c is obtained by:

$$q_c^t = \phi(W_c s_{t:t+c-1} + b_c) \tag{4}$$

where W_c is the convolution kernel and b_c is bias term. The feature after convolution is:

$$q_c = q_c^1 || q_c^2 || \cdots || q_c^{T+1-c} \tag{5}$$

Then we take max-pooling over all feature maps:

$$\tilde{q}_c = \max_t(q_c) = \max_t[q_c^1, q_c^2, \ldots, q_c^{T+1-c}], c \in \{1, 2, 3\} \tag{6}$$

After max-pooling, we concatenate three vectors into a long vector as our question feature q:

$$q = [\tilde{q}_1, \tilde{q}_2, \tilde{q}_3] \tag{7}$$

We conduct triple convolution operation with each window size as 1, 2, 3. With the convolution-pooling process, we can improve the interactions among different words and learn more meaningful phrase. Finally, we obtain more semantic question feature.

2.3   Attention with Semantic Consistency

Given the image regions feature and question feature, we design a attention with semantic consistency model to predict answer. In Fig. 2, to answer "what is the color of the bench?", we should first locate the object region, meanwhile, we hope that our found region is close with the question semantic information, which we bridge the semantic gap between image region and question with semantic cross correlation.

Based on the former two subsection, we have got the image region feature matrix V_I and the encoded question vector q. We first input them into a singe layer perceptron and then employ a softmax function to compute the image attention maps:

$$C = \phi((W_r V_I + b_r) \oplus (W_q q + b_q)) \tag{8}$$
$$P_I = softmax(W_P C + b_P) \tag{9}$$

where $V_I \in \mathbb{R}^{d \times L}$ and $q \in \mathbb{R}^d$, d is the dimension of vector and L is the number of image regions. We set $W_r, W_q \in \mathbb{R}^{k \times d}$ and $W_P \in \mathbb{R}^k$, k is attention dimension, then $C \in \mathbb{R}^{k \times L}$ and $P_I \in \mathbb{R}^L$, where P_I denotes the attention weights distribution of image regions. The $\oplus$ indicates element-add between a matrix and vector.

After we get the attention distribution, we compute weighted average $\bar{v}_I$ of image region feature v_i. And then we combine $\bar{v}_I$ with the question feature q to predict the answer since we have fused question semantic information with visual feature:

$$\bar{v}_I = \sum_i P_i v_i \tag{10}$$

$$h = \bar{v}_I + q \tag{11}$$

Following [13,34], we can also employ multiple attention mechanism. For example, for mth attention layer:

$$C^m = \phi((W_r^m V_I + b_r^m) \oplus (W_q^m h^{m-1} + b_q^m)) \tag{12}$$

$$P_I^m = softmax(W_P^m C^m + b_P^m) \tag{13}$$

where h^0 is initialized with q. Then we integrate the weighted average image feature $\bar{v}_I^m$ with previous question feature to generate a new question feature:

$$\bar{v}_I^m = \sum_i P_i^m v_i \tag{14}$$

$$h^m = \bar{v}_I^m + h^{m-1} \tag{15}$$

Semantic Cross Correlation. In order to reduce the semantic difference between two modalities $X \in \mathbb{R}^{d1}$ and $Y \in \mathbb{R}^{d2}$ (usually $d_1 \neq d_2$), we can employ a linear projection into them, respectively, so that the two modalities feature can be mapped into a high-level semantic common space:

$$R_1 : \mathbb{R}^{d1} \to \mathbb{R}^d \qquad R_2 : \mathbb{R}^{d2} \to \mathbb{R}^d \tag{16}$$

where $R_1 \in \mathbb{R}^{d \times d1}$, $R_2 \in \mathbb{R}^{d \times d2}$. To construct the semantic cross correlation between two modalities, we hope that their features are similar as far as possible in space $\mathbb{R}^d$:

$$R_1(X) \approx R_2(Y) \tag{17}$$

if the $d1 = d$, then the above equation can be rewrite as:

$$X \approx R_1^{-1} R_2(Y) = R(Y) \tag{18}$$

where $R = R_1^{-1} R_2$ is also a linear projection. Therefore, given semantic question feature q and image region weighted average feature $\bar{v}_I$, we can bridge the semantic gap between them with optimizing the following cross-correlation:

$$Loss_1 = \| q - R(\bar{v}_I) \|_2^2 \tag{19}$$

By minimize the above equation, the semantic consistency between question and image can be guaranteed, which makes our work more meaningful.

2.4 Answer Generation

Following the previous work, we regard our VQA model as a multi-class classification problem. After the multiple attention layer, we use the last output $\bar{v}_I^m$ which is input into a common latent space with question feature, and then fuse these two features to predict answer, next a *softmax* function is employed as:

$$f = R(\bar{v}_I^m) + q \tag{20}$$

$$p_{ans} = softmax(W_{ans}f + b_{ans}) \tag{21}$$

where $W_{ans} \in \mathbb{R}^{D \times d}$ is parameter and $b_{ans} \in \mathbb{R}^D$ is a bias term, D is the number of outputs which represents the number of answers. Because our model is to solve multi-class classification problem, then we can define the loss as:

$$Loss_2 = - \sum_{a_i \in \Omega} \mathbf{1}\{y = a_i\} \log p(y = a_i | q, \bar{v}_I; \theta) \tag{22}$$

where a_i is a answer from answers vocabulary Ω. $\mathbf{1}\{\cdot\}$ is a indicator function, when the $\{\cdot\}$ condition is true, the $\mathbf{1}\{\cdot\} = 1$ else 0. y is our predicted answer and θ denotes all parameters that need to be learned. Finally, we can formulate our problem by optimizing the following objection function:

$$L(\theta | \mathbb{D}) = \lambda Loss_1 + Loss_2 \tag{23}$$

the λ is a hyper-parameter that balance the two loss terms and $\mathbb{D}$ is training datasets.

3   Experiments

3.1   Datasets

We evaluate our model on two widely used datasets: COCO-QA [22] and VQA [1]. The images of these datasets are both from Microsoft COCO dataset [14].

COCO-QA. Toronto COCO-QA automatically generates question-answer (QA) pairs by analyzing syntactic structure of the original image captioning with Stanford Parser, then replacing the keywords from captioning. There are 78736 training QA pairs and 38948 testing pairs in this dataset. Four types of questions are included: *Object, Number, Color*, and*Location*. The answers in this dataset are all single-word.

VQA. This is the largest dataset so far. All questions and answers are annotated by human. There are 248349 training questions, 121512 validation questions and 244,302 testing questions in the dataset. For each image, three questions are related, and for each question, ten answers are provided by annotators. The questions in this dataset can be roughly divided into: *Yes/No, Number* and *Other*.

3.2   Evaluation Metrics

Since we regarded our VQA model as a classification problem, one simple way to evaluate VQA is accuracy, namely the predicted answer perfectly matches the groundtruth. However, under some circumstances, it is difficult to determine which answer is absolutely correct, for example "table" and "desk", "bike" and "bicycle". To solve this problem, one way is to use Wu-Palmer similarity (WUPS) score [31]. The WUPS score calculates the similarity between two words based on their longest common subsequence in a taxonomy tree, whose value ranges from 0 to 1. WUPS0.9 and WUPS0.0 are common used as evaluation metrics. There are slightly difference in evaluation for VQA dataset. [1] proposed a new evaluation metric:

$$ACC(\mathbf{ans}) = \min\{\frac{\#humans\,that\,said\,\mathbf{ans}\#}{3}, 1\} \qquad (24)$$

which means the answer is true as least three annotators agreed on the answer since there are ten answers for each question.

3.3 Experimental Settings

For image representation, we use the ResNet-152 to extract image feature. The parameter of the CNN is fixed. For question embedding, we set the number of three convolution filter as 512, respectively. The dimension of word embedding and all other hidden layers are set 512. We employ hyperbolic tangent function as our activation function. Following the [13,34], we set our attention layers with two layers. For COCO-QA dataset, there are 430 unique words and the vocabulary size is 10,158. For VQA, we use the top 1000 frequently used answers as output, which covers 86.53% of training and validation questions. The vocabulary size of this dataset is 14,771. All vocabulary words are extracted from training questions, and we add a additional character "UNK" for the words in testing questions that are not shown in training set. All questions are tokenized by using Python Natural Language Toolkit (NLTK).

In our experiments, we use Theano framework to train the model. We train our networks by stochastic gradient descent(SGD) with a learning momentum 0.9, weight-decay 0.0005 and initialized learning rate 0.05. All parameters are initialized with gaussian distribution. We set the batch size as 100 for COCO-QA and 128 for VQA since its a larger dataset. Parameters regularization and dropout strategy with dropout ratio 0.5 are employed. Our model is still end-to-end trainable though we do not finetune the ResNet.

3.4 Results and Analysis

Several recently proposed VQA methods are used as baseline for our model. For COCO-QA dataset, we compare our method with 2-VIS-BLSTM [22], IMG-CNN [18], ATT-VGG-SEG [5], SAN(2,CNN) [34], DPPnet [20], QRU [13], HYBRID [11], CO-Attention [16]. For the larger dataset VQA, since some models are not evaluated on official server, we only compare the published results: LSTM Q+I [1], ATT-VGG-SEG [5], SAN(2,CNN) [34], DPPnet [20], QRU [13], HYBRID [11], CO-Attention [16].

The experimental results on COCO-QA and VQA dataset are shown in Tables 1 and 2, respectively. From Table 1, we can see our basic model(without semantic consistency) has shown a pretty well accuracy since we employ the best deep CNN ResNet to extract the image features. The last of column of Table 1 demonstrate that attention with semantic consistency significantly

Table 1. Experiments results on COCO-QA dataset, in percentage.

Methods	Object	Number	Color	Location	Accuracy	WUPS0.9	WUPS0.0
2-VIS+BLSTM [22]	58.17	44.79	49.53	47.34	55.09	65.34	88.64
IMG-CNN [18]	-	-	-	-	58.40	68.50	89.67
ATT-VGG-SEG [5]	62.46	45.70	46.81	53.67	58.10	68.44	89.85
SAN(2, CNN) [34]	64.50	48.60	57.90	54.00	61.60	71.60	90.90
DPPnet [20]	-	-	-	-	61.19	70.84	90.61
QRU [13]	65.06	46.90	60.50	56.99	62.50	72.58	91.62
HYBRID [11]	-	-	-	-	63.18	73.14	91.32
CO-Attention [16]	68.00	51.00	62.90	58.80	65.40	75.10	92.00
Ours	66.52	48.28	61.30	57.18	63.76	73.47	91.38
Ours+semantic	67.06	48.89	63.48	57.19	64.55	74.06	91.52

Table 2. Results on VQA dataset. All results test on the official server.

Methods	test-dev				test-std
	Yes/No	Number	Other	All	All
LSTM Q+I [1]	80.50	36.77	43.08	57.75	58.16
ATT-VGG-SEG [5]	-	-	-	-	48.38
SAN(2, CNN) [34]	79.30	36.60	46.10	58.70	58.90
DPPnet [20]	80.71	37.24	41.69	57.22	57.36
QRU [13]	82.29	37.02	47.67	60.72	60.76
HYBRID [11]	80.47	37.50	46.72	59.57	60.06
CO-Attention [16]	79.70	38.70	51.70	61.80	62.10
Ours+semantic	81.40	35.74	49.20	60.96	61.16

Q: what is the batter swinging ?

A1: bat
A2: bat
GT: bat

Q: what is the man holding a snowboard on top of a snow covered ?

A1: mountain
A2: mountain
GT: hill

Q: what is the color of the cat ?

A1: black
A2: black
GT: black

Q: what is the color of the barns ?

A1: brown
A2: red
GT: red

Fig. 3. Example question-image pairs from COCO-QA dataset. "A1" indicates the answer predicted by the basic model, "A2" indicates the answer given by our best model(with semantic consistency), and "GT" represents the groundtruth answer. (Color figure online)

improve our model performance. We observe that our best model(with semantic consistency) outperform the 2-VIS+BLSTM, IMG-CNN, ATT-VGG-SEG, DPPnet, SAN(2,CNN), QRU and HYBRID by 17.17%, 10.53%, 11.1%, 5.59%, 4.79%, 3.28% and 2.17% in accuracy, respectively. Similar improvements trends can be also observed for evaluation metric WUPS0.9 and WUPS0.0. And especially for question type *color*, our model achieve the best performance among all VQA methods, which demonstrates our model learns semantic information between question and image well when there is a single object in image. Our semantic model also outperforms SAN by 3.97%, 0.06%, 9.35% and 5.91% in the different question types of *Object, Number, Color* and *Location*, respectively. Similarly with QRU, our best model improves by 3.07%, 4.24%, 4.93% and 0.04% in accuracy among different question types. Although SAN and QRU are similar with our works since we both employ multiple image attention step, they show inferior performance, which further demonstrates the semantic consistency plays an importance role in our model. An obviously defect for whole performance of our model is to show a little gap with CO-Attention [16] from Table 1. However, the model of co-attention is extremely complex since it employs multiple question and image attention. It computes attention distribution about image and question from three level, which has a large amount calculation.

After setting all hyper-parameters, we choose our best model (with semantic consistency) to test on VQA dataset that includes test-dev and test-standard partitions. Following [7], we divide the validation set into two halves. We employ the training set and one half of validation to train, and the remaining one half to valid in order to choose the best model. Then we evaluate our best model on VQA official server, and all results are shown in Table 2. In Table 2, our best model outperform most of VQA methods except CO-Attention. The overall results show that our model outperforms the LSTM Q+I method which proposes the VQA dataset by 5.16% and improves 1.12%, 14.21% in the question type of *Yes/No* and *Other*, respectively. However, our model shows worst performance on question type *Number* among all VQA methods. This maybe due to that our model cannot construct the question semantic cross correlation with image very well when there are multiple object in image. Our model outperforms QRU and HYBRID by 0.65% and 1.83% in test-standard overall. Both of the QRU and HYBRID model employ a pre-trained GRU network for skip-thought sentence embedding model which is trained in an encoder-decoder unsupervised manner on millions of corpus. Our model use a simple word embedding way and do not need to fine-tune the large GRU network. HYBRID model need analyze the datasets and count the number of different question types. QRU model first employs edge boxes to detect the object region and then extracts the object features by inputting it to CNN, which makes the model achieve the best performance on question type *Yes/No* among all VQA methods. However, this work makes the final results depend on the accuracy of object detection to some extent. On the contrary, we input the image to CNN and use the question text directly, and achieve a comparable results in the meantime.

Some sample question-image pairs from COCO-QA dataset are shown in Fig. 3. All questions and answers are represented as lower case. We can see the second example in Fig. 3, when asking "what is the man holding a snowboard on top of a snow covered?" and given answer "hill", our basic and best model predict the answer "mountain". From the classification accuracy aspect, we predict the wrong answers. As we know, the word meaning between "hill" and "mountain" is similar. Therefore, it is significant for us to employ WUPS measurement. The last example in Fig. 3, when determining the color of "barns", our basic model give the answer "brown" which is obviously the color of "cow", and our best model predicted the correct answer, which demonstrates that our model have a capability to distinguish the foreground and background feature.

4   Conclusion

In this paper, we proposed a new attention with semantic consistency model which is an end-to-end trainable neural network framework for VQA problem. By constructing the semantic cross correlation between image and question, we integrate the image and question features to predict answers. On the other hand, we conduct several comparative experiments to determine how much the semantic loss affect our model. Experimental results on two benchmark dataset COCO-QA and VQA demonstrate that our model outperform most of state-of-the-art VQA methods. In addition, we further analyze the strength and weakness of our model in detail through showing some question-image pairs.

Acknowledgment. This work was supported in part by the National Natural Science Foundation of China under Project 61502080, Project 61632007, and the Fundamental Research Funds for the Central Universities under Project ZYGX2016J085, Project ZYGX2014Z007.

References

1. Antol, S., Agrawal, A., Lu, J., Mitchell, M., Batra, D., Zitnick, C.L., Parikh, D.: VQA: visual question answering. In: ICCV, pp. 2425–2433 (2015)
2. Bahdanau, D., Cho, K., Bengio, Y.: Neural machine translation by jointly learning to align and translate. CoRR abs/1409.0473 (2014)
3. Berant, J., Liang, P.: Semantic parsing via paraphrasing. In: ACL, pp. 1415–1425 (2014)
4. Calixto, I., Liu, Q., Campbell, N.: Incorporating global visual features into attention-based neural machine translation. CoRR abs/1701.06521 (2017). http://arxiv.org/abs/1701.06521
5. Chen, K., Wang, J., Chen, L., Gao, H., Xu, W., Nevatia, R.: ABC-CNN: an attention based convolutional neural network for visual question answering. CoRR abs/1511.05960 (2015). http://arxiv.org/abs/1511.05960
6. Derczynski, L., Shaw, R., Solway, B., Wang, J.: Question answering against very-large text collections. CoRR abs/1304.7157 (2013)

7. Fang, H., Gupta, S., Iandola, F.N., Srivastava, R.K., Deng, L., Dollár, P., Gao, J., He, X., Mitchell, M., Platt, J.C., Zitnick, C.L., Zweig, G.: From captions to visual concepts and back. In: CVPR, pp. 1473–1482 (2015)

8. Graves, A.: Generating sequences with recurrent neural networks. CoRR abs/1308.0850 (2013)

9. He, K., Zhang, X., Ren, S., Sun, J.: Deep residual learning for image recognition. In: CVPR, pp. 770–778 (2016)

10. Hochreiter, S., Schmidhuber, J.: Long short-term memory. Neural Comput. 9(8), 1735–1780 (1997)

11. Kafle, K., Kanan, C.: Answer-type prediction for visual question answering. In: CVPR, pp. 4976–4984 (2016)

12. Krizhevsky, A., Sutskever, I., Hinton, G.E.: Imagenet classification with deep convolutional neural networks. In: NIPS, pp. 1106–1114 (2012)

13. Li, R., Jia, J.: Visual question answering with question representation update (QRU). In: NIPS, pp. 4655–4663 (2016)

14. Lin, T.-Y., Maire, M., Belongie, S., Hays, J., Perona, P., Ramanan, D., Dollár, P., Zitnick, C.L.: Microsoft COCO: common objects in context. In: Fleet, D., Pajdla, T., Schiele, B., Tuytelaars, T. (eds.) ECCV 2014. LNCS, vol. 8693, pp. 740–755. Springer, Cham (2014). doi:10.1007/978-3-319-10602-1_48

15. Lu, J., Xiong, C., Parikh, D., Socher, R.: Knowing when to look: Adaptive attention via A visual sentinel for image captioning. CoRR abs/1612.01887 (2016). http://arxiv.org/abs/1612.01887

16. Lu, J., Yang, J., Batra, D., Parikh, D.: Hierarchical question-image co-attention for visual question answering. In: NIPS, pp. 289–297 (2016)

17. Luong, T., Pham, H., Manning, C.D.: Effective approaches to attention-based neural machine translation. In: EMNLP, pp. 1412–1421 (2015)

18. Ma, L., Lu, Z., Li, H.: Learning to answer questions from image using convolutional neural network. In: AAAI, pp. 3567–3573 (2016)

19. Malinowski, M., Rohrbach, M., Fritz, M.: Ask your neurons: A neural-based approach to answering questions about images. In: ICCV, pp. 1–9 (2015)

20. Noh, H., Seo, P.H., Han, B.: Image question answering using convolutional neural network with dynamic parameter prediction. In: CVPR, pp. 30–38 (2016)

21. Pan, Z., Jin, P., Lei, J., Zhang, Y., Sun, X., Kwong, S.: Fast reference frame selection based on content similarity for low complexity HEVC encoder. J. Vis. Commun. Image Represent. 40, 516–524 (2016)

22. Ren, M., Kiros, R., Zemel, R.S.: Exploring models and data for image question answering. In: NIPS, pp. 2953–2961 (2015)

23. Simonyan, K., Zisserman, A.: Very deep convolutional networks for large-scale image recognition. CoRR abs/1409.1556 (2014). http://arxiv.org/abs/1409.1556

24. Song, J., Gao, L., Liu, L., Zhu, X., Sebe, N.: Quantization-based hashing: a general framework for scalable image and video retrieval. Pattern Recogn. (2017)

25. Song, J., Gao, L., Nie, F., Shen, H.T., Yan, Y., Sebe, N.: Optimized graph learning using partial tags and multiple features for image and video annotation. IEEE Trans. Image Process. 25(11), 4999–5011 (2016)

26. Song, J., Yang, Y., Huang, Z., Shen, H.T., Luo, J.: Effective multiple feature hashing for large-scale near-duplicate video retrieval. IEEE Trans. Multimedia 15(8), 1997–2008 (2013)

27. Szegedy, C., Liu, W., Jia, Y., Sermanet, P., Reed, S.E., Anguelov, D., Erhan, D., Vanhoucke, V., Rabinovich, A.: Going deeper with convolutions. In: CVPR, pp. 1–9 (2015)

28. Tian, Q., Chen, S.: Cross-heterogeneous-database age estimation through correlation representation learning. Neurocomputing **238**, 286–295 (2017)
29. Vinyals, O., Toshev, A., Bengio, S., Erhan, D.: Show and tell: A neural image caption generator. In: CVPR, pp. 3156–3164 (2015)
30. Wang, J., Zhang, T., Sebe, N., Shen, H.T., et al.: A survey on learning to hash. IEEE Trans. Pattern Anal. Mach. Intell. (2017)
31. Wu, Z., Palmer, M.: Verbs semantics and lexical selection. In: ACL, pp. 133–138 (1994)
32. Xiong, C., Merity, S., Socher, R.: Dynamic memory networks for visual and textual question answering. In: ICML, pp. 2397–2406 (2016)
33. Xu, K., Ba, J., Kiros, R., Cho, K., Courville, A.C., Salakhutdinov, R., Zemel, R.S., Bengio, Y.: Show, attend and tell: Neural image caption generation with visual attention. In: ICML, pp. 2048–2057 (2015)
34. Yang, Z., He, X., Gao, J., Deng, L., Smola, A.J.: Stacked attention networks for image question answering. In: CVPR, pp. 21–29 (2016)
35. Yao, T., Pan, Y., Li, Y., Qiu, Z., Mei, T.: Boosting image captioning with attributes. CoRR abs/1611.01646 (2016). http://arxiv.org/abs/1611.01646
36. You, Q., Jin, H., Wang, Z., Fang, C., Luo, J.: Image captioning with semantic attention. In: CVPR, pp. 4651–4659 (2016)
37. Yu, F., Koltun, V.: Multi-scale context aggregation by dilated convolutions. CoRR abs/1511.07122 (2015)

Deep Semantic Indexing Using Convolutional Localization Network with Region-Based Visual Attention for Image Database

Mingxing Zhang[1], Yang Yang[1(✉)], Hanwang Zhang[2], Yanli Ji[3], Ning Xie[1], and Heng Tao Shen[1]

[1] School of Computer Science and Engineering,
Center for Future Media, UESTC, Chengdu, China
superstar_zhang@hotmail.com, dlyyang@gmail.com, seanxiening@gmail.com,
shenhengtao@hotmail.com
[2] Department of Computer Science, Columbia University, New York City, USA
zhanghanwang@gmail.com
[3] School of Automation, Center for Future Media, UESTC, Chengdu, China
yanliji@uestc.edu.cn

Abstract. In this paper, we introduce a novel deep semantic indexing method, a.k.a. captioning, for image database. Our method can automatically generate a natural language caption describing an image as a semantic reference to index the image. Specifically, we use a convolutional localization network to generate a pool of region proposals from an image, and then leverage the visual attention mechanism to sequentially generate the meaningful language words. Compared with previous methods, our approach can efficiently generate compact captions, which can guarantee higher level of semantic indexing for image database. We evaluate our approach on two widely-used benchmark datasets: Flickr30K, and MS COCO. Experimental results across various evaluation metrics show the superiority of our approach as compared with other visual attention based approaches.

Keywords: Semantic indexing · Image database · Visual attention · Region proposals · Convolutional localization network

1 Introduction

With the proliferation of social networks, tremendous amount of image data have been created on the Web for sharing, self-expressing and distribution. Take Flickr as an example, a public picture sharing website, which received 1.8 million photos per day in average, from February to March 2012 [22]. Assuming the size of each photo is 1 megabytes (MB), it requires 1.8 terabytes (TB) storage every single day. Indeed, those large volumes of image dataset are precious resources for users to explore the human society, social events, public affairs, and so on [34,38]. However, how to effectively and efficiently manage and process such

© Springer International Publishing AG 2017
Z. Huang et al. (Eds.): ADC 2017, LNCS 10538, pp. 261–272, 2017.
DOI: 10.1007/978-3-319-68155-9_20

enormous database is still a challenging task for many real-life applications. To overcome this problem, we introduce a novel deep semantic indexing method for image database. Our method can efficiently generate a natural language caption to semantically index an image. By leveraging these compact captions to index image database, we may require less storage space and achieve faster processing speed. Moreover, because the captions contain high level of semantic meanings, our method can achieve preferable effects of semantic indexing compared with original visual data to enable more types of retrieval tasks.

Along with the progress of deep learning, recent work mainly focus on end-to-end image caption generation (from Convolutional Neural Network (CNN) encoding to Recurrent Neural Network (RNN) decoding) [6,17,21,31,33]. Specifically, this paradigm firstly encodes the image content using a deep CNN architecture, and then feeds the visual representation into a RNN network to sequentially generate the language words. All the parameters in such kind of models can be trained in an end-to-end manner. Thanks to the visual representation ability of CNN and the language modelling power of RNN, the CNN-RNN image captioners have shown promising results to some extent. However, encoding the whole image into a static representation may limit the generalization ability, because specific contents are inevitably ignored, thereby failing to comprehensively understand images in details.

Rather than attending to an individual image as a whole, humans can dynamically focus on different parts of an image as needed. This is an important mechanism in the human visual system which is usually called as visual attention [18]. As one of attention based approaches for image caption generation, Xu et al. [35] successfully incorporate the attention mechanism to improve the sentence quality of the aforementioned CNN-RNN paradigm. This approach dynamically selects the useful regions for generating a word. However, as Fig. 1.(a) shows, this kind of attention is based on fixed grid splits, which can hardly adapt to the content variations, thereby failing to recognize the semantic regions and understand the overall content in an image.

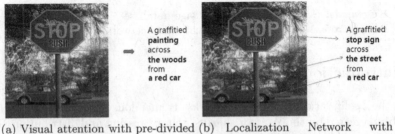

(a) Visual attention with pre-divided and fixed grid splits

(b) Localization Network with Region-based Visual Attention

Fig. 1. An example of the intention of our approach. Previous visual attention is based on pre-divided and fixed grid splits which cannot adapt to the content variations. We propose a Localization Network with Region-based Visual Attention (LocAtt) framework which can generate a semantic region for predicting next word.

Recently, Faster RCNN [26] achieves both high accuracy and efficiency for object detection. The core component in Faster RCNN is the Region Proposal Network (RPN), which can generate hundreds of semantic region proposals for an image. Inspired by the RPN, in this work, we propose a novel Localization Network with Region-based Visual Attention (LocAtt) framework to generate descriptive captions for semantic indexing in image database. As the first attempt, LocAtt combines the object localization and visual attention mechanism together into a unified end-to-end architecture. As shown in Fig. 1.(b), LocAtt generates a semantic region for the language module to focus on when predicting the next word. We conduct extensive experiments on benchmark datasets: Flickr30k [25], and MS COCO [19]. The results show that our approach outperforms other visual attention based approaches.

The rest of the paper is organized as follows. Section 2 briefly reviews related work and Sect. 3 details the proposed approach. Experimental results are reported in Sect. 4, followed by the conclusion in Sect. 5.

2 Related Work

Image Database Indexing. With the large scale of data and rapid expanding in image databases, automatic and efficient indexing techniques have become increasingly important for many real-life applications. Generally speaking, existing methods for image database indexing can be divided into two classes: (1) Text-base indexing [14,36] and (2) Content-based indexing [2,5,7,30,37]. Text-based methods index the images based on the metadata such as tags, keywords or text annotated with the image. This type of methods often result of ambiguity and inadequacy for image indexing, and also leads to irrelevant search results. As the development of computer vision techniques, content-based indexing have been widely explored. Content-based methods usually map the original image to a feature vector. However, visual feature can't be directly understood by users and there exists a large gap between visual feature space and human's cognition. In this paper, we introduce a novel deep semantic indexing method for image database. Our method can automatically generate a natural language caption describing an image as a reference to index the image, which can be regarded as the combination of text-base indexing and content-based indexing.

Neural Image Captioning. Recently, with the success of sequence to sequence learning using neural networks, such as [1,3,23,28], some methods based on neural networks for image captioning have been proposed. Similar to machine translation, those methods translate an image to a sentence. Kiros et al. [16] first developed a multi-model log-bilinear model using features from the images, setting the corner stone for caption generation with neural networks. After that, Kiros et al. [17] furthered their work by simultaneously realizing ranking and generation in a natural fashion. Mao et al. [21] took a next step by employing a recurrent neural language model instead of the original feed-forward one. Different from whose models which look the image at each time step,

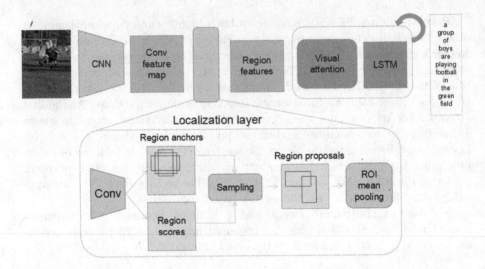

Fig. 2. The framework of our proposed LocAtt for image captioning. Our framework takes an image as input and simultaneously outputs the caption words as well as their corresponding semantic regions. Our framework is end-to-end in both training and testing stages.

Vinyals et al. [31] only input the image to the Long Shot Term Memory (LSTM) network at the beginning and then generate caption words sequentially.

Visual Attention. Inspired by the human ability of selectively concentrating on some information while ignoring others, the attention mechanism has widely used in many tasks, like machine translation [1,20], action classification [24,27], and image/video caption generation [35,39]. Rather than encoding the whole image into a static feature vector, attention based methods can dynamically focus on different parts of an image as needed for defferent tasks. For example, Xu et al. [35] proposed an attention framework that pre-divided the image into fixed grid splits. Sharma et al. [27] used an LSTM network with a soft attention module to focus on pre-divided local regions in each frame for action recognition. Those methods achieve impressive performance, however this pre-division is simple but can't adapt to the object variations and can't understand well what is happening in an image.

Object Detection. Because object detection is a fundamental process for various vision tasks, there has been a long line of works for it [9,11,26,29,32]. Among them, RCNN [11] laid the foundation for the region proposal based methods. However, how to effectively generate region proposals always plagued researchers [13]. Selective Search [29] prevailed the fields of object detection for a long time, combining superpixels in a greedy manner. EdgeBoxes [40] balanced the proposal quality and speed, contributing to less time consumed for searching compared with selective search. However, Faster RCNN [26] detector innovatively proposed a Region Proposal Network (RPN) which shared convolutional features with the detection networks, and thus formed one integrated pipeline for both region proposals generation and object detection.

3 Our Model

3.1 Overall Framework

The overall framework of our approach is depicted in Fig. 2. Our framework is designed to be end-to-end, which can simultaneously generate language words and their corresponding semantic regions. Generally, an image is first propagated through a CNN network to generate a convolutional feature map, which serves as the input of our localization layer. The localization layer generates several semantic region proposals, which are highly related to the image caption. Then the features of proposed regions are fed into the Region-based Visual Attention component, which sequentially selects a Region of Interest (RoI) and puts it into the language generating network (e.g., LSTM) to generate the next word.

3.2 Convolutional Localization Network

Our convolutional localization network consists of a convolutional network and a region localization layer. The convolutional network is based on the VGG16 network due to the excellent representative power for visual data. Specifically, we employ the first 13 convolutional layers with 3×3 kernel size and 4 interspersed max-pooling layers with 2×2 kernel size. The output of our convolutional network corresponds to the *con5_3* layer in the original VGG-16 network. The convolutional network encodes the input image of shape $3 \times W \times H$ to the convolution feature map of shape $512 \times \lfloor \frac{W}{16} \rfloor \times \lfloor \frac{H}{16} \rfloor$. The derived feature maps form the input to the region localization layer.

The localization layer is based on the Region Proposal Network (RPN), which aims to generate some region proposals for further processing. After receiving the feature map with shape $512 \times \lfloor \frac{W}{16} \rfloor \times \lfloor \frac{H}{16} \rfloor$ and certain internal processing, the localization layer outputs B identified semantic region proposals. The returned region proposals consist of 3 tensors, i.e., Region Coordinates, Region Scores and Region Features. Region Coordinates is a matrix of shape $B \times 4$ giving bounding box coordinates for each output region. Region Scores is a vector of length B giving the confidence score for the output regions. A higher score represents that this region proposal is more likely to be the corresponding region of interest ground truth. Region Features is a matrix of shape $B \times C$ giving features for the output regions. C is the feature dimension. In the following paragraphs, we describe the internal processing of our localization layer in details.

Convolutional Anchors. Our localization layer generates semantic region proposals by regressing offsets from a set of translation-invariant anchors, which is similar to Faster RCNN. Specifically, at each point in the $\lfloor \frac{W}{16} \rfloor \times \lfloor \frac{H}{16} \rfloor$ feature map, we consider 12 anchors centred on it with 3 different aspect ratios and 4 different spatial scales. In order to predict the region coordinates and scores corresponding to each anchor, we use a 3×3 convolution with 256 filters followed by a rectified linear non-linearity and a 1×1 convolution with 5 filters. After those computing, we get a 5 dimensional vector for each region with one for confidential score and others for offset regression.

Box Regression. Because the ground truth of region box is annotated using the pixel coordinates, we first need to project the location of each anchor in the $\lfloor \frac{W}{16} \rfloor \times \lfloor \frac{H}{16} \rfloor$ feature map back to the original $W \times H$ image field. Then we adopt the same parameterized regression from anchors to the region proposals as Fast RCNN [10] does. Given the center coordinate of anchor box (x_a, y_a), weight w_a, height h_a, and the offset regression of our model (t_x, t_y, t_w, t_h), we can compute the center coordinate of output region (x, y) as well as its weight w and height h as follows:

$$x = x_a + t_x w_a \qquad y = y_a + t_y h_a \qquad (1)$$

$$w = w_a exp(t_w) \qquad h = h_a exp(t_h) \qquad (2)$$

Box Selecting. Similar to the Fast RCNN [10], in order to save more details, we up-sample the original image to make the smaller side fixed to be 600 but we also restrain the max side not to exceed 1000. During up-sampling, we keep the image spatial ratio unchanged. For a typical image size 600×800, there exist nearly 20,000 anchor boxes.

At training time, we randomly sample 256 boxes from an image with half are the positive regions and half are the negatives. A region is positive if it has an intersection over union (IoU) of at least 0.7 with some ground-truth region or has the maximal IoU with each ground-truth region. A region is negative if it has $0 < \text{IoU} < 0.3$ with all ground-truth regions. At test time we apply the non maximum suppression (NMS) based on the predicted region proposal confidence scores to select the $B = 200$ most confident region proposals.

For each selected region proposal, we first project its bounding box to the top layer of the convolutional network and clip the corresponding feature maps on this layer. Then we apply the mean pooling for the clipped feature maps to generate the output region features.

3.3 Region-Based Visual Attention

The Region-based Visual Attention module in our framework is recurrent, which receives the hidden state h_{t-1} of the language generation model at previous time step and the features R of region proposals outputted from the region localization layer to determine which region should attend.

We use the *Hard attention* mechanism [35] to select the most appropriate one from region proposals to attend at each time step. Specifically, for each region proposal $R_i, i = 1 \cdots B$, the mechanism generates a positive weight α_i which can be interpreted as the probability of which the region should be attended for predicting the next word. The weight α_i of each region proposal is computed by the function f_{att} of a multi-layer perceptron conditioned on the previous hidden state h_{t-1}.

$$e_{ti} = f_{att}(R_i, h_{t-1}) \tag{3}$$

$$\alpha_{ti} = \frac{exp(e_{ti})}{\sum_{i=1}^{B} exp(e_{ti})} \tag{4}$$

Once the probabilities for all region proposals (which sum to one) are computed, our model samples one of them to focus on according to the multinomial distribution of their probabilities. So the region proposal with higher probability is more likely to be selected.

The reason we use the *Hard attention* mechanism rather than the *Soft attention* is that the generated region proposals are semantic and contain the necessary information we need to attend. So attending one of them which has the highest probability to be related to the next word prediction is enough. In addition, paying attention on one region proposal will not incorporate other useless information and thus is more benefit for word prediction.

3.4 Language Generation Model

We use a LSTM network [12] to produce the caption by generating one word at each time step conditioned on the feature of selected region, the previous hidden state and the previously generated words. Using $T_{s,t} : \mathbb{R}^s \rightarrow \mathbb{R}^t$ to denote a simple affine transformation with parameters that are learned, we have following equations:

$$\begin{pmatrix} i_t \\ f_t \\ o_t \\ g_t \end{pmatrix} = \begin{pmatrix} \sigma \\ \sigma \\ \sigma \\ tanh \end{pmatrix} T_{D+m+m,n} \begin{pmatrix} Ey_{t-1} \\ h_{t-1} \\ \hat{z}_t \end{pmatrix}$$

$$c_t = f_t \odot c_{t-1} + i_t \odot g_t$$

$$h_t = o_t \odot tanh(c_t)$$

here, i_t, f_t, c_t, o_t, h_t are the input gate, forget gate, memory, output gate and hidden state of LSTM, respectively. The vector $\hat{z}_t \in \mathbb{R}^C$ is the feature vector of the selected region, and keeps varying across different time steps. $E \in \mathbb{R}^{m \times K}$ is the word embedding matrix. m and n denote the dimensions of word embedding and LSTM memory respectively, σ and $\odot$ present the logistic sigmoid activation and element-wise multiplication respectively.

In order to initiate the memory and hidden state of LSTM, we extract the 4096-dimensional feature of whole image from $fc7$ layer in VGG16 network and then pass this feature through two separate MLPs respectively to obtain the initiations. We use the full image feature for initiations because we expect to give the LSTM network a quick overview of the image content at first.

3.5 Loss Function

At training time, our framework takes the annotated image captions and region bounding boxes as ground truth. Our model's loss comes from two modules: region localization module and language generation module. In the region localization module, we use the binary logistic loss for the confidences of sampled positive and negative regions. For bounding box regression, we use the smooth L1 loss for the parameterized coordinates of bounding box. In the language generation module, we use the cross-entropy loss for the caption word at each time-step. Our loss function can be formulated as follow:

$$L = - log(P(y|x)) + \lambda \left(\sum_i L_{cls} (p_i, p_i') + \sum_i p_i' L_{reg} (r_i, r_i') \right) + \beta \theta^2 \quad (5)$$

where p_i', r_i' is the ground truth for object/background and parametrized coordinates of bounding box respectively. p_i' is 1 if the anchor is positive and 0 otherwise. $\sum_i L_{cls} (p_i, p_i')$ is the binary logistic loss for the confidences of positive regions, $\sum_i L_{reg} (r_i, r_i')$ is the smooth L1 loss for bounding box regression. y is the ground truth of caption sentence and $- log(P(y|x))$ is the cross-entropy loss between the predicted sentence and ground truth. θ^2 is the weigh decay of model parameters. λ, β are the coefficients for the loss of region localization module and weight decay respectively.

4 Experiments

4.1 Datasets and Settings

We evaluate our approach on the popular Flickr30K and MS COCO datasets. The Flickr30K dataset has 31,783 images and the MS COCO dataset is more challenging with 123,287 images. Each image in both datasets has 5 or more captions which are manually annotated by different people. For each dataset, we construct a fixed size vocabulary which contains 9998 top frequency words appearing in the annotated captions along with another two words for *UNKOWN* and *END* tokens. We follow the same splits[1] which were commonly used in previous works [15,35].

In all experiments, the dimension of memory and hidden state of our LSTM caption generator is set to 1800. We set $\lambda = 0.1$ and $\beta = 0.0001$ in Eq. 5. At the training stage, we firstly optimize the parameters of convolutional localization network and LSTM caption generator module independently. Then we finetune all the parameters of our model together.

4.2 Quantitative Evaluation

For quantitative evaluation, we leverage the popular automatic metrics of BLEU _1,2,3,4 and METEOR which have correlation with human judgement. The

---

[1] https://github.com/karpathy/neuraltalk.

Table 1. Results of different methods on the MS COCO dataset.

Method	BLEU-1	BLEU-2	BLEU-3	BLEU-4	METEOR
BRNN [15]	64.2	45.1	30.4	20.3	-
Google NIC [31][a]	66.6	46.1	32.9	24.6	-
mutimodal-RNN [21]	67.0	49.0	35.0	25.0	-
LRCN [6]	62.8	44.2	30.4	21.0	-
MSR/CMU [4]	-	-	-	19.0	20.4
Visual concepts [8][a]	-	-	-	25.7	23.6
Toronto-Soft attention [35]	70.7	49.2	34.4	24.3	23.9
Toronto-Hard attention [35]	71.8	50.4	35.7	25.0	23.0
Ours	68.8	51.6	38.3	29.0	24.3

[a]indicates a different test data split.

Table 2. Results of different methods on the Flickr30K dataset.

Method	BLEU-1	BLEU-2	BLEU-3	BLEU-4	METEOR
Google NIC [31][a]	66.3	42.3	27.7	18.3	-
Log Bilinear [16]	60.3	38.0	25.4	17.1	16.9
Toronto-soft attention [35]	66.7	43.4	28.8	19.1	18.5
Toronto-Hard attention [35]	66.9	43.9	29.6	19.9	18.5
Ours	64.8	44.7	30.8	21.6	19.5

[a]indicates a different test data split.

BLEU metrics seek correlation at the corpus level, while the METEOR metric produces good correlation at the sentence or segment level. We use the standard Microsoft COCO Caption Evaluation toolkit[2] to compute those metrics.

Tables 1 and 2 show the scores of different metrics for different methods on the Flickr30K and MS COCO datasets respectively. From these results, we can find that: (1) Visual attention is helpful to improve the quality of generated sentence compared with non-attention methods. (2) Our model which takes semantic regions localization and visual attention can significantly improve the performance with respect to others. Particularly, our model outperforms the Toronto-Hard Attention model [35] using fixed region splits with 2.6 and 4.0 percentages of improvement for BLEU-3 and BLEU-4 metrics as well as 1.3% of improvement for METEOR metric. It is necessary to note that the BLEU-1 score of our model is poor than that of Toronto-Hard Attention model. However, the scores of other metrics of our model are better, especially the BLEU-4 which can better reflect the fluency and grammaticality for the generated sentence than the BLEU-1, and the METEOR which has more correlation with human judgement at the sentence or segment level than the BLEU metrics.

---

[2] https://github.com/tylin/coco-caption.

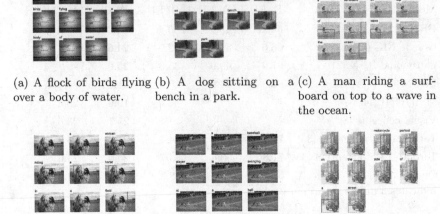

(a) A flock of birds flying over a body of water.

(b) A dog sitting on a bench in a park.

(c) A man riding a surfboard on top to a wave in the ocean.

(d) A woman riding a horse in a field.

(e) A baseball player is swinging at a ball.

(f) A motorcycle parked on the side of a street.

Fig. 3. Some captioning examples generated by our LocAtt framework. The words in the left top of pictures are generated by our model. The red bounding box in each picture is the corresponding semantic region which our model localized. Better zoom in to review. (Color figure online)

4.3   Visualization and Qualitative Analysis

In order to see the alignments between generated words and selected regions, as well as understand what our model has learned, we visualize some captioning examples in Fig. 3. From this Figure, we can see that our model achieves very strong alignments which correspond to human intuition except for some articles and prepositions such as *a*, *the*, *of*, which don't have any specific image content alignment themselves. What's more, our model not only learns the alignments for nouns that are corresponding to the objects in images, but also learns the alignments for some more semantic verbs. We can also notice that several mistakes of word prediction exists in our examples. However, in the case of predicting a wrong word, our model can still make suitable alignment and localize the right region for this word, which makes those mistakes can be understood by us.

5   Conclusion

In this paper, we introduce a novel deep semantic indexing method for image database. Our method combines semantic region localization and visual attention to automatically generate captions for image dataset indexing. Using those compact captions, less storage space is needed and reaches fast processing speed, as well as achieves preferable indexing performance compare with original visual data. We conduct a variety of experiments on two popular benchmarks and the results demonstrate our approach outperforms other visual attention based approaches and is also competitive with the sate-of-art approach.

Acknowledgement. This work was supported in part by the National Natural Science Foundation of China under Project 61572108 and Project 61502081, the National Thousand-Young-Talents Program of China, and the Fundamental Research Funds for the Central Universities under Project ZYGX2014Z007 and Project ZYGX2015J055.

References

1. Bahdanau, D., Cho, K., Bengio, Y.: Neural machine translation by jointly learning to align and translate. arXiv (2014)
2. Bhuiyan, A.: Content-based image retrieval for image indexing. IJACSA **6**(6), 71–79 (2015)
3. Bin, Y., Yang, Y., Shen, F., Xu, X., Shen, H.T.: Bidirectional long-short term memory for video description. In: ACM MM, pp. 436–440. ACM (2016)
4. Bhuiyan, A., Chen, X., Zitnick, C.L.: Learning a recurrent visual representation for image caption generation. arXiv (2014)
5. Chiueh, T.-C.: Content-based image indexing. In: VLDB, vol. 94, pp. 582–593 (1994)
6. Donahue, J., Hendricks, L.A., Guadarrama, S., Rohrbach, M., Venugopalan, S., Saenko, K., Darrell, T.: Long-term recurrent convolutional networks for visual recognition and description. In: CVPR, pp. 2625–2634 (2015)
7. Ewald, M.D.: Content-based image indexing and retrieval in an image database for technical domains. Trans. MLDM **2**(1), 3–22 (2009)
8. Fang, H., Gupta, S., Iandola, F., Srivastava, R.K., Deng, L., Dollár, P., Gao, J., He, X., Mitchell, M., Platt, J.C., et al.: From captions to visual concepts and back. In: CVPR, pp. 1473–1482 (2015)
9. Felzenszwalb, P.F., Girshick, R.B., McAllester, D., Ramanan, D.: Object detection with discriminatively trained part-based models. TPAMI **32**(9), 1627–1645 (2010)
10. Girshick, R.: Fast r-cnn. In: ICCV, pp. 1440–1448 (2015)
11. Girshick, R., Donahue, J., Darrell, T., Malik, J.: Rich feature hierarchies for accurate object detection and semantic segmentation. In: CVPR, pp. 580–587 (2014)
12. Hochreiter, S., Schmidhuber, J.: Long short-term memory. NC **9**(8), 1735–1780 (1997)
13. Hosang, J., Benenson, R., Dollr, P., Schiele, B.: What makes for effective detection proposals? TPAMI **38**(4), 814–830 (2016)
14. Hyvönen, E., Saarela, S., Styrman, A., Viljanen, K.: Ontology-based image retrieval. In: WWW (2003)
15. Karpathy, A., Li, F.-F.: Deep visual-semantic alignments for generating image descriptions. In: CVPR, pp. 3128–3137 (2015)
16. Kiros, R., Salakhutdinov, R., Zemel, R.S.: Multimodal neural language models. In: ICML, vol. 14, pp. 595–603 (2014)
17. Kiros, R., Salakhutdinov, R., Zemel, R.S.: Unifying visual-semantic embeddings with multimodal neural language models. arXiv (2014)
18. Koch, C., Ullman, S.: Shifts in selective visual attention: towards the underlying neural circuitry. In: Vaina, L.M. (ed.) MI. SYLI, vol. 188, pp. 115–141. Springer, Dordrecht (1987). doi:10.1007/978-94-009-3833-5_5

19. Lin, T.-Y., Maire, M., Belongie, S., Hays, J., Perona, P., Ramanan, D., Dollár, P., Zitnick, C.L.: Microsoft COCO: common objects in context. In: Fleet, D., Pajdla, T., Schiele, B., Tuytelaars, T. (eds.) ECCV 2014. LNCS, vol. 8693, pp. 740–755. Springer, Cham (2014). doi:10.1007/978-3-319-10602-1_48
20. Luong, M.-T., Pham, H., Manning, C.D.: Effective approaches to attention-based neural machine translation. arXiv (2015)
21. Mao, J., Xu, W., Yang, Y., Wang, J., Huang, Z., Yuille, A.: Deep captioning with multimodal recurrent neural networks (m-rnn). arXiv (2014)
22. Michel, F.: How many photos are uploaded to flickr every day and month? (2012). http://www.flickr.com/photos/franckmichel/6855169886/
23. Neubig, G.: Neural machine translation and sequence-to-sequence models: a tutorial. arXiv (2017)
24. Nguyen, T.V., Song, Z., Yan, S.: Stap: spatial-temporal attention-aware pooling for action recognition. TCSVT **25**(1), 77–86 (2015)
25. Plummer, B.A., Wang, L., Cervantes, C.M., Caicedo, J.C., Hockenmaier, J., Lazebnik, S.: Flickr30k entities: collecting region-to-phrase correspondences for richer image-to-sentence models. In: ICCV, pp. 2641–2649 (2015)
26. Ren, S., He, K., Girshick, R., Sun, J.: Faster r-cnn: towards real-time object detection with region proposal networks. In: NIPS, pp. 91–99 (2015)
27. Sharma, S., Kiros, R., Salakhutdinov, R.: Action recognition using visual attention. arXiv (2015)
28. Sutskever, I., Vinyals, O., Le, Q.V.: Sequence to sequence learning with neural networks. In: NIPS, pp. 3104–3112 (2014)
29. Uijlings, J.R.R., van de Sande, K.E.A., Gevers, T., Smeulders, A.W.M.: Selective search for object recognition. IJCV **104**(2), 154–171 (2013)
30. Vailaya, A., Figueiredo, M.A.T., Jain, A.K., Zhang, H.-J.: Image classification for content-based indexing. TIP **10**(1), 117–130 (2001)
31. Vinyals, O., Toshev, A., Bengio, S., Erhan, D.: Show and tell: a neural image caption generator. In: CVPR, pp. 3156–3164 (2015)
32. Viola, P., Jones, M.: Rapid object detection using a boosted cascade of simple features. In: CVPR, vol. 1, p. I-511. IEEE (2001)
33. Wang, C., Yang, H., Bartz, C., Meinel, C.: Image captioning with deep bidirectional LSTMs. In: ACM MM, pp. 988–997. ACM (2016)
34. Xindong, W., Zhu, X., Gongqing, W., Ding, W.: Data mining with big data. TKDE **26**(1), 97–107 (2014)
35. Xu, K., Ba, J., Kiros, R., Cho, K., Courville, A., Salakhutdinov, R., Zemel, R.S., Bengio, Y.: Show, attend and tell: neural image caption generation with visual attention. arXiv **2**(3), 5 (2015)
36. Yanai, K.: Image collector: an image-gathering system from the world-wide web employing keyword-based search engines. In: ICME (2001)
37. Yang, Y., Luo, Y., Chen, W., Shen, F., Shao, J., Shen, H.T.: Zero-shot hashing via transferring supervised knowledge. In: ACM MM, pp. 1286–1295. ACM (2016)
38. Yang, Y., Zha, Z.-J., Gao, Y., Zhu, X., Chua, T.-S.: Exploiting web images for semantic video indexing via robust sample-specific loss. TMM **16**(6), 1677–1689 (2014)
39. Yao, L., Torabi, A., Cho, K., Ballas, N., Pal, C., Larochelle, H., Courville, A.: Describing videos by exploiting temporal structure. In: ICCV, pp. 4507–4515 (2015)
40. Zitnick, C.L., Dollár, P.: Edge boxes: locating object proposals from edges. In: Fleet, D., Pajdla, T., Schiele, B., Tuytelaars, T. (eds.) ECCV 2014. LNCS, vol. 8693, pp. 391–405. Springer, Cham (2014). doi:10.1007/978-3-319-10602-1_26

Demo Papers

Real-Time Popularity Prediction on Instagram

Deming Chu[1], Zhitao Shen[2], Yu Zhang[3], Shiyu Yang[4(✉)], and Xuemin Lin[4]

[1] East China Normal University, Shanghai, China
ned_chu@qq.com
[2] Ant Financial Services Group, Shanghai, China
shenzt@gmail.com
[3] eBay Inc. Shanghai, Shanghai, China
webmaster@joe1cafe.com
[4] The University of New South Wales, Sydney, Australia
{yangs,lxue}@cse.unsw.edu.au

Abstract. Social network services have become a part of modern daily life. Despite explosive growth of social media, people only pay attention to a small fraction of them. Therefore, predicting the popularity of a post in social network becomes an important service and can benefit a series of important applications, such as advertisement delivery, load balancing and personalized recommendation etc. In this demonstration, we develop a real-time popularity prediction system based on user feedback e.g. count of likes. In the proposed system, we develop effective algorithms which utilize the temporal growth of user feedbacks to predict the popularity in real-time manner. Moreover, the system is easy to be adapted for a variety of social network platforms. Using datasets collected from Instagram, we show that the proposed system can perform effective prediction on popularity at early stage of post.

Keywords: Real-time · Predicting popularity · Social network

1 Introduction

With the popularity of mobile devices and lower bandwidth cost, people are more and more connected to each other through the Internet. People not only browse but also share and produce web contents, which leads to flood of information. At the same time, a small fraction of contents attract most of attention from public and bring most of flow out. The identification of potential popular content can help grasp pulse of flow. As a result, the service providers can make more profits out, advertisers can maximize their revenues through better advertisement placement and users can focus on most relevant information. Popularity prediction problem is clearly defined in [3]. And the popularity is usually characterized by number of user feedbacks e.g. count of likes. We also follow this convention in our demonstration. In most of the social networks, especially those focus on short contents e.g. Twitter and Instagram, a post will become popular in a short time (usually less than 24 h) [4]. Therefore, how to predict the popularity of a post just in a short time after it has been posted becomes increasingly important.

© Springer International Publishing AG 2017
Z. Huang et al. (Eds.): ADC 2017, LNCS 10538, pp. 275–279, 2017.
DOI: 10.1007/978-3-319-68155-9_21

Challenges. Due to the diversity and high update frequency property of a post in social network, it is difficult to effectively predict the popularity of a post and it is even harder when considering the real-time requirement. The main challenges are in two folds: Firstly, the popularity of a post is usually affected by many factors and most of them are either difficult to measure or changing frequently. Secondly, most posts loses attention from public days after publication, which means meaningful prediction must be made within several hours or even shorter.

In this demonstration, we propose a real-time prediction system for social network content. The system consists of two main components: back-end and front-end. In the back-end, crawlers continuously crawl target pages from social network platform and several regression model based algorithms are implemented to support prediction task. In terms of front-end, we build user interface upon Django[1] to visualize recent posts that are expected to be popular and individual prediction on single post as user input. The proposed system is evaluated by datasets which are crawled from Instagram[2]. Our main contributions are summarized as follows:

- A popularity prediction system with novel and effective prediction algorithms which can perform effective real-time prediction on popularity of social network content at early stage.
- A web interface that can present recent posts that are expected to be popular and individual prediction on a post in a user-friendly way.

2 System

Figure 1 demonstrates framework of our system which consists of two part: back-end part and front-end part. Most of work falls in the back-end part.

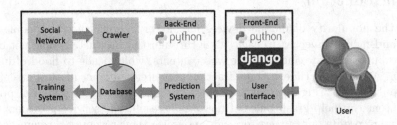

Fig. 1. System framework

---

[1] https://www.djangoproject.com/.
[2] https://www.instagram.com.

2.1   Back-End

Crawler. Crawler is built upon Scrapy[3] which is a popular crawler framework. For a given social platform, homepage of selected bloggers are used as target pages in advance. In order to predict popularity in real time, the crawler is designed to crawl target pages continuously. The gap between two queries on a certain blogger is no more than tens of minutes to gain sufficient data for prediction. Moreover, we save original data on first visit to a post and ignore data that doesn't change later on. As a result, we collect a set of original data of post and a set of update data of post.

Training and Prediction System. There are two important time in prediction problem: indication time t_i and reference time t_r. Prediction algorithm is running at t_i to predict popularity at t_r. In this demonstration, we take $t_r = 24h$, which means the system predict popularity 24 h after publication. The reference time is chosen based on definition of *effective shelf-life 90%*: time passed between its first visit and the time at which it has received a fraction 90% of the visits it will ever receive [1].

In this demonstration, we characterize popularity through number of user feedbacks, more specifically, count of likes. Regression-based model is built individually for each blogger upon one's past posts, using data point at t_i and data point at t_r. As popularity of post has strong association with its author, prediction can be well performed.

We propose three prediction algorithms, namely *Normalized Linear Regression, Normalized Linear Regression in Log Scale* and *Regression on Coefficient*. We also implement two algorithms as competitors, namely *Linear Regression* and *Linear Regression in Log Scale* are presented in [2]. Coefficient of each model are trained beforehand in training system, so that prediction system can handle queries in real-time. We introduce the idea of each algorithm as following:

- *Linear Regression.* Collect all history posts of a blogger and build linear regression model on popularity at t_i and popularity at t_r
- *Linear Regression in Log Scale.* This method is almost the same as linear regression, except that linear regression model is built in log scale.
- *Normalized Linear Regression.* Popularity at t_i is normalized based on publication time of post and estimated number of blogger's online followers.
- *Normalized Linear Regression in Log Scale.* Combination of linear regression in log scale and normalized linear regression.
- *Regression on Coefficient.* Fit the growth curve of popularity with power function kx^r where x is hours from publication. Using k and log of popularity at t_r to build linear regression model.

---

[3] https://scrapy.org/.

2.2   Front-End

Two scenarios are constructed in front-end part of system to present popularity prediction result in a friendly way. In the first scenario, top 10 posts that are most likely to be popular are collected from posts known to public in the last 24 h. In the second scenario, individual prediction on a certain post is presented in line chart.

3   Demonstration

3.1   Dataset

The proposed system are trained and evaluated using dataset which is collected from Instagram. Instagram is world's most popular photo/video sharing social network platform, where users could upload photographs and short videos, follow other users' feeds and share their feeling. Target bloggers are selected from two sources, 100 of them are from a top Instagram accounts list[4], the rest are randomly picked. In total, the dataset contains about 500 target bloggers, 100,000 posts and 100,000,000 updates. For a given post, gap between two updates is about several minutes.

3.2   User Interface

Figure 2 shows the first scenario, where top 10 recent posts that are most likely to be popular are presented. Details of the post are also shown in this scenario, including post time, details of blogger, estimated popularity and content of the post. Figure 3 shows the second scenario for individual post prediction. Users can choose blogger and post, then the historical prediction results of the popularity in 24 h is presented. In this line chart, x axis means time from publication. Meanwhile, y axis means popularity prediction result at x divided by the real popularity at 24 h after publication, in other words, relative error to ground-truth. The closer y to 1, the better our prediction algorithm is.

Fig. 2. Top 10 recent post

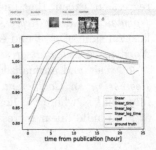

Fig. 3. Historical prediction

---

[4] http://zymanga.com/millionplus/.

Acknowledgement. The research is supported by the National Natural Science Foundation of China under Grant No. 61232006, 61672235, 61401155.

References

1. Castillo, C., El-Haddad, M., Pfeffer, J., Stempeck, M.: Characterizing the life cycle of online news stories using social media reactions. In: Proceedings of the 17th ACM Conference on Computer Supported Cooperative Work & Social Computing, pp. 211–223. ACM (2014)
2. Szabo, G., Huberman, B.A.: Predicting the popularity of online content. Commun. ACM **53**(8), 80–88 (2010)
3. Tatar, A., de Amorim, M.D., Fdida, S., Antoniadis, P.: A survey on predicting the popularity of web content. J. Internet Serv. Appl. **5**(1), 8 (2014)
4. Zaman, T., Fox, E.B., Bradlow, E.T., et al.: A bayesian approach for predicting the popularity of tweets. Ann. Appl. Stat. **8**(3), 1583–1611 (2014)

DSKQ: A System for Efficient Processing of Diversified Spatial-Keyword Query

Shanqing Jiang[1]([⊠]), Chengyuan Zhang[2], Ying Zhang[3], Wenjie Zhang[1],
Xuemin Lin[1,4], Muhammad Aamir Cheema[1,5], and Xiaoyang Wang[1]

[1] The University of New South Wales, Sydney, Australia
jiangsq91@gmail.com, {zhangw,lxue,xiaoyangw}@cse.unsw.edu.au
[2] Central South University, Changsha, China
cyzhang@csu.edu.cn
[3] University of Technology, Sydney, Australia
ying.zhang@uts.edu.au
[4] Shanghai Key Laboratory of Trustworthy Computing,
East China Normal University, Shanghai, China
[5] Clayton School of Information Technology,
Monash University, Melbourne, Australia
aamir.cheema@monash.edu

Abstract. With the rapid development of mobile portable devices and location positioning technologies, massive amount of *geo-textual* data are being generated by a huge number of web users on various social platforms, such as Facebook and Twitter. Meanwhile, *spatial-textual* objects that represent Point-of-interests (POIs, e.g., shops, cinema, hotel or restaurant) are increasing pervasively. Consequently, how to retrieve a set of objects that best matches the user's submitted spatial keyword query (SKQ) has been intensively studied by the research communities and commercial organisations. Existing works only focus on returning the nearest matching objects, although we observe that many real-life applications are now using diversification to enhance the quality of the query results. Thus, existing methods fail to solve the problem of diversified SKQ efficiently. In this demonstration, we introduce DSKQ, a diversified in-memory spatial-keyword query system, which considers both the textual relevance and the spatial diversity of the results processing on road network. We present a prototype of DSKQ which provides users with an application-based interface to explore the diversified spatial-keyword query system.

Keywords: Diversification · Spatial-keyword query · Boolean range query

1 Introduction

Extensive amount of *spatial-textual objects* that possess both text descriptions and geo-locations are now available thanks to the proliferation of geo-positioning technologies. As a consequence, there has been lot of commercial interest in

© Springer International Publishing AG 2017
Z. Huang et al. (Eds.): ADC 2017, LNCS 10538, pp. 280–284, 2017.
DOI: 10.1007/978-3-319-68155-9_22

spatial keyword query system since the last decade. The massive amount of available *spatial-textual* data enables users to retrieve a set of objects that best matches the user's submitted spatial keyword query [2] (i.e., SKQ, which includes a geographical location and a set of keywords), in terms of both spatial proximity to query location and textual relevance to query keywords. For instance, a user may want to find all nearby *restaurants* which serve both *steak* and *pancake*.

To make sense of retrieving *spatial-textual objects* and satisfy the increasing location-aware demand of users, it is critical to develop efficient system to support spatial keyword search on road networks, since most of the spatial-textual objects are located on predefined road networks and the computation cost of the road network distance is much higher than that in the Euclidean space. Moreover, it has been widely accepted [1] that the usefulness of a retrieved object depends not only on its relevance to the query (i.e., distance and keyword constraint) but also on other objects in the results. In many scenarios, users are more interested in retrieving more diversified results and are less likely to expect highly similar objects at the same time [5]. In [6], Zhang et. al. proposed a diversified spatial keyword query system on road network, which retrieves a set of the objects each of which contains all query keywords and is within certain distance to the query location in terms of road network distance, with those objects being spatially diversified, that is, the pair-wise distance (i.e., dissimilarity) between two objects in the result should be reasonably large.

Example. *Consider a businessman traveling in Sydney wants to select a 5-star hotel in CBD area with indoor swimming pool, meanwhile exploring local features around where he lives during his spare time. Since we don't know any preference of his lifestyle, a spatially diversified set of results will better meet his request by providing more possibles options.*

In this paper, we propose a novel system, named DSKQ, to demonstrate the diversified spatial keyword search queries in [6].

The system consists of two main components, one of which is the front-end graphic user interface (GUI) which is built for user to explore the query process in an interactive manner, the other is the back-end indexing structure built upon road network and spatial textual objects. The front-end GUI enables user to input a query by specify a query location, set of keywords and range, the system will visualise the diversified search result that satisfying the query constraints. As for the back-end, we develop an efficient signature-based inverted indexing technique as well as an efficient incremental network expansion algorithm for spatial keyword search on road networks.

Contributions. Our main contributions are summarised as follows:

- A novel indexing structure and network expansion algorithm are developed to efficiently precess diversified spatial keyword queries.
- An application based graphic user interface is designed to enable user to better explore the query precess.

2   DSKQ System

Figure 1 demonstrates our system architecture. There are two main components of our system: the back-end and the front-end.

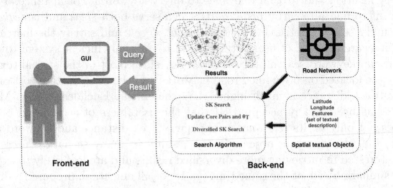

Fig. 1. System framework

Back-end. The back-end is where the road network and spatial textual objects index lies as well as the query algorithm processes. We use a structure that combines the R-tree and inverted file [6] to organise the road network and spatial textual objects in memory. We utilised the popular connectivity-clustered access method (CCAM) [4] to represent the road network which effectively organises the adjacent lists of the road nodes and enable us to take advantage of the query location and reduce the costs during the query processing. Nodes of the road network are indexed in a network R-tree [3] using their geo-location (i.e. latitude and longitude) as key. Road network edges are represented as adjacent lists of a nodes, in which the end-node and weight (i.e. network distance) are stored. We also build a R-tree to organise the minimal bounding rectangles (MBR) of the edges, and, associate object to its corresponding edge by utilising the MBR of the edge in a branch-and-bound fashion.

Front-end. The front-end of the system is an application interface implemented under the Java Swing Framework which can be used to explore the DSKQ procedure. The interface display the road network with Google Map[1] and user specified query location and query range. After a query is being issued, the system will highlight the diversified result set (i.e., spatial textual objects each contains all the query keyword and within the road network distance from query location and, the pair-wise distance are maximised) in the map for user to explore.

3   Demonstration

Figure 2 shows the application GUI of our system. Figure 2(a) is the initial application interface, where the road network and, user's geo-location is displayed

---

[1] http://maps.google.com.

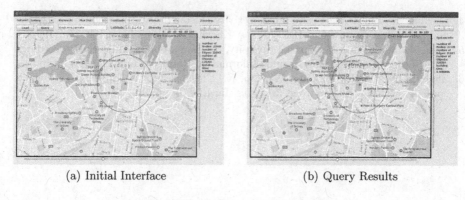

(a) Initial Interface   (b) Query Results

Fig. 2. System interface and query results

and user can issue a query to retrieve the nearby interested objects. For example, a user issues a query with location *Sydney* (latitude: -33.874412, longitude: 151.211418), a set of keywords *steak, wine* and *pancake* and number of retrieving results *4*. The query range is set to *1 km* as default, which can also be adjusted according to user's preference. After submitting this query to our system, all the retrieved objects will be highlighted with their corresponding textual description in Fig. 2(b). User can also zoom and drag the map to explore other interested regions.

4 Conclusion

In this paper, we present DSKQ, a diversified spatial-keyword query system on road network which retrieves a set of diversified *spatial-textual objects* that satisfies the query constraints (i.e., spacial and keyword constraints). A signature-based inverted indexing is built on road network and effective network expansion algorithm as well as pruning techniques are developed to accelerate query processing. Moreover, an application based graphic user interface is also developed to enable users to interactively explore the query process and examine the returned results.

References

1. Angel, A., Koudas, N.: Efficient diversity-aware search. In: Proceedings of the 2011 ACM SIGMOD International Conference on Management of Data, SIGMOD 2011, pp. 781–792. ACM, New York (2011)
2. Chen, L., Cong, G., Jensen, C.S., Wu, D.: Spatial keyword query processing: an experimental evaluation. In: Proceedings of the VLDB Endowment, vol. 6, pp. 217–228. VLDB Endowment (2013)
3. Papadias, D., Zhang, J., Mamoulis, N., Tao, Y.: Query processing in spatial network databases. In: Proceedings of the 29th International Conference on Very Large Data Bases, vol. 29, pp. 802–813. VLDB Endowment (2003)

4. Shekhar, S., Liu, D.R.: Ccam: a connectivity-clustered access method for networks and network computations. IEEE Trans. Knowl. Data Eng. **9**(1), 102–119 (1997)
5. Tang, J., Sanderson, M.: Spatial diversity, do users appreciate it? In: Proceedings of the 6th Workshop on Geographic Information Retrieval, p. 22. ACM (2010)
6. Zhang, C., Zhang, Y., Zhang, W., Lin, X., Cheema, M.A., Wang, X.: Diversified spatial keyword search on road networks. In: EDBT, pp. 367–378 (2014)

Author Index

Ahmad, Sabbir 3
Albarrak, Abdullah M. 45
Ali, Mohammed Eunus 3

Bao, Zhifeng 29
Boroujeni, Forough Rezaei 138

Cao, Liangfu 248
Cheema, Muhammad Aamir 280
Chu, Deming 275

Duong, Chi Thang 125

Gao, Lianli 248

He, Shiyuan 98
Hu, Mengqiu 98
Hua, Wen 71
Huang, Xiu 217

Indrawan-Santiago, Maria 17

Ji, Yanli 261
Jiang, Shanqing 280

Kamal, Rafi 3

Lan, Hai 233
Lee, Sharon X. 178
Li, Jingjing 193
Li, Lingxiao 17
Li, Xuelong 98
Li, Zhihui 110, 193
Lin, Xuemin 275, 280
Liu, Luyao 110
Liu, Yiqun 205
Liu, Zheng 165
Lu, Ke 193
Luo, Cheng 205

Ma, Shaoping 205
Mao, Jiaxin 205

Naeem, M. Asif 59
Nguyen, Kim Tung 59
Nguyen, Quoc Viet Hung 125

Peng, Yuwei 233

Qi, Jianzhong 3

Rao, Weixiong 85
Ruan, Boyu 71

Scheuermann, Peter 3
Shao, Zhou 17
Sharaf, Mohamed A. 45
Shen, Fumin 98, 217
Shen, Heng Tao 98, 217, 248, 261
Shen, Zhitao 275
Shi, Wei 165
Siuly, Siuly 151
Song, Jingkuan 248
Stantic, Bela 125, 138

Taniar, David 17
Tanin, Egemen 3

Wang, Hua 151
Wang, Sen 125, 138
Wang, Sheng 29
Wang, Xiaoyang 280
Weber, Gerald 59

Xie, Ning 217, 261
Xu, Xing 248

Yadamjav, Munkh-Erdene 29
Yang, Shiyu 275
Yang, Yang 98, 217, 261
Yang, Zihao 217
Ye, Guo 98
Yu, Jeffrey Xu 165

Yuan, Mingxuan   85
Yuan, Pengcheng   85

Zarei, Roozbeh   151
Zeng, Jia   85
Zhang, Bang   29
Zhang, Chengyuan   280
Zhang, Hanwang   261
Zhang, Min   205
Zhang, Mingxing   261

Zhang, Ruiyuan   71
Zhang, Wenjie   280
Zhang, Yanchun   151
Zhang, Ying   280
Zhang, Yu   275
Zhao, Qinpei   85
Zheng, Weiguo   165
Zheng, Yukun   205
Zhou, Xiaofang   71
Zhu, Lei   110, 193

Printed in the United States
By Bookmasters